AF598055

Circuit Design for Reliability

Ricardo Reis • Yu Cao • Gilson Wirth
Editors

Circuit Design for Reliability

Editors
Ricardo Reis
Instituto de Informática
Universidade Federal do Rio
Grande do Sul (UFRGS)
Porto Alegre, Rio Grande do Sul, Brazil

Yu Cao
School of ECEE
Arizona State University
Tempe, USA

Gilson Wirth
Electrical Engineering Department
Universidade Federal do Rio
Grande do Sul (UFRGS)
Porto Alegre
Rio Grande do Sul
Brazil

ISBN 978-1-4614-4077-2 ISBN 978-1-4614-4078-9 (eBook)
DOI 10.1007/978-1-4614-4078-9
Springer New York Heidelberg Dordrecht London

Library of Congress Control Number: 2014952728

Printed on acid-free paper

Springer is part of Springer Science+Business Media (www.springer.com)

Contents

Chapter 1
Introduction

Ricardo Reis, Yu Cao, and Gilson Wirth

The scaling of CMOS technology to the nanometer regime inevitability increases reliability concerns, profoundly impacting all aspects of circuit performance and posing a fundamental challenge to future IC design. These reliability concerns arise from many different sources, and become more severe with continuous scaling.

VLSI design in the late CMOS era is then driven by an increasing challenge to cope with unreliable components at the device and circuit levels. Early approaches focused only on technology improvement, but ignoring these effects in the design process causes an excessive amount of over-margining, with negative impact in both cost and performance of the final product. At the device and circuit levels, physical understanding, modeling, detection, and successful design techniques for leading reliability mechanisms are vitally important, not only to current robust design practice, but also to the prediction and management of system reliability over the full product life cycle. It is critical to understand, simulate, and mitigate their impact during the stage of physical and circuit design. This book presents physical understanding, modeling and simulation, on-chip characterization, layout solutions, and design techniques that are effective to enhance the reliability of various circuit units.

We start by addressing the physical mechanisms of reliability, focusing on aging and noise. Aging is among the critical reliability issues facing present and future deeply downscaled CMOS devices, and the major concern is the so-called Bias Temperature Instability (BTI). The elementary definitions and experimental

---

R. Reis • G. Wirth
Electrical Engineering Department, Universidade Federal do Rio Grande do Sul (UFRGS), Porto Alegre, Rio Grande do Sul, Brazil

Y. Cao (✉)
School of ECEE, Arizona State University, Tempe, AZ, USA
e-mail: ycao@asu.edu

R. Reis et al. (eds.), *Circuit Design for Reliability*,
DOI 10.1007/978-1-4614-4078-9_1

observations of BTI are first briefly reviewed. Afterwards it is shown how the understanding of gate oxide defect properties can be used to explain BTI and its variability, and possible technological solutions for both nFETs and pFETs are discussed. The charge trapping phenomena that plays a dominant role in BTI also dominates low-frequency noise. A modeling and circuit simulation approach that focuses on operation conditions relevant for digital and analog design, including large signal AC operation, is presented. The modeling and simulation approach aims at helping circuit designers to cope with this reliability issue.

Next, the modeling and characterization is addressed. Any modeling effort needs a solid base to be built on. Models are based on accurate measurement methodologies, which contribute to understanding the mechanisms behind the device degradation. In this book, authors survey characterization methods for the purpose of device modeling that includes the effects of temporal parameter degradation, such as BTI and noise. Particular focus is placed on the approaches that utilize on-chip circuits to enhance the accuracy and fidelity of the measurements. Modeling must focus on real circuit operation conditions, such as Dynamic Voltage Scaling (DVS), or biasing found in mixed signal designs where the stress waveform is much more random. Modeling solutions suitable to allow aging simulation under all dynamic stress conditions are presented.

Variability in electrical parameters, such as variability in threshold voltage, is also known to be a major reliability concern. Among the sources of variability is random dopant fluctuation (RDF). We discuss how RDF can induce significant variation in saturation (on) current and transconductance degradation—two key metrics for benchmark performance of digital and analog integrated circuits.

An analysis of ongoing scaling related trends of integrated circuits is also presented. For this purpose, authors discuss major trends observed in process technology, as well as in memory and logic designs. While newer technology styles, such as FinFETs provide necessary advances to enable further technology scaling, we can observe that there exists also a trend for reduced resilience with ongoing scaling. Our hope is that with the roadmap presented in this book, which concentrates on FinFET-based designs, we are able to sensitize the research community to present and near future problems associated with this technology. Additionally, different architectures for SRAM memories are analyzed, to demonstrate the different design choices available and tradeoffs that have to be made.

Finally, the last chapters present design techniques for reliability. FinFETs have emerged as alternatives to conventional bulk MOSFETs in scaled technologies due to superior gate control of the channel, lower short channel effects and higher scalability. However, width quantization in FinFETs constrains the design space of FinFET-based circuits, especially SRAMs in which transistor sizing is critical for the circuit robustness. The adverse effects of width quantization can be mitigated by appropriate device-circuit co-design of FinFET-based memories. Some of such techniques are described, with an emphasis on the device-circuit interactions associated with each methodology. The impact of different technology options in FinFETs like gate-underlap, fin orientation, fin height, gate workfunction and independent control of the gates on the stability, power and performance of 6T SRAMs is discussed.

Electromigration is an aging effect that affects metal wires (interconnections). In this book, authors briefly introduce physical foundations of electromigration (EM). We discuss physical parameters affecting EM wire lifetime and we introduce some background related to the existing EM physical simulators. In the work presented in this book for EM physical simulation, authors adopt the atomic concentration balance-based model. We discuss the simulation setup and results. We present VEMA, a variation-aware electromigration (EM) analysis tool for power grid wires. The tool considers process variations caused by the chemical–mechanical polishing (CMP) and edge placement error (EPE). VEMA is a full-chip EM analysis tool. It performs EM lifetime calculation, and analyzes process variation effects on EM reliability and reports variation tolerances of EM-sensitive power grid wires.

Integrated circuits subject to ionizing radiation are sensitive to transient faults caused by the interaction of ionizing particles with the semiconductor. At ground level, the main source of radiation are neutrons that interact with the material generating secondary particles that can ionize the silicon, provoking transient upset in circuits fabricated in nanometer technology. The interaction of the ionizing radiation with the transistor may provoke transient and permanent effects. Different fault tolerance techniques can be applied to FPGAs according to their type of configuration technology, architecture and target operating environment. A chapter is dedicated to present a set of fault mitigation techniques for SRAM, FLASH and ANTIFUSE-based FPGAs. A test methodology to characterize those FPGA under radiation is also presented. Results from neutron-induced faults are presented and discussed.

The relevance of power consumption and leakage currents is also increasing. Although several Design-for-Low-Power and Design-for-Variability options are already available in modern EDA suites, the contrasting nature of the two metrics makes their integration extremely challenging. Most of the approaches used to compensate and/or mitigate circuit variability (e.g., Dynamic Voltage Scaling and Adaptive Body Biasing) are, in fact, intrinsically power inefficient, as they exploit the concept of redundancy, which is known to originate power overhead. In this book, we introduce possible solutions for concurrent leakage minimization and variability compensation. More specifically, authors propose Power-Gating as a mean for simultaneously controlling static power consumption and mitigating the effects induced by two of the most insidious sources of variability, namely, electrical parameter variations and Aging due to BTI.

High-performance clock network design has been a challenge for many years due to the drastically increasing effect of process variability. In addition, tight power budgets have lowered supply voltage levels, which make designs more sensitive to noise. Together, variability and noise present a colossal challenge to clock designers in order to meet timing, yield, and power simultaneously. This chapter discusses the different strategies that designers use to ameliorate variability and noise problems in clock network design.

The project of this book was born during the successful IEEE Circuits and Systems Society Summer School on Physical Design of Reliable Circuits that was held in Porto Alegre, Brazil, from January 12 to 15, 2010. Some chapters of the book are based on the talks realized during the school and other well-known invited authors wrote other chapters.

It is our hope that this book will provide the vitally important physical understanding and design techniques for state of the art and future technologies, ranging from technology modeling, fault detection and analysis, circuit hardening, and reliability management in different circuit units.

Chapter 2
Recent Trends in Bias Temperature Instability

B. Kaczer, T. Grasser, J. Franco, M. Toledano-Luque, J. Roussel, M. Cho, E. Simoen, and G. Groeseneken

Abstract The paradigm shifts occurring in the past few years in our understanding of BTI are reviewed. Among the most significant ones is the shift from perceiving NBTI in terms of the Reaction-Diffusion model to analyzing BTI with the tools originally developed for describing low-frequency noise. This includes the interpretation of the time, temperature, voltage, and duty cycle dependences of BTI. It is further demonstrated that a wealth of information about defect properties can be obtained from deeply-scaled devices, and that this information can allow projection of variability issues of future deeply downscaled CMOS devices. The chapter is concluded by showing the most promising technological solutions to alleviate both PBTI and NBTI.

1 Introduction

Among the critical reliability issues facing present and future deeply downscaled CMOS devices is the so-called Bias Temperature Instability (BTI). While BTI in nFET devices was generally ascribed to charge trapping in the high-k portion of the gate oxide, the interpretation of BTI in pFET devices still generates controversy [1–7]. This phenomenon in pFET devices has been previously described by diffusion of hydrogen from and back into substrate/gate oxide interface states [8]. The so-called Reaction–Diffusion model based on this assumption is still popular, especially in the design community [9], despite being inconsistent with some crucial observations [10].

In this chapter, the elementary definitions and experimental observations of BTI are first briefly reviewed. One of the most intriguing properties of BTI—the lack

B. Kaczer (✉) • J. Franco • M. Toledano-Luque • J. Roussel • M. Cho • E. Simoen • G. Groeseneken
imec, Leuven, Belgium
e-mail: kaczer@imec.be

T. Grasser
TU Wien, Wien, Austria

G. Groeseneken
KU, Leuven, Belgium

R. Reis et al. (eds.), *Circuit Design for Reliability*,
DOI 10.1007/978-1-4614-4078-9_2

of characteristic time scale, especially in pFETs—is then argued to point to a *dispersion* in the underlying mechanism [5, 7, 11]. In CMOS technologies, a response on many time scales is typical for low-frequency noise [and its manifestation as Random Telegraph Noise (RTN) in deeply-scaled devices], suggesting that (the principal component of) BTI is in fact caused by the same defects [12, 13]. The link between BTI and low-frequency noise is then further developed—it is shown that many properties of gate oxide defects can be directly extracted from BTI relaxation measurements in deeply-scaled devices [14] and the noise-inspired model capable to fully describe these properties is reviewed. Afterwards it is shown how the understanding of gate oxide defect properties can be used to explain variability of BTI in deeply-scaled technologies, as well as possible technological solutions for both nFETs and pFETs.

2 Brief Overview of BTI

BTI is a consequence of charging of defect states in the gate oxide and at its interface [2]. The defects could be both pre-existing or generated during device operation. The trapped charge results in a shift of the device parameters, such as its threshold voltage V_{th}, channel mobility, transconductance, and subthreshold slope, and generally a decrease of the FET's drive current. The name is derived from the phenomenon being strongly accelerated by *temperature* T and gate *bias* V_G. BTI in n-channel FET devices, which are typically biased in circuits at positive V_G, is referred to as Positive BTI (PBTI), while negative BTI (NBTI) takes place in p-channel FETs. Constant V_G stress bias is often referred to as static, or "DC" BTI, while periodically interrupted V_G stress is called "AC", or dynamic BTI.

3 Static BTI

Figure 2.1 illustrates the typical gradual shift of pFET threshold voltage ΔV_{th} during accelerated stress at elevated T [5]. The stress data are typically measured at several V_G's and ΔV_{th} is extrapolated to 10 years at the circuit operating voltage V_{DD} (or $V_{DD} + 10\ \%$). The extrapolated ΔV_{th} must be below a given value (typically 30 or 50 mV) for the technology to qualify.

This simple extrapolation procedure is, however, complicated by ΔV_{th} decreasing *immediately* after the stress bias is removed, as illustrated in Fig. 2.1 [15]. As will be discussed henceforth, this *recovery*, or *relaxation*, component R typically proceeds simultaneously on many time scales, making it difficult to determine its beginning or end and thus separating it from the final *non-recoverable*, or *permanent*, component P [2, 6]. This ΔV_{th} relaxation is thus a crucial problem for BTI measurement, interpretation, and extrapolation.

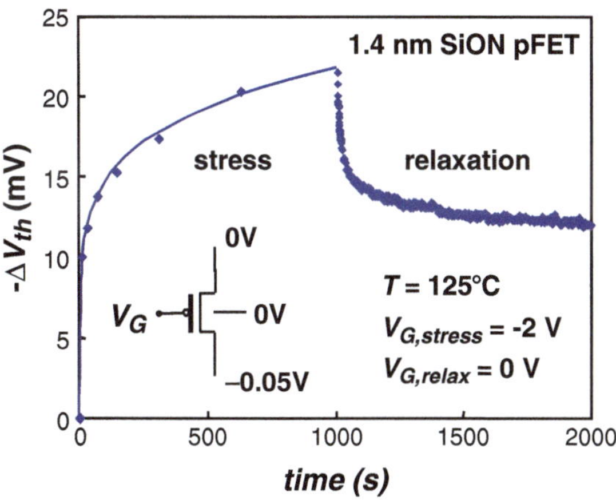

Fig. 2.1 Shift in pFET threshold voltage is observed during negative gate bias stress. When the stress bias is removed, a recovery of the effect is seen (note: $V_{th} \sim 0$ V for this device). Inset illustrates the biases applied at FET terminals during the BTI measurement

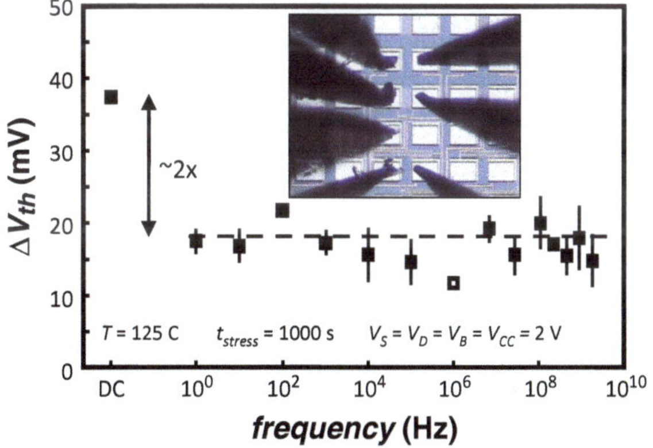

Fig. 2.2 The V_{th} shifts due to 50 % AC unipolar NBTI stress in pFETs are seen independent of frequency in the entire frequency range of 1 Hz–2 GHz. The V_{th} shift of the corresponding DC NBTI stress obtained on an identical device is shown for comparison. Inset: Micrograph of the *on-chip* circuit for the DC and AC BTI measurements consisting of a ring oscillator, a frequency divider, a buffer, a pass-gate-based multiplexer, and the device under test [17]

4 Dynamic BTI

In many CMOS applications, such as logic, the majority of the FETs are constantly switched and thus exposed to *dynamic* stress [16]. Figure 2.2 documents that NBTI is present at frequencies up to the GHz range, i.e., there does not appear to be any "cut-off" time constant of the degradation mechanism above ~1 ns [17]. Furthermore, the AC bias signal reduces BTI with respect to the DC stress. This provides some additional reliability margin, which can be factored in during the application design phase [9].

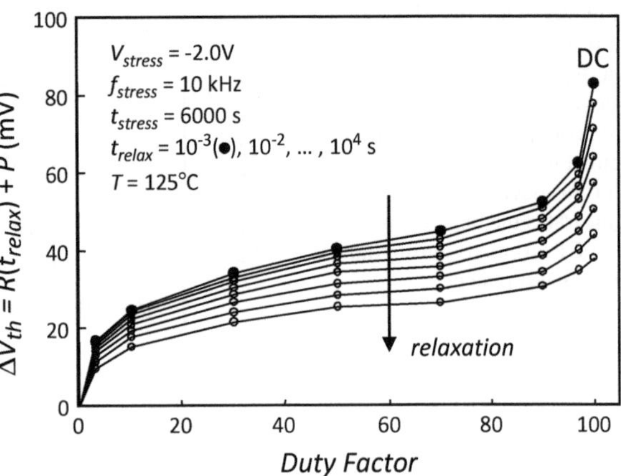

Fig. 2.3 Total degradation ΔV_{th} after 6000 s of unipolar NBTI stress shows a distinctive dependence on the duty factor *DF*. In particular, a weak dependence, or a "plateau" between ~10 and ~90 % is observed, complemented by rapid ΔV_{th} increase for the outermost *DF* values. Data at different relaxation times are shown

In an arbitrary FET of an arbitrary digital circuit, the average probability of a signal being high can vary between 0 and 100 %. The dependence of BTI on the *duty cycle* (called *duty factor* or *DF* here) thus needs to be studied. A NBTI ΔV_{th}-*DF* dependence with an inflection point around *DF* ~50 %, first reported in [17], is shown in Fig. 2.3 [12].

5 Similarity Between BTI Relaxation and Low-Frequency Noise

Long, log(t)-like behavior of ΔV_{th} without a characteristic time scale is typically observed in both the initial portion of NBTI degradation [13, 18] and the recovery phase. Figure 2.4 illustrates that the *rate of degradation* $\mathrm{d}\Delta V_{th}/\mathrm{d}t_{relax}$ [7] extracted from the log(t_{relax})-like ΔV_{th} NBTI relaxation transient after even a *very short*, 0.1 s stress, follows $1/t_{relax}$ for over seven decades. Such behavior is a signature of states with discharging time constants covering as many decades [19].

Incidentally, *superposition of states with widely distributed time scales* is the standard explanation of the 1/f noise spectra [20], which are clearly observed in our pFETs (Fig. 2.4). This obvious similarity leads us to argue that the same states with widely distributed time scales in fact play a fundamental role in both NBTI and noise measurements.

6 Semi-quantitative Model for BTI Relaxation

In order to visualize this common property it is beneficial to consider an equivalent circuit representing states with widely distributed time scales [19]. Note that in either NBTI relaxation or 1/f noise measurements, no maximum or minimum "cut-

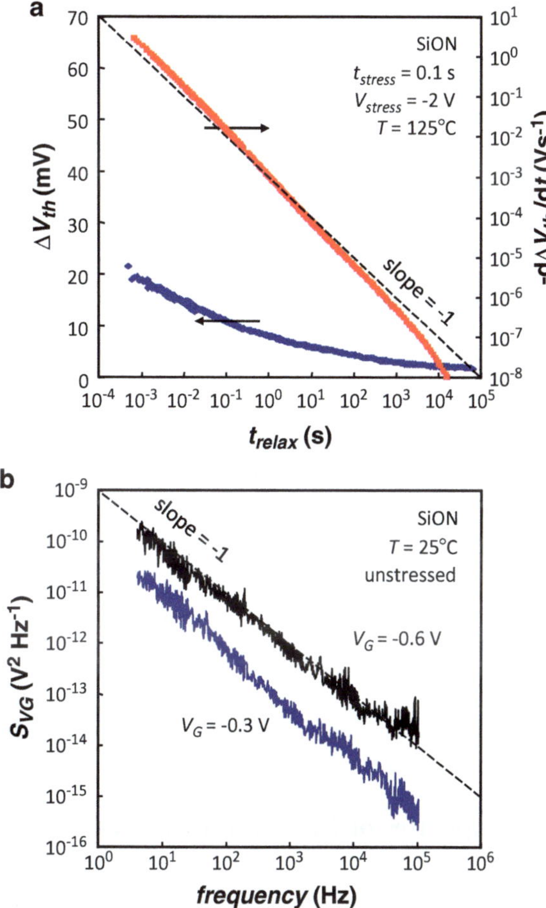

Fig. 2.4 (**a**) A characteristic long, log-like ΔV_{th} relaxation trace is observed after even short (pulse-like) NBTI stress. The *rate* of recovery $\mathrm{d}\Delta V_{th}/\mathrm{d}t_{relax}$ following $\sim 1/t_{relax}$ for $\sim$7 decades is a signature of states with discharging time constants covering as many decades. (**b**) Gate-referred noise spectra measured on the same (unstressed) devices show clear 1/f dependence, routinely explained by a superposition of states with widely distributed time scales

off" times are typically observed [12]. For the sake of simplicity it is therefore assumed here that the time constants are log-uniformly distributed from times much shorter than the switching time of a pFET to very long, corresponding to the lifetime of a CMOS application. Such states with widely distributed time scales are then represented by "RC" elements in Fig. 2.5 with the total FET ΔV_{th} being proportional to the sum of voltages ("occupancies") on all capacitors. For the sake of simplicity, it is assumed that all RC elements have the same weight and can be partially occupied, which emulates the behavior of a large-area device. Most properties of the recoverable component can be reproduced when the ohmic resistors in Fig. 2.5 are replaced with a *non-linear* component (simulated by two diodes with different parameters, see Fig. 2.5), which emulates different charging (i.e., capture) and discharging (i.e., emission) time constants of each defect [21]. Such a circuit correctly reproduces *DF* (Fig. 2.6, cf. Fig. 2.3) and also the log-like relaxation (Fig. 2.4a) and the log-like initial phase of stress (not shown) [19].

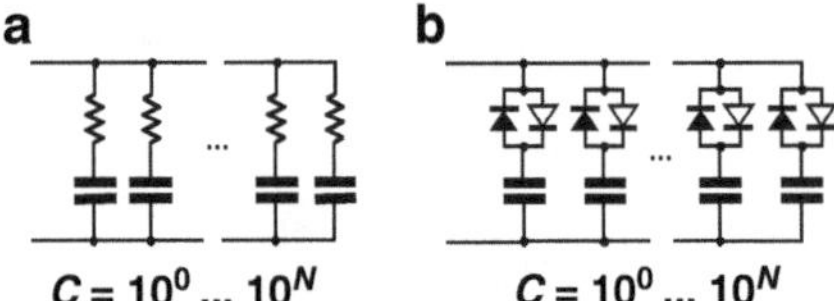

Fig. 2.5 (**a**) An equivalent circuit with exponentially increasing capacitances used to emulate defect states with widely distributed time scales, such as those active in low-frequency noise. (**b**) The same circuit modified to account for charging (i.e., capture) and discharging (i.e., emission) time constants being voltage dependent, represented by asymmetric diodes. The sum of voltages on capacitors is assumed to be proportional to FET ΔV_{th}

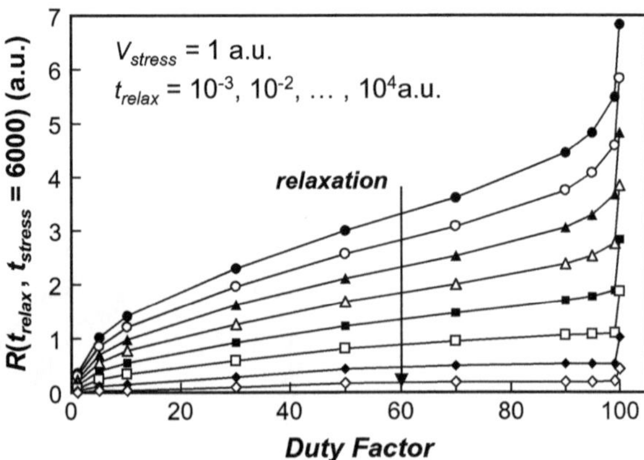

Fig. 2.6 The plateau in *DF* dependence of *R* is also qualitatively well reproduced by the equivalent circuit in Fig. 2.5, as is the decrease with increasing relaxation time (cf. Fig. 2.3, which, however, shows the sum of *R* and *P*)

7 Properties of Individual Defects

Figure 2.7 shows two typical ΔV_{th} relaxation transients following positive V_G stress on a single 70×90 nm^2 nMOSFET (i.e., corresponding to PBTI). Conversely to the continuous relaxation curves obtained on large devices, a quantized ΔV_{th} transient is observed in the deeply-scaled devices. In such devices, the relaxation is observed to proceed in discrete voltage steps, with each step corresponding to discharging of a *single* oxide defect [12, 22, 23]. Upon repeated perturbation, each defect shows up in the relaxation trace with a characteristic "fingerprint" consisting of its discharge, or *emission* time, and its voltage step [14].

Figure 2.8 shows the two-dimensional histogram of the heights and the emission times of the steps when the experiment was repeated 70 times at the same stressing and relaxing condition as in Fig. 2.7 [14]. In Fig. 2.8, four clusters are clearly formed that correspond to four active defects in the time window of the experimental setup.

The emission times of each defect are stochastically distributed and follow an exponential distribution. This allows us to determine the average emission time τ_e. The capture time of each trap can be obtained by varying the stress (i.e., charging) time from 240 ms down to 2 ms. The intensity of the cluster decreases with reducing

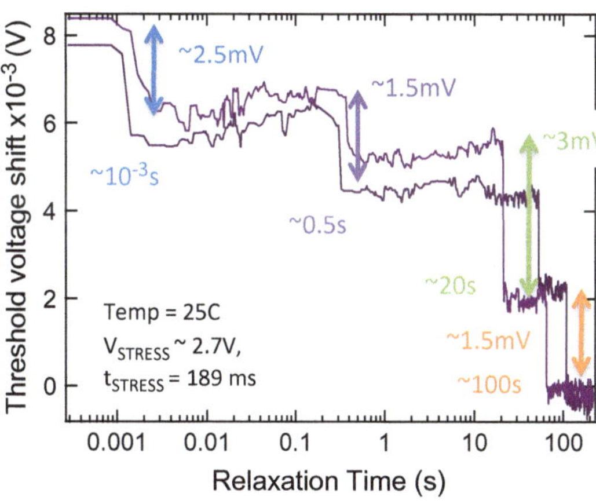

Fig. 2.7 Characteristic ΔV_{th} transients of a single $70 \times 90\ nm^2$ 1 nm-SiO_2/1.8 nm-HfSiO nMOSFET device stressed at 25 °C and $V_G = 2.8$ V for 184 ms. Four discrete drops are observed indicating the existence of four active traps at this stress condition

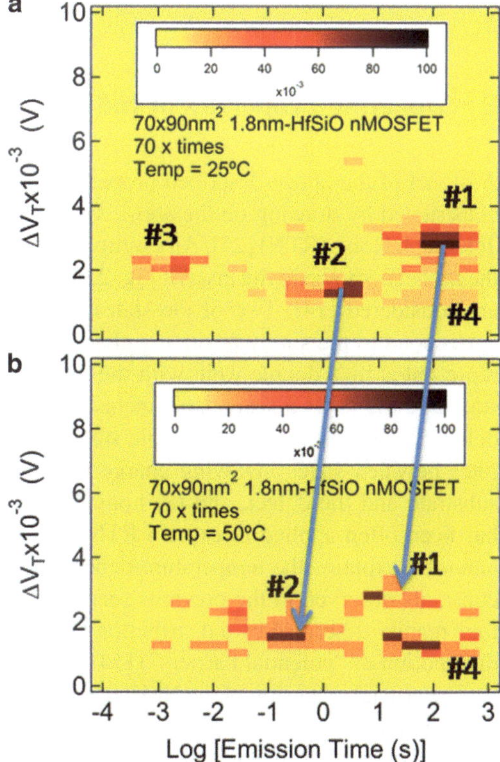

Fig. 2.8 Two-dimensional histograms (TDDS spectra) of the heights and emission times of the steps extracted from 70 ΔV_{th} transients of the particular device of Fig. 2.7 at (**a**) 25 °C and (**b**) 50 °C. Four clusters are formed that shift horizontally to shorter emission times with increasing temperature. Note that trap #3 disappears from the experimental window at 50 °C

stress time when the characteristic capture time is in the range of the stress time. The fit of the intensity to $P_c = 1 - \exp(-t_{stress}/\tau_c)$ lets us calculate the average capture time τ_c. This technique is known as Time Dependent Defect Spectroscopy (TDDS) [14].

In Fig. 2.8, an identical experiment was repeated at 50 °C on the same device. Note the large horizontal shift of the clusters to shorter emission times with only a 25 °C temperature increase. The Arrhenius plots of the emission and capture times obtained at T from 10 to 50 °C (not shown) provide activation energies of 0.48 eV for emission and 0.25 eV for capture. Similarly thermally activated capture and emission times are also observed in both nFET and pFET (i.e., corresponding to NBTI) with conventional SiO_2 gate oxide [14, 23, 24]. One can therefore conclude for all these cases that *both emission and capture in both electron and hole gate oxide traps are without a doubt thermally activated processes.* This experimental fact is incompatible with direct elastic tunneling theories widely used in different oxide trap characterization techniques and calculations. Consequently, a new model that takes into account this thermal dependence has to be considered.

8 Modeling Properties of Individual Defects

A model of the above-described properties of individual gate oxide defects can be constructed by drawing on the above similarities with low-frequency and Random Telegraph Noise (RTN) [25]. An example of the configuration coordinate diagram of the model is shown in the inset of Fig. 2.9. Four different configurations of the defect are considered [14]. Two of the states are electrically neutral while two of them correspond to the singly positively charged state. In each charge state the defect is represented by a double well, with the first of the two states being the equilibrium state and the other a secondary (metastable) minimum. The time dynamics of the defect can be described by a simple stochastic Markov process. Broadly, transition rates between states involving charge transfer assume (1) tunneling between the substrate and the defect, and (2) nonradiative multiphonon (NMP) theory, which has been often applied to explain RTN [26, 27]. Introduction of the NMP theory naturally explains the temperature dependence of both capture and emission time constants observed in the previous section. The wide distribution of time scales is then readily described by a distribution of the overlaps of the potential wells (i.e., a distribution of "potential barriers") [14].

The crucial extension of the NMP theory is the assumption of the relative position of the potential wells changing with gate bias [14], quite naturally introducing the required strong V_G dependence. As documented in Fig. 2.9, the model successfully describes the bias as well as the temperature dependences of the characteristic time constants. It is also noted that, contrary to techniques for the analysis of RTN, which only allow monitoring the defect behavior in a rather narrow time window, TDDS can be used to study the defects capture and emission times over an extremely wide range.

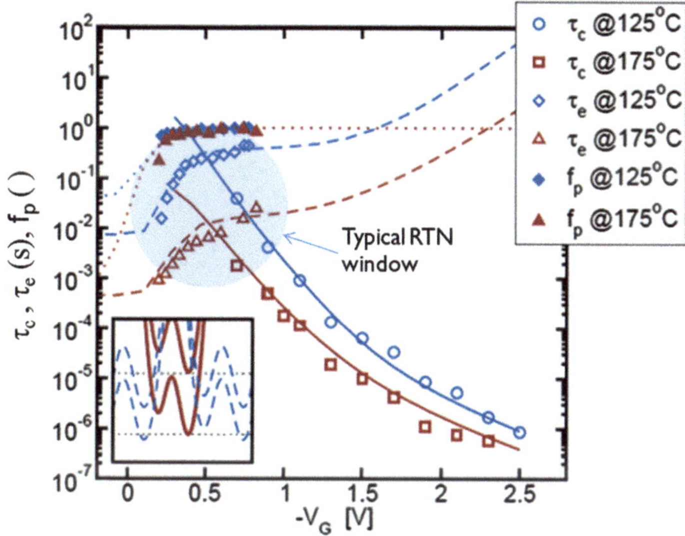

Fig. 2.9 Simulated capture and emission time constants (*lines*) compared with to the experimental TDDS values obtained on SiO_2 pFETs (*symbols*) during NBTI stressing at 125 and 175 °C and varying V_G. The experimental occupation probability of the charged state f_p is also indicated. The configuration coordinate diagram is shown in the inset (*dashed line*: neutral defect state; *solid line*: charged state potential)

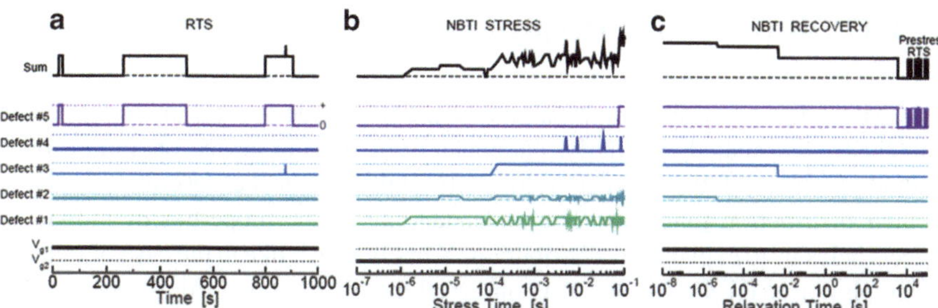

Fig. 2.10 Simulated RTN, stress, and recovery behavior of a nano-scale device using a stochastic solution algorithm of the proposed model. (**a**) At the threshold voltage ($Vg1$), the RTN is dominated by defect #5 with the occasional contribution from defect #3. Defects #1, #2, and #4 remain positively charged within the 'simulation/experimental' window. (**b**) During stress ($Vg2$), the capture times are dramatically reduced by the higher (more negative) gate voltage and the defects #3 and #5 become predominantly positively charged ($\tau c \ll \tau e$). Defects #1, #2, and #4 start producing RTN. (**c**) During recovery (back at $Vg1$), trapped charge is subsequently lost and the dynamic equilibrium behavior is gradually restored

It has been previously argued that the phenomenon called NBTI relaxation in pFET devices is in fact just a different facet of the well-known low-frequency noise in these devices. While the low-frequency noise corresponds to the channel/gate dielectrics system being in the state of dynamic equilibrium, NBTI relaxation corresponds to the perturbed system returning to this equilibrium [28]. Figure 2.10

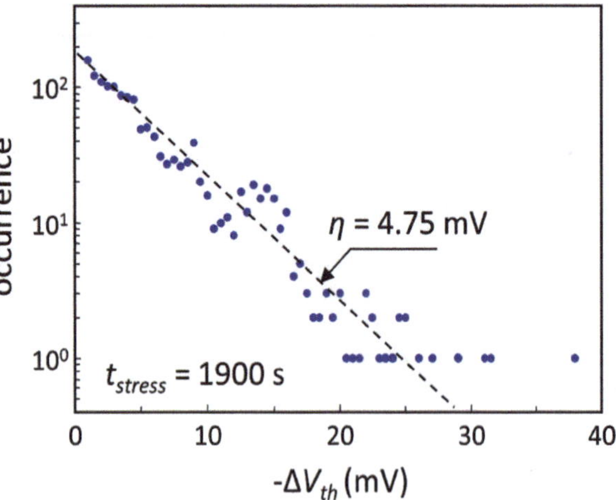

Fig. 2.11 Histogram of NBTI transient individual step heights measured on 72 devices shows a clear exponential distribution. The average V_{th} shift η corresponding to a single carrier discharge is 4.75 ± 0.30 mV in the pFETs with metallurgic length $L = 35$ nm, width $W = 90$ nm, and HfO_2 dielectrics with EOT $= 0.8$ nm

then illustrates this concept on a simulated example of a deeply scaled pFET containing only five active defects [23]. In particular it shows that the same defects can be responsible both for RTN as well as the NBTI relaxation and the (initial phase of) NBTI stress.

9 BTI Distribution in Deeply-Scaled FETs

As CMOS devices scale toward atomic dimensions, device parameters become statistically distributed. Similarly, parameter *shifts* during device operation, once studied in terms of the average value only, will have to be described in terms of their distribution functions. The understanding of the properties of individual defects helps us to explain this distribution. Namely, much like in the case of random telegraph noise (RTN) [29, 30], it is observed the distribution of down-steps ΔV_{th} due to *individual* discharging events to be *exponentially* distributed (Fig. 2.11). The exponential distribution of single-charge ΔV_{th} can be understood if non-uniformities in the pFET channel due to random dopant fluctuations (RDF) are considered [28–30]. A *single* discharging event in many devices routinely exceeded 15 mV, *and in several devices exceeded 30 mV*, the NBTI lifetime criterion presently used by some groups. For comparison, ΔV_{th} of less than 2 mV would be expected based on a simple charge sheet approximation. The large observed step height amplitude is due to the aggressively scaled dimensions of the pFETs used [28, 31].

Assuming the lateral locations of the trapped charges are uncorrelated, the overall ΔV_{th} distribution can be readily expressed as a convolution of individual exponential distributions [28, 31]. An actual population of stressed devices will

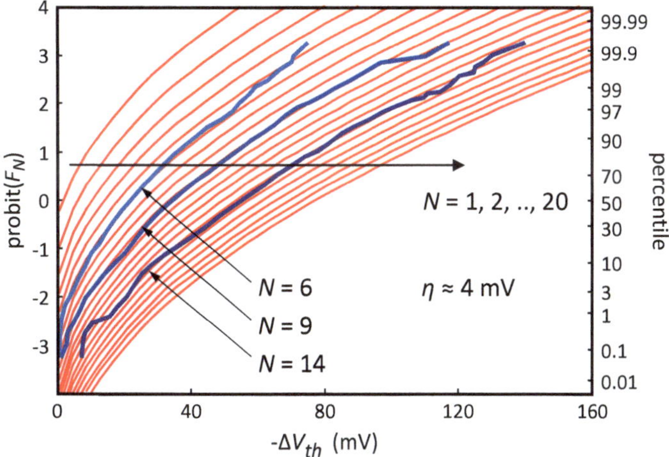

Fig. 2.12 Equation (2.1) in a probit plot rescaled to fit experimental distributions from Fig. 2.10 of Ref. [22], with the corresponding values of N and η readily extracted

consist of devices with a *different* number n of oxide defects in each device, with n being Poisson distributed [12, 22, 28]. The *total* ΔV_{th} distribution can be therefore obtained as

$$F_N\left(\Delta V_{th}, \eta\right) = \sum_{n=0}^{\infty} \frac{e^{-N} N^n}{n!} \left[1 - \frac{n\,\Gamma\left(n, \Delta V_{th}/n\right)}{n!}\right], \tag{2.1}$$

where N is the mean number of defects in the FET gate oxide and is related to the oxide trap (surface) density N_{ot} as $N = W\,L\,N_{ot}$. The CDF is plotted in Fig. 2.12 for several values of N. For comparison, measured total ΔV_{th} distributions for three different stress times from Ref. [22] are excellently fitted by the derived analytical description.

The advantage of describing the total ΔV_{th} distribution in terms of Eq. (2.1) is its relative simplicity and tangibility of the variables. The analytical description allows, among other things, to calculate NBTI threshold voltage shifts in an unlimited population of devices, a feat practically impossible through device simulations.

10 Technological Solutions

Once the underlying BTI mechanisms are understood, the defect properties can be modified to beneficial ends. Below, two possible technological solutions for both PBTI and NBTI are discussed.

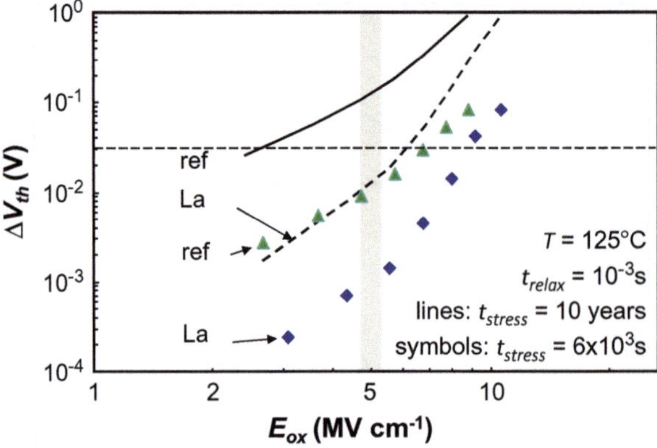

Fig. 2.13 A significant reduction of PBTI threshold voltage shift is observed in planar nFETs with La passivation ("La") over the reference stack ("ref") without passivation. Simplified power-law projection to 10 years shows passivated stack having sufficient reliability ($\Delta V_{th} < 30$ mV) at ~5 MV/cm operating field

11 Improving PBTI with Rare-Earth Incorporation

PBTI was considered a minor problem in technologies based on SiO_2. It arose as a reliability issue when high-k materials were incorporated into the gate stack. However, when rare earths were introduced to adjust the nFET initial threshold voltage, this issue was mitigated, as can be seen in Fig. 2.13. A significant reduction of PBTI is observed in planar nFETs with lanthanum with respect to a lanthanum-free reference [32].

Positive BTI in nFETs with high-k materials like HfO_2 has been linked to oxygen vacancies, which produce a defect level in the upper part of the oxide band gap. Group III elements compensate unpaired electrons around the oxygen vacancy in HfO_2 and the defects are "passivated" by being pushed up toward the conduction band minimum [33]. Such states are not easily accessible to nFET channel electrons, resulting in the significant reduction of negative charge capture in the stack and hence the reduction of PBTI.

12 Improving NBTI in High-Mobility SiGe pFETs

Reduction of gate stack EOT, which is one of the most efficient ways to improve FET performance, enhances NBTI due to increased oxide electric field. As a consequence, 10 year lifetime can be guaranteed for sub-1 nm EOT Si pFETs only at gate overdrive voltages far below the expected operating voltages (Fig. 2.14).

Another way to improve FET performance is the use of high-mobility substrates, such as buried-channel SiGe [34]. Because of the valence band offset between the

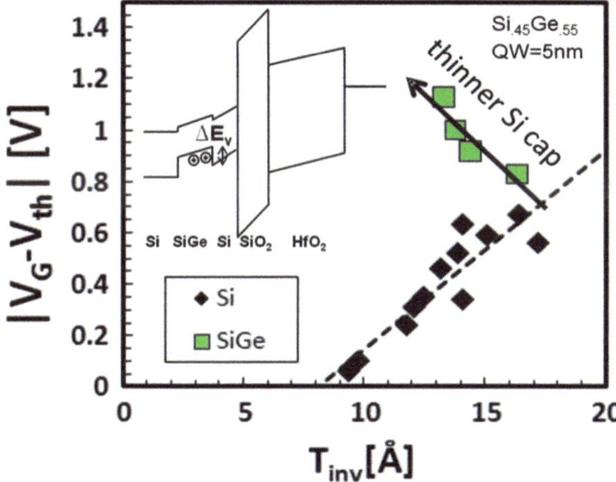

Fig. 2.14 Plot of the operating overdrive voltage $|V_G - V_{th}|$ for 10 year lifetime assuming a 30 mV threshold voltage shift criterion vs. the inversion capacitance equivalent thickness T_{inv}, for Si channel devices with different processing used as a reference, and for SiGe pFETs. For low T_{inv}, Si devices $|V_G - V_{th}|$ is below the expected operating voltage. In contrast to that, optimized SiGe devices show improved lifetime. *Inset*: gate-stack band diagram in inversion. Si cap acts as a barrier ΔE_V for holes

SiGe and the Si cap (see inset of Fig. 2.14), inversion channel holes are confined in the SiGe layer, which therefore acts as a quantum well (QW) for holes. The Si cap lowers the inversion capacitance as compared to the accumulation capacitance. For these devices it is therefore necessary to report the capacitance-equivalent thickness in inversion (T_{inv}, evaluated at $V_G = V_{th} - 0.6$ V) which will be affected by the thickness of the Si cap [35].

As can be seen from Fig. 2.14, SiGe-based device gate stacks significantly increase operating gate overdrive while still guaranteeing 10 year device lifetime and at the moment seem to be the only solution to the NBTI issue for sub-1 nm EOT devices. It has been recently observed that both increasing the Ge content in the channel as well as increasing the SiGe QW thickness reduces NBTI. Most intriguingly, a *reduction* of Si cap thickness also diminishes NBTI [35]. The most likely hypothesis explaining all three trends appears to be the energetic decoupling of the buried channel and the gate oxide defects [36].

Conclusions

In this chapter some of the shifts occurring in the past few years in our understanding of BTI were reviewed. Among the most significant ones is the shift from perceiving NBTI in terms of the Reaction–Diffusion model to analyzing BTI with the tools originally developed for describing low-frequency

(continued)

noise. This includes the interpretation of the time, temperature, voltage, and duty cycle dependences of BTI. It was further demonstrated that a wealth of information about defect properties can be obtained from deeply-scaled devices, and that this information can allow interpretation of variability issues of future deeply downscaled CMOS devices. This theme was complemented by showing the most promising technological solutions to alleviate both PBTI and NBTI.

References

1. J. H. Stathis and S. Zafar, Microelectronics Reliab. **46**, 270 (2006).
2. V. Huard, M. Denais, C. Parthasarathy, *Microelectronics Reliab.***46**, 1 (2006).
3. D. K. Schroder, Microelectronics Reliab. **47**, 841 (2007).
4. B. Kaczer, R. Degraeve, V. Arkhipov, N. Collaert, G. Groeseneken, M. Goodwin, as discussed at SISC, San Diego, CA, 2006.
5. B. Kaczer, V. Arkhipov, R. Degraeve, N. Collaert, G. Groeseneken, and M. Goodwin, *Proc. Int. Reliab. Phys. Symp.*, 381 (2005); *Appl. Phys. Lett.***86**, 143506 (2005).
6. T. Grasser, B. Kaczer, P. Hehenberger, W. Goes, R. O'Connor, H. Reisinger, W. Gustin, and C. Schlunder, *Int. Electron Devices Meeting Tech. Dig.*, 801 (2007).
7. A. Kerber, K. Maitra, A. Majumdar, M. Hargrove, R. J. Carter, and E. A. Cartier, *IEEE T. Electron Dev.***55**, 3175 (2008).
8. M. A. Alam, *Int. Electron Devices Meeting Tech. Dig.*, 345 (2003).
9. S. V. Kumar, C. H. Kim, S. S. Sapatnekar, *IEEE/ACM International Conference on Computer-Aided Design ICCAD'06*, 493 (2006).
10. T. Grasser, B. Kaczer, W. Goes, H. Reisinger, Th. Aichinger, Ph. Hehenberger, P.-J. Wagner, F. Schanovsky, J. Franco, Ph. Roussel, and M. Nelhiebel, to be presented at *Int. Electron Devices Meeting* 2010.
11. T. Grasser, B. Kaczer, and W. Goes, *Proc. Int. Reliab. Phys. Symp,* 28 (2008).
12. B. Kaczer, T. Grasser, J. Martin-Martinez, E. Simoen, M. Aoulaiche, Ph. J. Roussel, and G. Groeseneken, *Proc. Int. Rel. Phys. Symp.*, 55, 2009.
13. T. Grasser, B. Kaczer, W. Goes, Th. Aichinger, Ph. Hehenberger, and M. Nelhiebel, "A Two-Stage Model for Negative Bias Temperature Instability", *Proc. Int. Reliab. Phys. Symp.*, 2009.
14. T. Grasser, H. Reisinger, P.-J. Wagner, F. Schanovsky, W. Goes, and B. Kaczer, *Proc. Int. Rel. Phys. Symp.*, 16 (2010).
15. S. Rangan, N. Mielke, and E. C. C. Yeh, *Int. Electron Devices Meeting Tech. Digest*, 341 (2003).
16. G. Chen, K. Y. Chuah, M. F. Li, D. S. H. Chan, C. H. Ang, J. Z. Zheng, Y. Jin, and D. L. Kwong,*Proc. Int. Reliab. Phys. Symp.*, 196 (2003).
17. R. Fernández, B. Kaczer, A. Nackaerts, S. Demuynck, R. Rodríguez, M. Nafría, G. Groeseneken, *Int. Electron Devices Meeting Tech. Dig.*. 1 (2006).
18. H. Reisinger, O. Blank, W. Heinrigs, A. Mühlhoff, W. Gustin, and C. Schlünder, *Proc. Int. Reliab. Phys. Symp.*, 448 (2006).
19. B. Kaczer, T. Grasser, Ph. J. Rousse, J. Martin-Martinez, R. O'Connor, B. J. O'Sullivan, G. Groeseneken, *Proc. Int. Reliab. Phys. Symp.*, 20 (2008).
20. E. Milotti, arXiv:physics/0204033v1.
21. H. Reisinger, T. Grasser, W. Gustin, and C. Schlünder, *Proc. Int. Reliab. Phys. Symp.*, 1 (2010).

22. V. Huard, C. Parthasarathy, C. Guerin, T. Valentin, E. Pion, M. Mammasse, N. Planes and L. Camus, *Proc. Int. Reliab. Phys. Symp,* 289 (2008).
23. T. Grasser, H. Reisinger, W. Goes, Th. Aichinger, Ph. Hehenberger, P.-J. Wagner, M. Nelhiebel, J. Franco, and B. Kaczer, *Int. Electron Devices Meeting Tech. Dig.*, 729 (2009).
24. M. Toledano-Luque, B. Kaczer, Ph. Roussel, M.J. Cho, T. Grasser, and G. Groeseneken, *presented at WoDiM 2010.*
25. M. J. Uren, M. J. Kirton, and S. Collins, *Phys. Rev. B***37**, 8346 (1988).
26. M. J. Kirton and M. J. Uren, *Adv. Phys.***38**, 367 (1989).
27. A. Palma et al., *Phys. Rev. B***56**, 9565 (1997).
28. B. Kaczer, T. Grasser, Ph. J. Roussel, J. Franco, R. Degraeve, L.-A. Ragnarsson, E. Simoen, G. Groeseneken, H. Reisinger, *Proc. Int. Reliab. Phys. Symp.*, 26 (2010).
29. A. Asenov, R. Balasubramaniam, A. R. Brown, J. H. Davies, *IEEE T. Electron Dev.***50**, 839 (2003).
30. A. Ghetti, C. M. Compagnoni, A. S. Spinelli, A. Visconti, *IEEE T. Electron Dev.***56**, 1746 (2009).
31. B. Kaczer, Ph. J. Roussel, T. Grasser, and G. Groeseneken, *IEEE Electron Dev. Lett.***31**, 411 (2010).
32. B. Kaczer, A. Veloso, M. Aoulaiche, and G. Groeseneken, Microelectronic Engineering, **86**(7-9), 1894 (2009).
33. D. Liu and J. Robertson, *Appl. Phys. Lett.* 94, 042904 (2009).
34. N. Collaert, P. Verheyen, K. De Meyer, R. Loo and M. Caymax, *IEEE T. Nanotech***1**, 190 (2002).
35. J. Franco, B. Kaczer, M. Cho, G. Eneman, G. Groeseneken, and T. Grasser, *Proc. Int. Rel. Phys. Symp.*, 1082 (2010).
36. J. Franco, B. Kaczer, G. Eneman, J. Mitard, A. Stesmans, V. Afanas'ev, T. Kauerauf, Ph. J. Roussel, M. Toledano-Luque, M. Cho, R. Degraeve, T. Grasser, L.-Å. Ragnarsson, L. Witters, J. Tseng, S. Takeoka, W.-E. Wang, T.Y. Hoffmann, and G. Groeseneken, *Int. Electron Devices Meeting* 2010.

Chapter 3
Charge Trapping Phenomena in MOSFETS: From Noise to Bias Temperature Instability

Gilson Wirth and Roberto da Silva

Abstract Charge trapping phenomena is known to be a major reliability concern in modern MOSFETS, dominating low-frequency noise behavior and playing a significant role in aging effects such as Bias Temperature Instability (BTI). In this chapter we address this reliability issue.

We start by discussing MOSFET low-frequency (LF) noise, which is known to be dominated by charge capture and emission by defects (traps) close to the semiconductor–dielectric interface. Standard (LF) noise models used today (e.g. BSIM and PSP) do not properly model noise behavior under large signal excitation. A circuit level modeling and simulation approach, valid at both DC and large signal (AC) biasing, is presented.

The role of charge trapping and de-trapping in BTI (Bias Temperature Instability) is also discussed and modeled.

Mutual relation between the different reliability phenomena (LF noise, BTI and RDF) is also studied. For instance, random dopant fluctuations (RDF) may exacerbate the impact of BTI and LF noise on circuit performance.

1 Introduction

The major goal of this chapter is to compile and critically discuss recent work performed by the authors on modeling of charge trapping and de-trapping phenomena in nanometer scale CMOS devices [1–6, 16–18]. The role of charge trapping and de-trapping in both low-frequency noise and bias temperature instability (BTI) is discussed.

G. Wirth (✉)
Electrical Engineering Department, Universidade Federal do Rio Grande do Sul (UFRGS), Porto Alegre, Rio Grande do Sul, Brazil
e-mail: wirth@ece.ufrgs.br

R. da Silva
Physics Institute, UFRGS, Porto Alegre, Brazil

R. Reis et al. (eds.), *Circuit Design for Reliability*,
DOI 10.1007/978-1-4614-4078-9_3

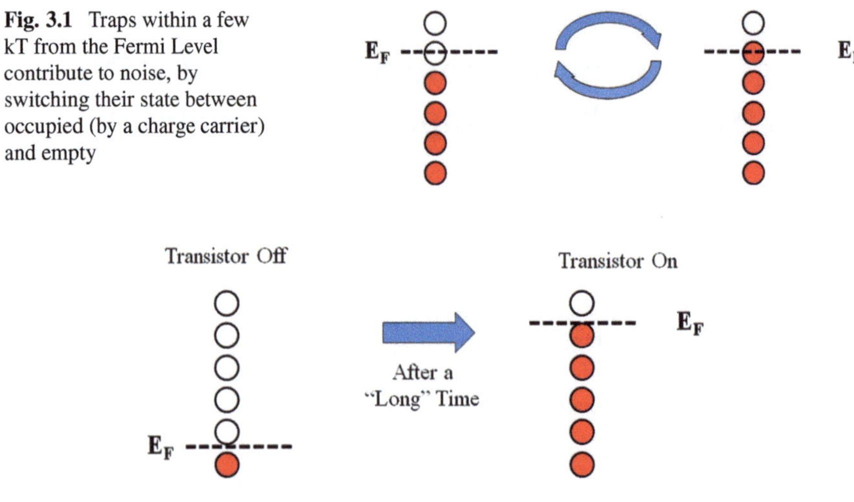

Fig. 3.1 Traps within a few kT from the Fermi Level contribute to noise, by switching their state between occupied (by a charge carrier) and empty

Fig. 3.2 If a transistor is abruptly turned on, the Fermi level abruptly changes and trap occupation probability follows the Fermi level (E_F) change. Charge trapping/de-trapping is not an instantaneous event. It is governed by characteristic time constants, and the trap occupation reflects the new Fermi level after an elapsed time. This figure depicts the situation for electron trapping in the channel of a NMOS transistor

Charge trapping and de-trapping at localized states (charge traps) at the interface or in the gate dielectric is a significant reliability issue for CMOS applications. It is known to be a source of low-frequency noise, besides playing a role in Bias Temperature Instability (BTI).

Traps that contribute to noise are the ones that keep switching their state between occupied and empty, as depicted in Fig. 3.1. These are the traps with occupation probability close to 50 %, which means that their capture and emission times are similar.

Traps that contribute to BTI are the ones that have a high probability to stay occupied after a charge trapping event, as depicted in Fig. 3.2. These are the traps with occupation probability close to 100 %, which means that their capture time is much shorter than the emission time.

Figure 3.3 depicts the mechanisms leading to BTI and low-frequency noise.

If a constant bias is applied to a Metal–Insulator–Semiconductor system, only the localized states (traps) with energy close to the Fermi level show significant activity, i.e., change their states between occupied and empty, by capture and subsequent emission of charge carriers from the channel region. This may be considered a steady state condition, where the *occupation probability* of a trap is time independent, i.e., a constant that does not change with time. This mechanism is known to originate low-frequency noise.

However, if the bias point changes abruptly, as is common in e.g. digital CMOS applications, the trap occupation probability abruptly changes. For instance, if the

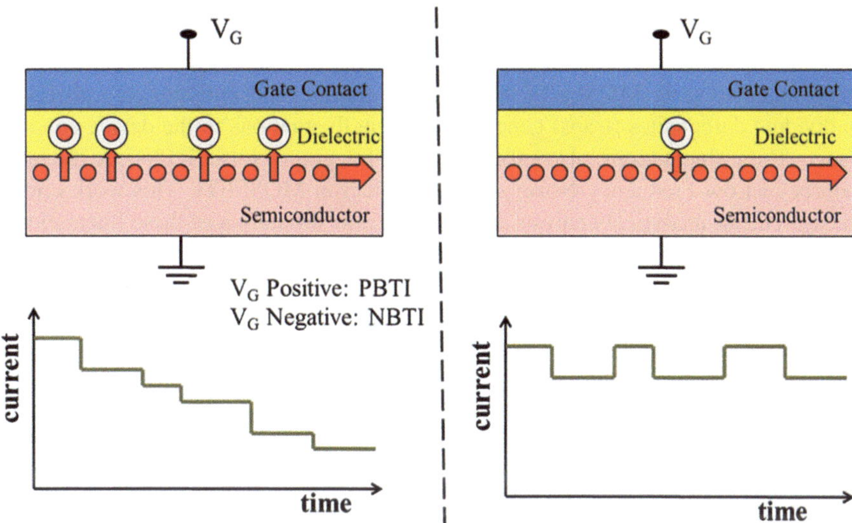

Fig. 3.3 (*Left hand side*) Traps that contribute to NBTI are the ones that stay occupied after an capture event occurs. This leads to a degradation of transistor on current over time. (*Right hand side*) Traps that contribute to noise are the ones that keep switching their state over time, exchanging charge carriers with the channel inversion layer. This leads to Random Telegraph Noise in the device current

gate voltage abruptly changes the Fermi level in such a way that trap occupation probability is increased, the rate at which charge carries are captured abruptly becomes larger than the rate at which carriers are emitted, and the number of trapped charge increases over time. Traps change their occupation state according to their characteristic time constant, meaning that the number of trapped charge does not instantaneously reflect the new occupation probability. The faster traps (with shorter capture time constants) become filled first, while the slowest traps take longer to become filled.

Each trap that becomes occupied degrades the channel conductivity, decreasing the device current. For nanometer scale devices, this current decrease is seen to occur in discrete steps, each step being related to the capture of a single channel carrier (in large area device it becomes hard to clearly identify the discrete steps). Since the dynamics of this occupation depends on the bias point and temperature, it may lead to bias temperature instability (BTI).

The first sections of this chapter cover the modeling of low-frequency noise, while the last section covers the modeling of BTI.

2 Charge Trapping Events as a Source of Noise

In nanometer scale MOSFETs the alternate capture and emission of carriers at individual defect sites (traps) generates discrete fluctuations in the device conductance. These fluctuations, also called Random Telegraph Noise (RTN), are the main source of Low-Frequency Noise in deep-submicron MOSFETs. This work covers measurement, analytical analysis and Monte Carlo simulation of these fluctuations.

The low-frequency noise model is based on device physics parameters which cause statistical variation in low-frequency noise behavior of individual devices. It includes detailed consideration of statistical effects for distribution of number of traps per device, the trap energy distribution, trap location and its device bias dependent noise contributions [1, 2]. Microscopic discrete quantities are used in model derivation, and analytical equations for the statistical parameters are provided.

In many practical applications the MOS device is not biased at steady state, but periodically switched. This operation regime is called cyclo-stationary excitation. The modeling approach adopted allows analysis of noise both at steady state operation as well as under cyclo-stationary excitation. Noise behavior under cyclo-stationary excitation is the main focus of this work.

First the autocorrelation of the RTN signal is calculated, and then the Wiener–Khinchin formula applied, leading to an analytical formulation for the RTN spectrum due to a single trap. After the evaluation of the power spectrum due to a single trap, the noise behavior resulting from the combined effect of all traps found in a device is derived. The modeling approach is valid for steady state as well as for any periodic excitation signal, and allows the derivation of relevant statistical parameters. Square wave excitation is used as a case study, to explore noise behavior in detail.

It is shown that since RTN-amplitude depends on the bias point strong variations of noise performance may appear not only between devices, but also for a single device operated under different bias conditions.

If the trap energy distribution in the band-gap is a convex curve (e.g. “U”-shaped), the model here presented yields a reduction in LF-noise under cyclo-stationary excitation when the device is biased between strong and weak inversion or even slight accumulation, which is in agreement to our experimental results [3, 7], and experimental results found in the literature [8–11]. However, while the average noise power is seen to decrease, the variability (normalized standard deviation) of noise power increases, as shown by the mathematical derivation, and experimental observations provided here.

3 Power Spectrum of the RTN Noise due to a Single Trap

The origin of RTN noise is the alternate capture and emission of charge carriers at discrete trap levels near the Si–SiO_2 interface. Figure 3.4 depicts the cross section of an n-channel MOSFET through the location of the interface trap. The influence of the traps on the electrical current flowing through the channel is twofold. On the one hand, the occupation of a trap changes the number of free carriers in the inversion layer. On the other hand, a charged trap state has an influence on the local mobility near to its position due to Coulomb scattering. If the MOSFET biasing is kept constant, a stationary RTN is observed at the terminals of the device as a discrete fluctuation in electrical current, δI_d being the amplitude of the current fluctuation, as shown in the inset of Fig. 3.5. The average high current time corresponds to the electron capture time constant (τ_c), and the average low current time corresponds to the emission time constant (τ_e).

The power spectrum of a RTN fluctuation can be evaluated by calculating the auto-covariance of the signal, and then applying the Wiener–Khinchin formula [12]. The power spectrum is then the Fourier transform of the auto-covariance, and is a Lorentzian, as given by Eq. (3.1) and depicted in Fig. 3.5.

$$S(\omega) = \frac{\delta^2}{\pi} \cdot \frac{\beta}{(1+\beta)^2} \cdot \frac{1}{\omega_0} \cdot \frac{1}{1+(\omega/\omega_0)^2} \tag{3.1}$$

where δ is the RTN amplitude, ω_0 is the angular corner frequency and β is equal to τ_c/τ_e. The amplitude of the RTN fluctuation induced by a trap depends on trap position along the channel. It may be exacerbated by factors such as random dopant fluctuations (RDF). For nanometer scale channel lengths (L), traps near the source end of the channel can cause significant mobility fluctuations and surface potential

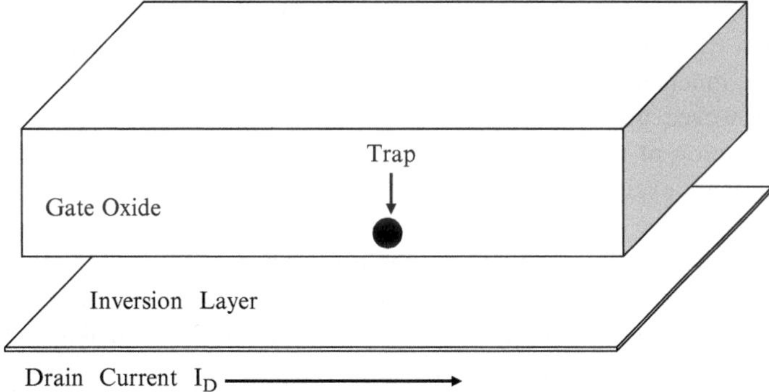

Fig. 3.4 Schematic cross section of the inversion layer of a MOS transistor through the location of an interface trap. If the trap is electrically charged the inversion layer is disturbed by the trap, affecting the drain current I_D. The trap not only affects the number of free carriers in the inversion layer but is also a source of electrical charge carrier scattering

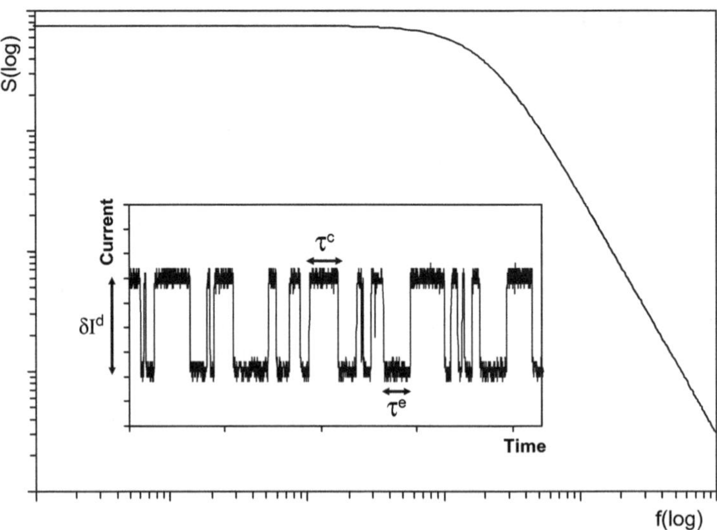

Fig. 3.5 Time and frequency domain representation of a stationary random telegraph noise (RTN). In frequency domain, the power spectrum of a RTN is a Lorentzian. In time domain discrete fluctuations are observed in the drain current, where τ_c is the average time in the high current state, which corresponds to the state where the trap is electrically neutral (empty). τ_e is the average time in the low current state, which corresponds to the state where the trap is electrically charged. δI_d is the amplitude of the current fluctuation

fluctuations, hindering electrical current flow and enhancing RTN amplitude (δ). A trap positioned near the source junction may be able to create a repulsive Coulomb blockade well surrounding the trap, hindering trap injection from the source into the channel. Thus, a significant mobility fluctuation will be added to carrier number fluctuation for carrier electrons trapped near the source side [17]. Another parameter that may impact RTN amplitude is trap position into the dielectric.

Equation (3.1) assumes that the capture and emission time constants (τ_c and τ_e) are time independent, constant values. This is true for constant (time invariant) biasing. However, under cyclo-stationary excitation the applied bias voltage is a periodic function of time, and $\tau_c(t)$ and $\tau_e(t)$ become periodic functions of time. Cyclo-stationary excitation is depicted in Fig. 3.6.

In order to derive the power spectrum under cyclo-stationary excitation, we did follow the methodology originally proposed by Machlup [12] for stationary RTN. A RTN is considered to be a purely random signal, which may be in one of two states, called 1 and 0. If the signal is in state 1, the probability of making a transition to 0 in a short time dt is assumed to be $dt/\tau_c(t)$. If the signal is in state 0, the probability of making a transition to 1 is assumed to be $dt/\tau_e(t)$. In this form, the state 1 is related to the empty trap, i.e., high current state in Fig. 3.5, while the state 0 is related to the occupied trap, i.e., low current state.

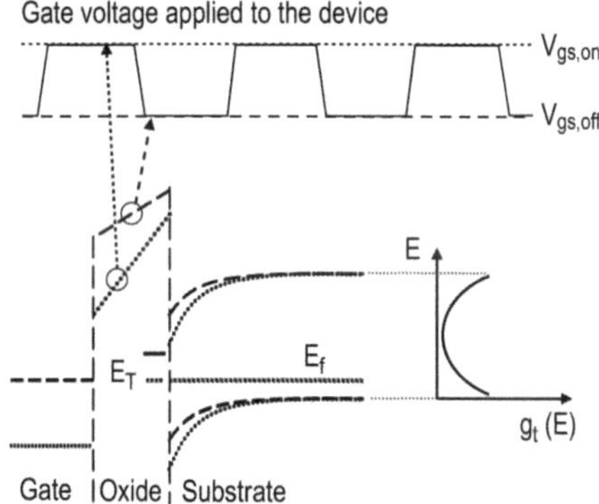

Fig. 3.6 Energy band diagram of MOSFET with noise relevant traps during two different phases (*dashed* and *dotted*) of square wave gate voltage biasing as shown on top of the figure. Note that traps close to Fermi level both in '*on*' and '*off*' state can contribute to the noise. $g_t(E)$ shows a U-shaped trap density

In order to derive the low frequency noise spectrum of cyclo-stationary RTN, we first calculate the autocorrelation of the RTN, and then apply the Wiener–Khinchin formula to obtain the spectrum. Let the RTN signal be *x*(*t*).

The autocorrelation is then given by

$$A(s) = < x(t) \cdot x(t+s) >_{\text{average}} = P(x(t) = 1) \cdot P_{11}(s) \tag{3.2}$$

where $P_{11}(s)$ is the probability of an even number of transitions in time *s*, given we start in state 1. $P(x(t) = 1)$, the probability of being in state 1 at time *t*, is given by

$$P(x(t) = 1) = \frac{\frac{1}{T}\int_0^T \frac{1}{\tau_e(t)} dt}{\frac{1}{T}\int_0^T \frac{1}{\tau_e(t)} dt + \frac{1}{T}\int_0^T \frac{1}{\tau_c(t)} dt} = \frac{<\frac{1}{\tau_e(t)}>}{<\frac{1}{\tau_e(t)}> + <\frac{1}{\tau_c(t)}>} \tag{3.3}$$

where *T* is the period of the cyclo-stationary excitation, and symbol <•> is an abbreviation for $(1/T) \int_0^T \bullet dt$, the time average value.

The autocorrelation can then be calculated as

$$A(s) = \frac{<\frac{1}{\tau_e(t)}>}{<\frac{1}{\tau_e(t)}> + <\frac{1}{\tau_c(t)}>} \cdot \left(1 + \int_0^s e^{\int_0^x \left(<\frac{1}{\tau_c(y)}> + <\frac{1}{\tau_e(y)}>\right) dy} \frac{1}{\tau_e(x)} dx\right) e^{-\int_0^s \left(<\frac{1}{\tau_c(y)}> + <\frac{1}{\tau_e(y)}>\right) dy} \tag{3.4}$$

This formulation for the autocorrelation is a generalization of the Machlup formula, and is valid for any kind of periodic excitation and any frequency. If τ_c and τ_e become constant (independent of time), Eq. (3.4) becomes equal to (7) in [12].

4 Approximation for Excitation Frequencies Higher than the Noise Frequency

The case of interest for most practical applications is for the noise at frequencies below the frequency of the cyclo-stationary excitation signal.

If the probability of a trap to switch state during one period T of the cyclo-stationary excitation signal is very small, a simplification may be done in the calculation of the autocorrelation. This case corresponds to the limit where $\tau_c(t)$ and $\tau_e(t)$ are much larger than the period T, leading to small transition probabilities $T/\tau_c(t)$ and $T/\tau_e(t)$.

In this case we can write, without loss of generality, that $s = nT$, where n is a positive integer (1, 2, 3, …). In this situation we have

$$\int_0^{s}\left(\frac{1}{\tau_e(y)}+\frac{1}{\tau_c(y)}\right)dy=\sum_{i=0}^{n-1}\int_{iT}^{(i+1)T}\frac{1}{\tau_e(y)}dy+\sum_{i=0}^{n-1}\int_{iT}^{(i+1)T}\frac{1}{\tau_c(y)}dy = nT\left(<\frac{1}{\tau_e(t)}>+<\frac{1}{\tau_c(t)}>\right) \tag{3.5}$$

leading to

$$A(s)=\frac{<\frac{1}{\tau_e(t)}>}{<\frac{1}{\tau_e(t)}>+<\frac{1}{\tau_c(t)}>}\left[1-\frac{T<\frac{1}{\tau_e(t)}>}{e^{T\left(<\frac{1}{\tau_e(t)}>+<\frac{1}{\tau_c(t)}>\right)}-1}\right]e^{-s\left(<\frac{1}{\tau_e(t)}>+<\frac{1}{\tau_c(t)}>\right)}+ + \frac{<\frac{1}{\tau_e(t)}>}{<\frac{1}{\tau_e(t)}>+<\frac{1}{\tau_c(t)}>}\,\frac{T<\frac{1}{\tau_e(t)}>}{e^{T\left(<\frac{1}{\tau_e(t)}>+<\frac{1}{\tau_c(t)}>\right)}-1} \tag{3.6}$$

For small values of T a Taylor expansion may be employed, leading to

$$A(s)=\frac{<\frac{1}{\tau_e(t)}>}{<\frac{1}{\tau_e(t)}>+<\frac{1}{\tau_c(t)}>}\left[1-\frac{<\frac{1}{\tau_e(t)}>}{<\frac{1}{\tau_e(t)}>+<\frac{1}{\tau_c(t)}>}\right]e^{-s\left(<\frac{1}{\tau_e(t)}>+<\frac{1}{\tau_c(t)}>\right)}\ldots+ + \frac{<\frac{1}{\tau_e(t)}>^2}{\left(<\frac{1}{\tau_e(t)}>+<\frac{1}{\tau_c(t)}>\right)^2} \tag{3.7}$$

This means that, if $\tau_c(t)$ and $\tau_e(t)$ are much larger than the period T of the excitation signal, the values of $1/\tau_c(t)$ and $1/\tau_e(t)$ in the integrals of (3.4) are equivalent to their time averages $<1/\tau_c(t)>$ and $<1/\tau_e(t)>$.

The power spectrum $S_i(\omega)$, due to a single trap (the i-th trap), is the calculated as the Fourier transform of the autocorrelation, leading to

$$S_i(\omega)=\frac{1}{2\pi}\int_{-\infty}^{\infty}-A(s)e^{i\omega s}ds \tag{3.8}$$

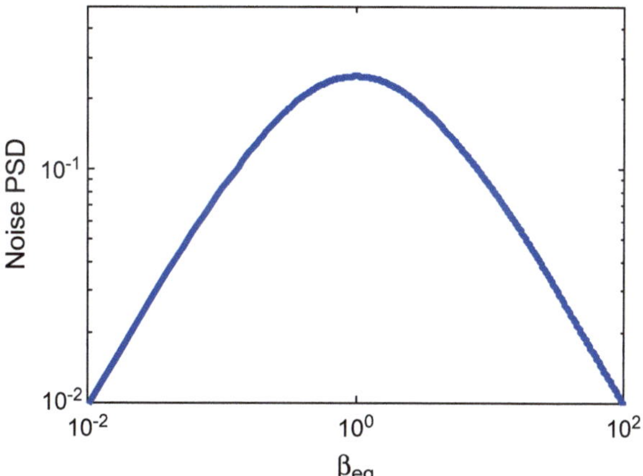

Fig. 3.7 Noise Power Spectral Density as a function of β_{eq}. For $\beta_{eq} = 1$, i.e., $<1/\tau_e> = <1/\tau_c>$, the noise power reaches its maximum

which is evaluated as being

$$S_i(\omega) = \frac{\delta_i^2}{\pi} \cdot \frac{\beta_{eq}}{\left(1+\beta_{eq}\right)^2} \cdot \frac{1}{\omega_i} \cdot \frac{1}{1+(\omega/\omega_i)^2} \tag{3.9}$$

Here δ_i determines the trap's voltage amplitude and ω_i the angular frequency, and

$$\beta_{eq} = <1/\tau_e(t)>/<1/\tau_c(t)> \tag{3.10}$$

Figure 3.7 depicts the behavior of noise power as a function of β_{eq}.

The cyclo-stationary noise spectrum is still Lorentzian, with angular corner frequency ω_i given by

$$\omega_i = <\frac{1}{\tau_c(t)}> + <\frac{1}{\tau_e(t)}> \tag{3.11}$$

This takes us to the conclusion that making a RTN signal cyclo-stationary leads to a Lorentzian spectrum with corner frequency equal to the sum of the inverse time average values of the capture and emission times. Please note that for stationary RTN the corner frequency is equal to the sum of the inverse values of the constant capture and emission times. Please refer to equation 9 in [12].

The result for this limit is valid for any kind of periodic excitation. This is the limit studied in [11, 13]. However, for this limit, we obtain the same result in a much simpler derivation than in [11, 13], and without making any further assumption or simplification. The single assumption is that the transition probabilities $dt/\tau_e(t)$ and $dt/\tau_c(t)$ are much smaller than the excitation period T.

Equations (3.11) and (3.9) are a generalization of the Machlup formulation for cyclo-stationary RTN with excitation frequency higher than the noise frequency.

The Machlup equations for the auto-correlation and power spectrum are recovered if we consider $<1/\tau_e(t)> = 1/\tau_e$ and $<1/\tau_c(t)> = 1/\tau_c$, i.e., constant, not time dependent values.

5 Average Power Spectrum of the RTN Noise due to the Ensemble of Traps

The noise behavior of a device results from the combined effect of all traps found in the device. The noise power spectrum may then be written as the summation of the contribution of each one of the N_{tr} traps found in the device. The observation that traps are Poisson distributed results in an average noise spectral density given by

$$<S> = \sum_{N_{tr}=0}^{\infty} \sum_{i=1}^{N_{tr}} <S_i> \frac{e^{-N} N_{tr}^{N}}{N_{tr}!} = N <S_i> \tag{3.12}$$

Here $<S_i>$ is the average noise contribution of a trap, N_{tr} is actual number of traps in a particular device, and N is average number of traps in an ensemble of devices.

As depicted in Fig. 3.8, if the corner frequencies of the Lorentzians corresponding to power spectral density (PSD) of different traps are equally spaced on a log scale, the summation of the power spectrum due to all traps leads to 1/f noise. In time domain, this corresponds to the situation where the charge trap characteristic time constants are uniformly distributed on a log scale.

The equations above are valid for any waveform of the periodic excitation signal. In order to explore the noise behavior in detail and allow comparison to experimental results, a case of particular interest for practical applications will be studied. It is square wave excitation.

6 Square Wave Excitation

Square wave excitation is chosen as case study, because of its interest in practical applications and because of the availability of experimental data. Comparison of model results to relevant experimental data is performed in the last section of this work.

Under square wave excitation, the bias voltage abruptly alternates between two states, called *on* and *off*. Please note that the state names *on* and *off* do not imply that the device has necessarily to be periodically turned on and off. The names refer to two distinct states, with different gate bias. With periodically changing gate bias, the Fermi level becomes a periodic function of time $E_F(t)$. This implies that the Fermi level alternates between two levels: E_{on} being the Fermi level during the *on* state and E_{off} being the Fermi level during the *off* state. The duty cycle α is the fraction

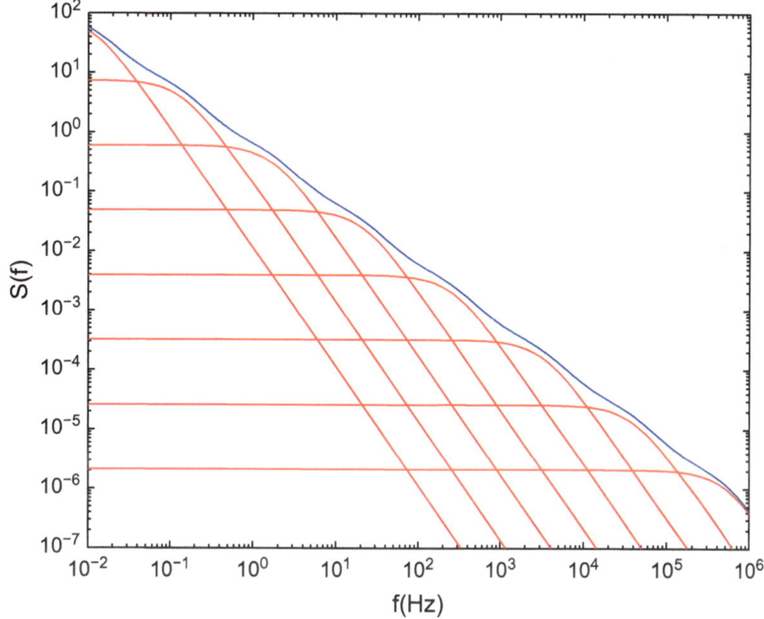

Fig. 3.8 If the corner frequencies of the Lorentzians (*red lines*) corresponding to different traps are equally spaced on a log scale, the summation of the power spectrum due to all traps leads to 1/f noise (*blue line*)

of the period T in which the device is in the *on* state. The capture and emission time constants of a trap are affected by the Fermi level.

For square wave cyclo-stationary excitation with duty cycle α, the time averaged capture and emission time constants may be written as

$$<1/\tau_c> = \left(\alpha/\tau_{c,on} + (1-\alpha)/\tau_{c,off}\right) \tag{3.13}$$

$$<1/\tau_e> = \left(\alpha/\tau_{e,on} + (1-\alpha)/\tau_{e,off}\right) \tag{3.14}$$

$$\beta_{eq} = \psi\left(E_{on}, E_{off}, \alpha\right) e^{2E_t/k_BT} \tag{3.15}$$

with

$$\psi\left(E_{on}, E_{off}, \alpha\right) = \frac{\alpha e^{-E_{on}/k_BT} + (1-\alpha)\, e^{-E_{off}/k_BT}}{\alpha e^{E_{on}/k_BT} + (1-\alpha)\, e^{E_{off}/k_BT}} \tag{3.16}$$

Assuming statistical independence of the random variables, the average noise of a transistor <*S*> is then given by:

$$<S> = \frac{N_{dec} < \delta^2 >}{\pi W L \omega} \int_{E_v}^{E_c} \frac{\psi e^{2E_t/k_B T}}{\left(1 + \psi e^{2E_t/k_B T}\right)^2} g\left(E_t\right) dE_t \qquad (3.17)$$

where W is the channel width and L the channel length of the transistor, and N_{dec} is the trap density per unit area and frequency decade. Please see (3.1) for a detailed explanation of these parameters. This equation again looks similar to the DC noise behavior (see [1]). In case the cyclo-stationary period T is short compared to the trap's time constants their time averages and the trap numbers close to the two Fermi levels determine the noise level. At frequencies significantly higher than the excitation frequency the noise is given by DC theory.

Equation (3.17) above clearly relates the noise reduction to the distribution of traps over energy. From the above equation, it follows that if the distribution of traps in energy over the bandgap is uniform, <S> is expected to remain approximately constant under cyclo-stationary excitation. In this case, the peak of the noise contribution of a trap always probes the same trap density, since the trap density around to the Fermi level is always the same. Note that for the peak of the noise contribution of a trap occurs for $E_T = E_F$. If the trap density is uniform over the bandgap, $g(E_t)$ corresponding to the peak of β_{eq} is always the same.

Monte Carlo (MC) simulations are performed in time domain. The RTN of each trap is evaluated, and then the power spectral density is calculated. The Monte Carlo simulation show very good agreement to the analytical results, as well as to our experimental results, as shown in the last section of this work.

In Fig. 3.9 the results of Eq. 3.17 show noise reduction as a function of the Fermi levels in the 'on' and 'off' state at frequencies lower than the excitation period.

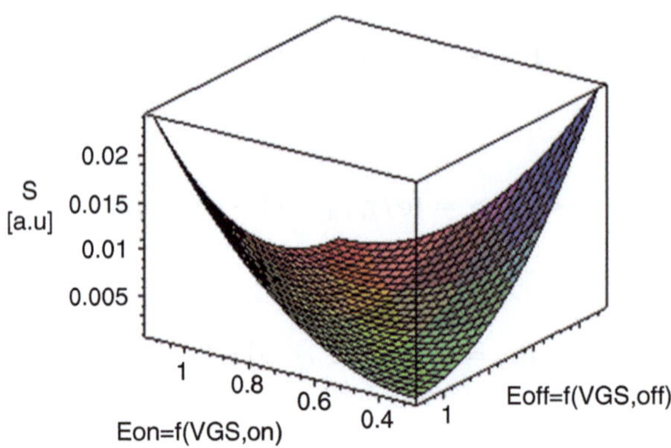

Fig. 3.9 Noise reduction as a function of the Fermi level in the 'on' and 'off' states. The noise power S is evaluated according to Eq. (3.17), considering the U-shaped trap density given by Eq. (3.18). The same parameter $a = 11$ is used in all figures and evaluations performed in this work. Note that the noise reduction is larger if the biasing levels are symmetrical in relation to the center of the U-shaped trap density

It can be seen that a higher noise reduction can be expected if the biasing levels are symmetrical in relation to the center of the U-shaped trap density. In this case, the traps that contribute most to the noise power are the ones close to the center of the bandgap, where the density is lowest.

In good agreement to the works in [11, 13], the trap energy distribution $g(E_t)$ is key for explaining experimentally observed findings. A parabolic U-shape trap density function is assumed here:

$$g\left(E_t\right) = aE_t^2 - a\left(E_c - E_v\right)E_t + k \tag{3.18}$$

where a is a fitting parameter. The integral of $g(E_t)$ from E_v to E_c is normalized to one with

$$k = \frac{1}{E_c - E_v}\left(\frac{a}{6}\left(E_c^3 - E_v^3\right) + \frac{a}{2}\left(E_c^2 E_v - E_v^2 E_c\right) + 1\right) \tag{3.19}$$

Figure 3.10 depicts a U-shape trap density.

7 Variability in the Power Spectrum of the RTN Noise due to the Ensemble of Traps

Since the microscopic approach maintains the statistically relevant parameters, the model proposed here also allows the modeling of the statistically relevant parameters. As derived in [2], the normalized standard deviation of noise performance under constant biasing is

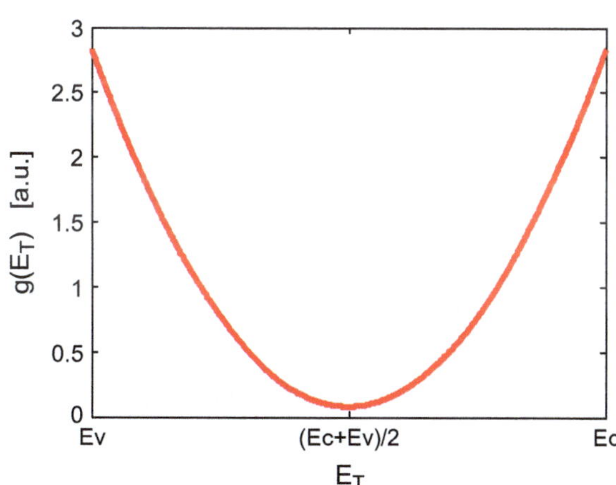

Fig. 3.10 U-shaped trap density. The trap density is higher close to the valance (E_V) and conduction (E_C) bands, and lower in the center of the bangap

$$\frac{\sigma_{np}}{<np_{BW}>} = \frac{2}{\pi} \frac{1}{\sqrt{N_{dec}WL}} \sqrt{\frac{<A^4>}{<A^2>^2}} \frac{b}{\left(\frac{f_H}{f_L}\right)^c} \tag{3.20}$$

Here f_H and f_L are the lower and upper boundaries of the bandwidth of interest in a given circuit design, respectively, and b and c are constants, with $b = 0.74$ and $c = 0.05$ [2].

A similar formulation can also be derived for cyclo-stationary operation. The standard deviation of noise power is given by:

$$\sigma_S = \sqrt{<S^2> - <S>^2} \tag{3.21}$$

Hence, in addition to $<S>^2$ based on Eq. (3.17), it is necessary to evaluate $<S^2>$. Again following a similar approach to [12], the resulting normalized standard deviation under square wave excitation is given by:

$$\frac{\sigma_S}{<S>} = \frac{\sqrt{\frac{<\delta^4>}{<\delta^2>^2}}}{\sqrt{N_{dec}WL}} \frac{\left(\int_{E_v}^{E_c} \frac{\psi^2 e^{4E_t/k_B T}}{\left(1 + \psi e^{2E_t/k_B T}\right)^4} g(E_t)\, dE_t \right)^{\frac{1}{2}}}{\int_{E_v}^{E_c} \frac{\psi e^{2E_t/k_B T}}{\left(1 + \psi e^{2E_t/k_B T}\right)^2} g(E_t)\, dE_t} \tag{3.22}$$

Figure 3.11 shows the normalized standard deviation as a function of the Fermi levels in the 'on' and 'off' state for the U-shaped trap density given by (3.18). The normalized standard deviation becomes higher when the traps that contribute most to noise are the ones close to the center of the band gap. In this case the normalized standard deviation increases even though the average noise power decreases.

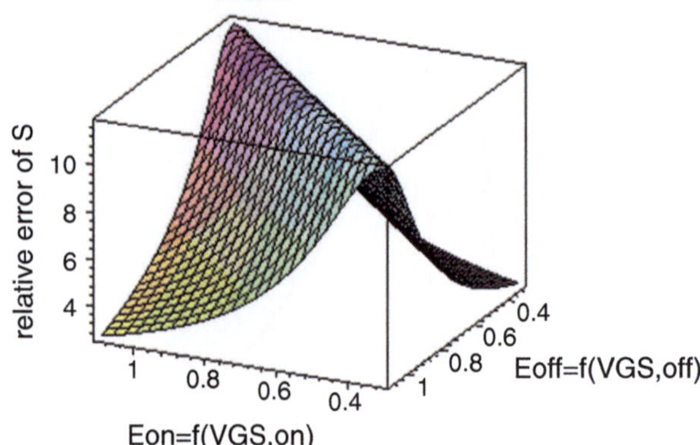

Fig. 3.11 Normalized standard deviation of noise performance as given by Eq. (3.22). An increase in normalized standard deviation is seen under cyclo-stationary excitation, especially for conditions where the average cyclo-stationary noise reduction is high

8 Experimental Results

The low frequency noise was experimentally investigated using short and long channel NMOS devices from a 45 nm technology. Technology details are described in [15]. The DC and large signal noise was measured using a semi-automatic prober setup using a BTA noise system, and the results are described in [3]. A square wave with 50 % duty cycle ($\alpha = 0.5$) was applied to the gate using a waveform generator HP33120A for measuring cyclo-stationary noise. Two types of cyclo-stationary measurements were performed, one with $V_{GS,OFF} = 0$ V and $V_{GS,ON} = V_T$, and the other with $V_{GS,OFF} = V_T$ and $V_{GS,\,ON} = V_{DD}$. Note, that the effect of modulation for a signal modulated by a square wave with 50 % duty cycle is a reduction of noise power (in baseband) by a factor of four.

After the standard modulation theory, for square wave excitation, switching operation can be represented by the multiplication of the noise with a square-wave signal with 50 % duty cycle, *m*(*t*), as follows [19]:

$$\mathrm{m}\,(\mathrm{t}) = \frac{1}{2} + \frac{2}{\pi}\sin\omega_{sw}t + \frac{2}{3\pi}\sin 3\omega_{sw}t + \frac{2}{5\pi}\sin 5\omega_{sw}t + \ldots$$

In frequency domain this corresponds to a convolution of the noise PSD with a spectrum with delta functions at dc, ω_{sw}, $3\omega_{sw}$, $5\omega_{sw}$, and so forth. The dc-term determines the resulting noise power in baseband, which is $(1/2)^2$ or -6 dB, if compared to the original noise power.

However, the measurements results did show a much larger noise reduction. The noise reduction was larger than a factor of ten, and the mean value of the LF-noise under DC conditions was seen to follow 1/f behavior, but with deviations at some frequencies, showing Lorentzian components [3]. For cyclo-stationary conditions these deviations were more pronounced in agreement to our theory. Also in good agreement to our theory, the noise reduction factor was found to be almost constant if only measured results below the excitation frequency are considered.

To compare theory and experiment, the average number of traps for all possible energetic levels in the bandgap were extracted from average noise behavior. The model comparisons are based on the total number of traps and the assumption of a U-shaped trap density with $a = 11$ (see Eq. 3.18).

Experimental results did show that whereas the average LF noise is decreasing under cyclo-stationary conditions, the normalized standard deviation of noise power increases. As shown in Fig. 3.12, this again is in good agreement to the model presented here. In Fig. 3.12, the model values for mean and upper spec in cyclo-stationary conditions are given by Eqs. 3.17 and 3.22, respectively. Here the upper boundary for the DC noise is evaluated using the model presented in [1].

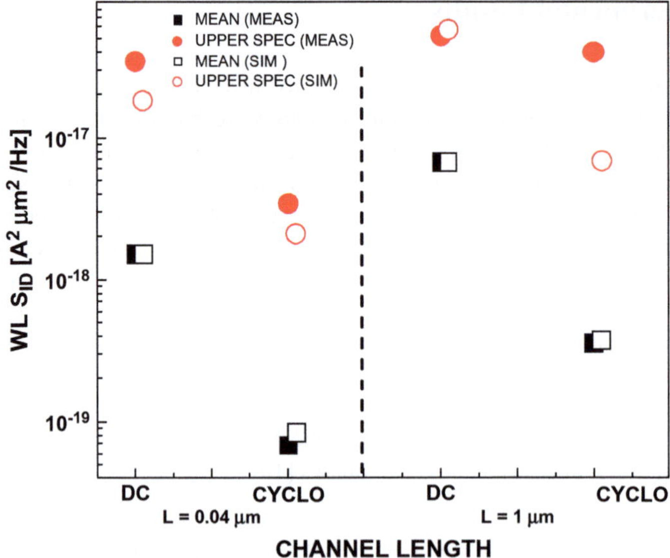

Fig. 3.12 Area normalized average and $+3\sigma$ worst case noise behavior under DC and CS conditions. All data are taken at 10 Hz noise frequency. *Open symbols* are data from simulation and *closed symbols* are measured data reported in [3]

9 The Charge Trapping Component of Bias Temperature Instability

Above we discussed the low-frequency noise originated from trapping and de-trapping at localized states (charge traps) at the interface or in the gate dielectric of transistors. Here we discuss the role of charge trapping in Bias Temperature Instability (BTI). It is widely accepted that charge trapping plays a role in the threshold voltage shifts (ΔV_T) produced by Bias Temperature stress. It was reported that a significant fraction of the threshold voltage shift is recovered spontaneously once the Bias Temperature stress is removed [20–22]. Although first observed decades ago, the phenomenon still remains controversial in both experimental and theoretical terms.

There is a vast literature on Negative Bias Temperature Instability (NBTI) in PMOSFETs, where most models are based on reaction diffusion. Reaction diffusion models involve breaking of Hydrogen-Silicon bonds at the Silicon-Gate Dielectric interface, related to the trapping of inversion layer holes, with the release and diffusion of a hydrogenic species. Recovery (relaxation) is assumed to occur with re-bonding of the Hydrogen-Silicon bonds, i.e., anneal out the interface traps [20]. Although reaction diffusion models have been very useful and successful, some aspects of NBTI are difficult to be fully explained in a reaction diffusion framework, as for instance the fast recovery which occurs if bias stress is removed [20]. Other phenomena, such as charge trapping, therefore have to be also considered in the NBTI mechanism [22, 23]. Clear steps caused by single trapping or de-trapping events were seen in experimental works, showing the discrete nature of V_T shifts, [22–24].

The discussion here presented will not focus on the origin and nature of the charge traps. Detailing lattice dynamics, tunneling mechanisms, or determining the physical location of trapping centers is also not the focus of this work. This work focuses instead on the charge trapping statistics (stochastic capture and emission events) that contribute to degradation of transistor on-current over time (BTI). The basic assumptions made in the modeling of BTI here presented are the same ones as in our work presented above, on the modeling of low-frequency noise: 1) charge trapping and de-trapping are stochastic events, governed by characteristic time constants, which are uniformly distributed on a log-scale; 2) the number of traps is assumed to be Poisson distributed, and the parameter of the Poisson distribution is assumed to be constant over the time interval of interest; 3) trap energy distribution is assumed to be U-shaped (this last assumption is key to explain the AC-Behavior), and 4) the amplitude of the fluctuation induced by a single trap is a random variable.

In this section we develop a theoretical analysis to describe the density of occupied traps as a function of bias, temperature and time, aiming to understand and model the charge trapping component of the aging (degradation) process that occurs in MOSFETs.

The same equation (model) applies to the degradation process, i.e., increase in number of occupied traps, as well as to the recovery process, i.e., decrease in number of occupied traps.

The capture and emission of charge carriers by a trap are described as simple Poisson processes governed by rates τ_c and τ_e, where the capture occurs with probability $p_{01}(dt) = dt/\tau_c$ and emission occurs with probability $p_{10}(dt) = dt/\tau_e$. State 1 stands for the occupied trap, while state 0 stands for the empty trap. τ_c and τ_e are then the average residence time in states 0 and 1, respectively.

We start by evaluating the device degradation process, which is described by the average number of occupied traps at time *t*, which we denote $<n(t)>$. First we write the equation for the probability of a particular trap, which is initially empty (state 0), to remain in the same state after an elapsed time *t*. We denote this probability as $P_{00}(t)$. This probability can be calculated observing that:

$$P_{01}\,(t + dt) = P_{01}(t)p_{11}(dt) + P_{00}(t)p_{01}(dt) \tag{3.23}$$

where $p_{11}(dt) = 1 - p_{10}(dt) = 1 - dt/\tau_e$ and $P_{00}(t) = 1 - P_{01}(t)$. This leads to a simple differential equation. The solution of this differential equation is [18]:

$$P_{01}(t) = \left[1 - \exp\left(-t/\tau_{eq}\right)\right]\tau_e/\left(\tau_c + \tau_e\right) \tag{3.24}$$

where $1/\tau_{eq} = 1/\tau_c + 1/\tau_e$. A similar evaluation can be performed for $P_{11}(t)$, leading to

$$P_{11}(t) = \left[\tau_e + \tau_c \exp\left(-t/\tau_{eq}\right)\right] / \left(\tau_c + \tau_e\right) \tag{3.25}$$

For a device which has N_{tr} traps, we can write the number of occupied traps at time t, $n(t)$, as being $n(t) = \sum_{i=1}^{Ntr} \theta_i(t)$, where $\theta_i(t)$ can assume the values 0 or 1. We can then evaluate the average number of traps occupied at time t as being:

$$\langle n(t)\rangle = \sum\nolimits_{Ntr=0}^{\infty} \frac{N^{Ntr} e^{-N}}{N_{tr!}} \sum\nolimits_{k=0}^{Ntr} \mathrm{k\,P\,(k,t)} \tag{3.26}$$

where $N^{Ntr} e^{-N}/N_{tr}!$ is the probability that N_{tr} traps are found in a device, i.e., the number of traps is assumed to be Poisson distributed, with parameter $<N_{tr}> = N$. While P(k, t) is the probability that k traps are occupied at time t, with $k = 0 \ldots N_{tr}$. The averaging must be performed over the successive stochastic capture and emission events of a trap, and over the number of traps of the ensemble of devices.

The average threshold voltage fluctuation is then obtained by multiplying (3.26) by $<\delta>$, which is the average fluctuation due to a single trap. This leads to:

$$< \Delta V_T(t) > = < \delta > \langle n(t)\rangle \tag{3.27}$$

Since traps have different time constants, which are statistically independent and identically distributed random variables, for considering P(k, t) in Eq. (3.26) we evaluate the average probability over different traps by writing a binomial. In this case

$$\overline{\mathrm{P\,(k,t)}} = \binom{\mathrm{Ntr}}{\mathrm{k}} \overline{\mathrm{P_{01}}\left(\tau_c, \tau_e, t\right)}^{\mathrm{k}}\, \overline{\mathrm{P_{00}}\left(\tau_c, \tau_e, t\right)}^{\mathrm{Ntr-k}} \tag{3.28}$$

where $\overline{\mathrm{P_{01}}\left(\tau_c, \tau_e, t\right)} = \iint d\tau_c d\tau_e \mathrm{P_{01}}\left(\tau_c, \tau_e, t\right) \mathrm{f}\left(\tau_c, \tau_e\right)$ and $\overline{\mathrm{P_{00}}\left(\tau_c, \tau_e, t\right)} = \iint d\tau_c d\tau_e \mathrm{P_{00}}\left(\tau_c, \tau_e, t\right) \mathrm{f}\left(\tau_c, \tau_e\right)$, with $f(\tau_c, \tau_e)$ being the joint probability function of time constants τ_c and τ_e over all traps.

τ_c and τ_e are random variables, whose values depend on temperature and bias point, and which we assume to follow [14]:

$$\tau_c = 10^{\mathrm{p}}\left[1 + \exp\left(-\mathrm{q}\right)\right] \tag{3.29}$$

$$\tau_e = 10^{\mathrm{p}}\left[1 + \exp\left(+\mathrm{q}\right)\right] \tag{3.30}$$

where $p \in [p_{min}, p_{max}]$. Please note that p_{min} and p_{max} define the times constants of the fastest and the slowest trap, respectively. This also limits the time interval in which the model here proposed is valid, in a similar way as in the analysis of low-frequency noise made above. Since p is assumed to be uniformly distributed, the charge trap characteristic time constants are uniformly distributed on a log scale. Again, this is the same assumption done in the low-frequency noise analysis above. The variable q is given by $q = (E_T - E_F)/k_BT \in [(E_V - E_F)/k_BT, (E_C - E_F)/k_BT]$, where E_C is the conduction band edge, while E_V is the valence band edge. E_T is the energy level of the trap. Consequently, τ_c and τ_e are temperature dependent. The assumption of the existence of individual defects with a very wide distribution of time constants is also in line with recent NBTI data [20–24]. The assumption of a single p, governing bothτ_c and τ_e, may be restrictive and deserves deeper investigation in future works, both theoretical and experimental. Nevertheless, we did run numerical analysis (Monte Carlo simulations) assuming independent p values for capture and emission. The BTI behavior is observed to stay essentially the same (no qualitative deviation on the BTI behavior is observed), as long as the p values remain uniformly distributed.

For the evaluation of Eq. (3.26), we consider Eqs. (3.28), (3.29) and (3.30). In evaluating the average over the different random variables, we separate the average over number of traps

$$\langle(\cdot)\rangle = \sum_{Ntr=0}^{\infty} (\cdot) \frac{N^{Ntr} e^{-N}}{N_{tr}!}$$

and time constants

$$\overline{(\cdot)} = \iint d\tau_c d\tau_e \, (\cdot) \, \mathrm{f}(\tau_c, \tau_e)$$

Please remember that the time constants τ_c and τ_e depend on Fermi level, trap energy level, and temperature, as given by (3.29) and (3.30). Evaluation the averages in this manner facilitates the analytical analysis.

This leads to:

$$\overline{\langle n(t)\rangle} = \overline{\mathrm{P}_{01}(\tau_c, \tau_e, \mathrm{t})} \sum_{Ntr=0}^{\infty} \frac{N^{Ntr} e^{-N}}{N_{tr}!} N_{tr} = N \, \overline{\mathrm{P}_{01}(\tau_c, \tau_e, \mathrm{t})}$$

$$= \frac{N}{ln10\,(p_{max} - p_{min})} \left(\int_{Ev}^{Ec} \frac{g(E_T)\, dE_T}{1 + e^{-(E_T - E_F)/k_BT}} \right) \left(\int_{10^{-p_{min}} t}^{10^{-p_{max}} t} \frac{(e^{-u} - 1)}{u} du \right) \tag{3.31}$$

where $g(E_T)$ describes the trap energy distribution in the band-gap, and in the second integral a change of variable was made, p = −log (u/t), $dp = -du$ / (u ln 10). N is the average number of traps found in a device.

In Eq. (3.31), the first integral contains the Fermi level and temperature dependence, while the second integral contains the time dependence. This means that this equation has the time dependence separated from the Fermi level (i.e., bias point, since the Fermi level is defined by the bias point) and temperature dependence. Hence, the model predicts that for measurements carried out at different temperatures, there are scaling factors that can be used as a multiplicative coefficients for the threshold voltage shift, making the curves for different temperatures to overlap (at all measured times). The same applies for measurements carried out at different Fermi levels, i.e., different stress voltages (bias points). Note that the Fermi level is a function of the applied voltage. Furthermore, since Eq. (3.31) is valid for both stress and recovery phases, it implies that the temperature and voltage dependence (scaling factor) during stress and during recovery is the same. This is very relevant and in agreement to experimental data, as discussed below.

Equation (3.31) can be evaluated numerically, leading to:

$$\langle n(t) \rangle \sim \varphi\,(T, E_F)\,(A + B\ log(t)) \tag{3.32}$$

where A and B are constants, and the last term clearly shows that the time evolution of number of occupied traps shows a log(t) behavior. The term $\varphi(T, E_F)$ describes the temperature and Fermi level dependence. Please note that (3.32) above is of the same form as equation (4) in [23], which was empirically written, as an approximation for the experimentally observed behavior.

The actual form of the term $\varphi(T, E_F)$ depends on the trap energy distribution in the band-gap $g(E_T)$. $g(E_T)$ is usually found to be a convex curve (e.g. U-shaped); see, for instance, [14, 15]. In this work, a U-shape trap density function is investigated. As in the case of low-frequency noise analysis for abruptly changing gate bias, the trap energy distribution $g(E_T)$ is key for explaining experimentally observed findings.

Equation (3.32) models both stress and recovery phases. Hence, both stress and recovery phases of BTI are described by the same Eq. (3.32). If the density of initially occupied traps is lower than the value expected for the bias point, the number of occupied traps increases logarithmically. This corresponds to stress phase of BTI. On the other hand, if the density of initially occupied traps is higher than the value expected for the bias point, the density of occupied states decreases logarithmically. This corresponds to the recovery phase of BTI. Besides analytical analysis and evaluation, we did run Monte Carlo simulations to confirm this behavior (Monte Carlo simulations were performed starting from different numbers of initially occupied traps). Some of the results are presented in Fig. 3.13.

The model here presented considers that the number of traps is constant, i.e., there is no trap generation or annihilation. For stress or recovery over long time intervals, there may be generation of traps during the stress phase, or annihilation of traps (annealing) during the recovery phase. In this case the average number of traps $<N_{tr}>$ in Eq. (3.26) becomes a function of time, $<N_{tr}(t)>$. If the number

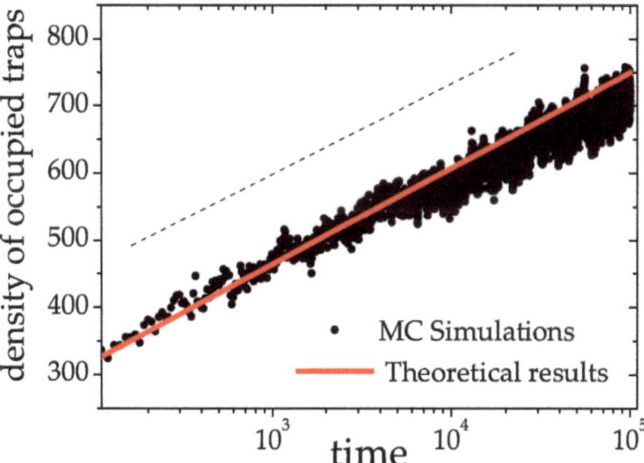

Fig. 3.13 The *full line* depicts the evolution of the density of occupied traps obtained by numerical integration of Eq. (3.31). The *points* correspond to Monte Carlo simulations performed under the same conditions

or traps increases with time during the stress phase, the evolution of number of occupied traps may become faster than log(t) as time evolves, as experimentally observed in many works, where the time dependence is found to follow a power law for long stress times [20, 21]. If a proper equation for $<N_{tr}(t)>$ is available, the modeling approach here presented can be used to evaluate this behavior. Modeling N_{tr} as a function of time is out of the scope of this work. This work focus on the log(t) behavior observed for recovery and short stress times, as discussed in the experimental session below.

For numerical analysis and Monte Carlo simulations performed in this work ($p_{max}-p_{min}$) was chosen to be 7, i.e., it is assumed that p is uniformly distributed over seven decades. The behavior does not depend on p_{max} and p_{min}, as long as the time being considered is much longer than the shortest time constant, and much shorter than longest time constant.

Now we discuss the situation that we call cyclo-stationary excitation, where the gate bias of the device is periodically switched.

In square wave cyclo-stationary excitation, the bias voltage abruptly alternates between two states, called *on* and *off*. In the context of this work, the *on* state refers to the state where stress is applied, and *off* refers to the state where no stress is applied. The fraction of the period T in which the device is in the *on* state is α (i.e., the duty cycle). Figure 3.14 depicts this situation.

With periodically changing gate bias, the Fermi level becomes a periodic function of time $E_F(t)$. As seen from Eqs. (3.23) to (3.28), the parameters that govern the charge trapping behavior are the capture and emission time constants, τ_c and τ_e, respectively. The behavior of these time constants under cyclo-stationary was already discussed above in the context of Low-Frequency Noise. It has been shown that as long as the cyclo-stationary period time T is short compared to the trap's time constants, the behavior of each trap is governed by an effective capture ($\tau_{c,eff}$) and emission ($\tau_{e,eff}$) time constant. The effective values are evaluated as being the time average values over one excitation period:

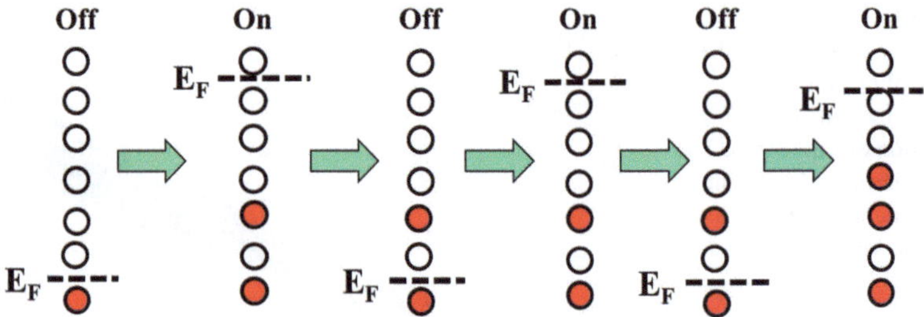

Fig. 3.14 This figure depicts the behavior of slow traps under switched bias. If the gate bias is periodically alternated between *on* and *off*, the Fermi Level (E_F) changes accordingly. If trap time constants are much longer than the *on* (stress) and *off* (recovery) cycle, traps are not able to follow device bias, and act according to equivalent time constants. For these traps, which are too slow to follow bias point change, the behavior is described by the equivalent time constants given by Eqs. (3.13) and (3.14)

$$\frac{1}{\tau_{c,eff}} = \frac{1}{T}\int_0^T \frac{1}{\tau_c} dt \tag{3.33}$$

$$\frac{1}{\tau_{e,eff}} = \frac{1}{T}\int_0^T \frac{1}{\tau_e} dt$$

Under square wave excitation the effective time constants become a simple function of the duty cycle α, as given by Eqs (3.13) and (3.14) above.

This formulation shows that there is no dependence of the effective time constants on the frequency of the cyclo-stationary excitation, as long as the excitation period T is short compared to the trap time constants.

Using the formulation for the equivalent time constants, Eq. (3.28) becomes

$$\overline{\mathrm{P}(\mathrm{k}, \mathrm{t})} = \binom{\mathrm{Ntr}}{\mathrm{k}} \overline{\mathrm{P}_{01}(\tau_{\mathrm{c,\,eff}}, \tau_{\mathrm{e,eff}}, \mathrm{t})^{\mathrm{k}} \mathrm{P}_{00}(\tau_{\mathrm{c,eff}}, \tau_{\mathrm{e,eff}}, \mathrm{t})^{\mathrm{Ntr-k}}} \tag{3.34}$$

Because the Fermi level is now a function of time, it is not possible to derive an analytical closed form expression as Eq. (3.31). However, numerical analysis, as well as Monte Carlo simulations, show that also under cyclo-stationary excitation the behavior has the form of Eq. (3.32), i.e., there is one term that describes the log(t) behavior, multiplied by another term that describes the temperature and Fermi level dependences. Furthermore, it can be shown below that a simple equation to write an effective Fermi level may be used, leading to a simple model, suitable for circuit simulation purposes.

Figure 3.15 shows the results of MC simulation where the switching bias is fast if compared to the trap time constants of all traps considered in the simulation. In this simulation, time constants of all traps are at least one order of magnitude longer than the period of the square wave excitation. In this figure, the dotted

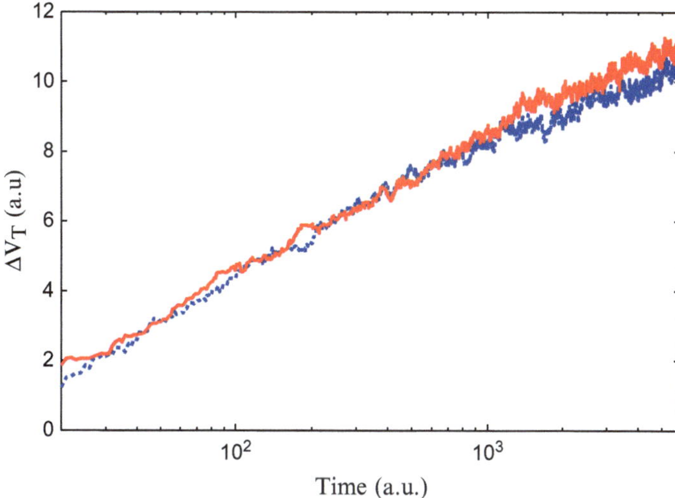

Fig. 3.15 Behavior of slow traps under square wave excitation. Steady increase of ΔV_T is seem, causing a BTI degradation which follows a log(t) behavior. *Dotted (blue) line* is detailed Monte Carlo simulation, considering usual time constants. *Solid (red) line* is Monte Carlo simulation using equivalent time constants, as given by Eqs. (3.13) and (3.14)

(blue) line is the result of the detailed MC simulation of the square wave excitation, applying the usual time constants during the stress (*on*) and recovery (*off*) phases. The solid (red) line is MC simulation using equivalent time constants, as given by Eqs. (3.13) and (3.14). In this simulation the same equivalent time constants are applied during the whole simulation. In other words, it is similar to a DC simulation (no switching bias), where the equivalent time constants are applied during the whole simulation time. Slow traps lead to a degradation component which steadily increases with time, with a log(t) behavior. In this MC simulation all traps are assumed to be initially empty. Note that the slow traps do not follow the switching bias. Steady increase in V_T is seen. These MC simulations confirm that the behavior is determined by the equivalent time constants, as given by (3.13) and (3.14).

Equation (3.34) can be solved numerically, but this approach is not practical for circuit simulation purposes. A simpler approach, considering and equivalent Fermi level $E_{F,eff}$, which is a function of duty cycle, can fit the experimental data. In this approach, empirically obtained after insights from extensive Monte Carlo simulations and numerical analysis, $E_{F,eff}$ is written as the time average of the Fermi level. For the case of square wave excitation with duty cycle α we have:

$$E_{F,eff}(\alpha) = \alpha E_{F,on} + (1-\alpha)\ E_{F,off} \tag{3.35}$$

With this empirical approach, the same Eq. (3.31) can model both steady bias and cyclo-stationary excitation. Monte Carlo numerical analysis confirms that the empirical formulation fits the numerical solution of (3.34).

Conclusion

This chapter discussed the role of charge trapping in low-frequency noise and bias temperature instability (BTI). A statistical model for the charge trapping phenomena in MOS devices was presented. The model is based on discrete microscopic device physics parameters. Besides evaluating the average behavior, the model here proposed allows the derivation of statistical relevant parameters. The model is applied to study the random telegraph noise (RTN) and Bias Temperature Instability (BTI).

The impact of charge trapping is modeled as momentary changes in threshold voltage (or drain current). In analog circuits, this may lead to low-frequency noise and be a source of jitter in oscillators. In digital circuits it may lead to aging effects and transient effects, since circuit behavior may change due to transient V_T fluctuations between two logic operations of a digital circuit.

The modeling approach focuses on operation conditions relevant for digital and analog design, including large signal AC operation.

It is shown that if the density of traps in the band-gap is a convex curve ("U"-shaped) a reduction in noise power may be achieved under cyclo-stationary excitation. However, the variability in noise behavior (normalized standard deviation of noise power) is shown to increase. This increase in variability will be another challenge to analog and RF circuit designs in deep sub-micron CMOS technologies.

Regarding BTI, an analytical model for both stress and recovery phases of BTI is presented. Furthermore, the model properly describes device behavior under periodic switching, also called AC-BTI or cyclo-stationary operation. It is shown that a universal logarithmic law describes the time dependence of charge trapping in both stress and recovery phases, and that the time dependence may be separated from the temperature and bias point dependence.

Acknowledgments Gilson Wirth and Roberto da Silva thank CNPq and FAPERGS for research grants that partially support the work reported here.

References

1. G Wirth, J Koh, R da Silva, R Thewes and R Brederlow, "Modeling of Statistical Low-Frequency Noise of Deep-Submicron MOSFETs.", *IEEE Trans. Electron Dev.*, **52**, p. 1576–1588 (2005).
2. G Wirth, R da Silva and R Brederlow. "Statistical Model for the Circuit Bandwidth Dependence of Low-Frequency Noise in Deep-Submicrometer MOSFETs", *IEEE Trans. Electron Dev.*, **54**, p. 340–345 (2007).

3. G Wirth, R da Silva, P Srinivasan, J Krick and R Brederlow. "Statistical model for MOSFET low-frequency noise under cyclo-stationary conditions" *Int Electron Dev Meeting - IEDM 2009*, p. 30.5.1-4, (2009).
4. G Wirth and R da Silva, "Low-Frequency Noise Spectrum of Cyclo-Stationary Random Telegraph Signals", *Electrical Eng.*, **90**, p. 435–41 (2008).
5. R da Silva, G Wirth and L Brusamarello. "An appropriate model for the noise power spectrum produced by traps at the Si SiO interface: a study of the influence of a time-dependent Fermi level. Journal of Statistical Mechanics" *Journal of Statistical Mechanics. Theory and Experiment*, **2008**, p. P10015 (2008).
6. R Brederlow, J Koh, G Wirth, R da Silva, M Tiebout, R Thewes. "Low Frequency Noise Considerations for CMOS Analog Circuit Design" *Proceedings of the 2005 International Conf on Noise and Fluctuations (ICNF)*. p. 703–708 (2005).
7. R Brederlow, J Koh and R Thewes. "A physics-based low frequency noise model for MOSFETs under periodic large signal excitation" *Solid-State El.*, **50**, p. 668–73 (2006).
8. I Bloom and Y Nemirovsky. "1/f noise reduction of metal-oxide-semiconductor transistors by cycling from inversion to accumulation" *Appl Phys Lett*, **58**, p.1664–6 (1991).
9. B Dierickx and E Simoen. "The decrease of "random telegraph signal"noise in metal-oxide-semiconductor field-effect transistors when cycled from inversion to accumulation". *J Appl Phys*, **71**, p. 2028–2029 (1992).
10. M Ertürk, T Xia and W Clark. "Gate voltage dependence of MOSFET 1/f noise statistics". *IEEE Electron Dev Let.*, **28**, p. 812–814 (2007).
11. A van der Wel, E Klumperink, E Hoekstra and B Nauta. "Relating random telegraph signal noise in metal-oxide-semiconductor transistors to interface trap energy distribution" *Appl Phys Lett.*, **87**, p. 183507 (2005).
12. S Machlup, "Noise in Semiconductors: Spectrum of a Two-Parameter Random Signal" *J Appl Phys*, **35**, p. 341–343 (1954).
13. A Roy and C Enz, "Analytical Modeling of Large-Signal Cyclo-Stationary Low-Frequency Noise with Arbitrary Periodic Input" *IEEE Trans. Electron Dev.*, **54**, p.2537-2545 (2007).
14. M Kirton and M Uren, "Noise in solid-state microstructures: A new perspective on individual defects, interface states and low-frequency (1/f) noise" *Adv. in Physics*, **38**, p. 367–468 (1989).
15. Ekbote, S.; Benaissa, K.; Obradovic, B.; Liu, S.; Shichijo, H.; Hou, F.; Blythe, T.; Houston, T.W.; Martin, S.; Taylor, R.; Singh, A.; Yang, H.; Baldwin, G, "45nm Low-Power CMOS SoC Technology with Aggressive Reduction of Random Variation for SRAM and Analog Transistors" *2008 VLSI Tech. Symp.*, p. 160–161 (2008).
16. Wirth, Gilson I. ; da Silva, Roberto ; Kaczer, Ben. "Statistical Model for MOSFET Bias Temperature Instability Component Due to Charge Trapping". *IEEE Transactions on Electron Devices*, p. 2743–2751 (2011).
17. Ashraf, Nabil ; Vasileska, Dragica ; Wirth, Gilson ; Srinivasan, P. "Accurate Model for the Threshold Voltage Fluctuation Estimation in 45-nm Channel Length MOSFET Devices in the Presence of Random Traps and Random Dopants". *IEEE Electron Device Letters*, p. 1044–1046 (2011).
18. Camargo, Vinícius V. A. ; Ashraf, Nabil ; BRUSAMARELLO, Lucas ; Vasileska, Dragica ; Wirth, Gilson . "Impact of RDF and RTS on the performance of SRAM cells". *Journal of Computational Electronics*, p. 122–127, (2010).
19. E Klumperink, S Gierkink, A van der Wel, and B Nauta; "Reducing MOSFET 1/f noise and power consumption by switched biasing ," *Solid-State Circuits, IEEE Journal of* , vol.35, no.7, pp.994-1001 (2000).
20. D. K. Schroder, "Negative bias temperature instability: What do we understand?", Microelectron. Reliab., vol. 47 no. 6, pp. 841–852, (2007).
21. S. E. Rauch, "Review and reexamination of reliability effects related to NBTI statistical variations," IEEE Trans. Device Mater. Rel., vol. 7, no. 4, pp. 524–530, (2007).

22. B. Kaczer, T. Grasser, J. Martin-Martinez, E. Simoen, M. Aoulaiche, Ph. J. Roussel, and G. Groeseneken, "NBTI from the Perspective of Defect States with Widely Distributed Times," Proc. Int. Rel. Phys. Symp., p. 55 (2009).
23. T. Grasser and B. Kaczer, "Evidence That Two Tightly Coupled Mechanisms Are Responsible for Negative Bias Temperature Instability in Oxynitride MOSFETs", *IEEE Trans. on Electron Dev.*, vol. 56, pp. 1056–1062 (2009).
24. Ang, D.; Teo, Z.; Ho, T.; Ng, C.; , "Reassessing the mechanisms of negative-bias temperature instability by repetitive stress/relaxation experiments," *Device and Materials Reliability, IEEE Transactions on*, vol. 11 , no. 1, pp. 19 – 34 (2011).

Chapter 4
Atomistic Simulations on Reliability

Dragica Vasileska and Nabil Ashraf

Abstract Discrete impurity effects in terms of their statistical variations in number and position in the inversion and depletion region of a MOSFET, as the gate length is aggressively scaled, have recently been researched as a major cause of reliability degradation observed in intra-die and die-to-die threshold voltage variation on the same chip resulting in significant variation in saturation drive (on) current and transconductance degradation—two key metrics for benchmark performance of digital and analog integrated circuits. In the following chapter, the authors have highlighted the random dopant fluctuation (RDF) based Ensemble Monte Carlo (EMC) device simulation study conducted by the Computational Electronics (COMPUTEL) research group of Arizona State University. In addition to RDF, random number and position of interface traps lying close to Si:SiO_2 interface engender additional concerns leading to enhanced experimentally observed fluctuations in drain current and threshold voltage. In this context, the authors of this chapter present novel EMC based simulation studies on trap induced random telegraph noise (RTN) responsible for statistical fluctuation pattern observed in threshold voltage, its standard deviation and drive current in saturation for 45 nm gate length MOSFET device. From the observed simulation results and their analysis, it can be cogently projected that with continued scaling in gate length and width, RTN effect will eventually supersede as a major reliability bottleneck over the typical RDF phenomenon. The fluctuation patterns observed by EMC simulation outcomes for both drain current and threshold voltage have been analyzed and explained from analytical device physics perspectives.

1 Introduction

The transistor mismatch due to random variations in process parameters has become one of the major issues in deep-submicrometer technology. The term transistor mismatch refers to the fact that supposedly identical transistors at the design phase

---

D. Vasileska (✉) • N. Ashraf
Department of Electrical, Computer and Energy Engineering, Arizona State University, Tempe, AZ 85287-5706, USA
e-mail: vasileska@asu.edu; nashraf@asu.edu

R. Reis et al. (eds.), *Circuit Design for Reliability*,
DOI 10.1007/978-1-4614-4078-9_4

come out as distinct devices after manufacturing due to process variations. The extrinsic factors include variation in the implantation dose and energy, oxidation and annealing temperatures, etc. On the other hand, the intrinsic factors include variations due to channel dopant number, interface, and fixed oxide charge. Both the extrinsic and the intrinsic factors influence the transistor output performance and yield. The transistor mismatch effect will become worse in the future due to demanding requirement on process tolerance. Hence, there is a direct need to develop a very good understanding on the impact of variation in a given unit process on resulting variation in a given circuit parameter. This will enable the process engineering and manufacturing team to define appropriate process monitor and control criteria for all the unit processes involved. This will also help during the technology development phase to select a particular process integration scheme that could minimize the mismatch among various available options. In this book chapter we first discuss the impact of discrete impurity effects on critical device parameters (Sect. 2). In Sect. 3 we elaborate on the influence of the random telegraph noise fluctuations on the variation in the threshold voltage and device on current. Conclusions from this work are presented in "Conclusions" section.

2 Discrete Impurity Effects

2.1 Some General Considerations

Statistical fluctuations of the channel dopant number were predicted by Keyes [1] as a fundamental physical limitation of MOSFETs down-scaling. Entering into the nanometer regime results in a decreasing number of channel impurities whose random distribution leads to significant fluctuations of the threshold voltage and off-state leakage current. These effects are likely to induce serious problems on the operation and performances of logical and analog circuits.

It has been experimentally verified by Mizuno and co-workers [2] that threshold voltage fluctuations are mainly caused by random fluctuations of the number of dopant atoms (RDF) and other contributions such as fluctuations of the oxide thickness are comparably very small. It follows from these remarks that impurities cannot be considered anymore using the continuum doping model in advanced semiconductor device modeling but the precise location of each individual impurity within a full Coulomb interaction picture must be taken into account.

2.2 Drift-Diffusion Simulations of Discrete Impurity Effects

To illustrate the importance of the discrete impurity effects on the threshold voltage and the off-state current, in 1996 the group from ASU simulated a number of MOSFET device structures [3]. The corresponding potential profile under equilibrium

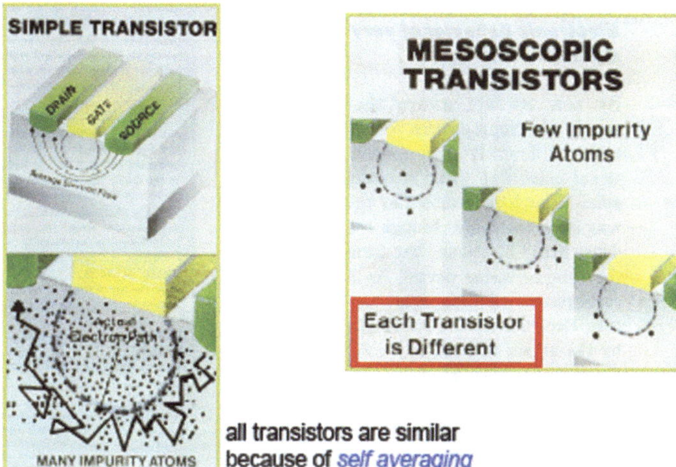

Fig. 4.1 The role of discrete impurities in large devices (*left panel*) and in mesoscopic devices (*right panel*)

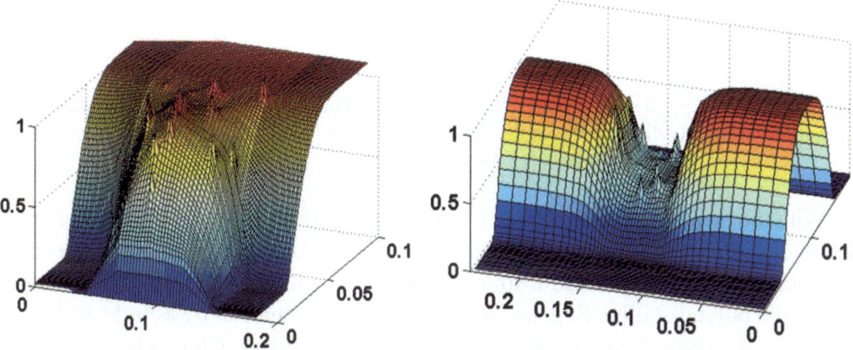

Fig. 4.2 *Top panel*: Potential profile in a cut through the depth of the device. *Bottom panel*: Potential profile in the semiconductor in a plane parallel to the semiconductor/oxide interface

conditions along the depth and in the plane parallel to the semiconductor/oxide interface is shown in the top and bottom panels of Fig. 4.2. In our analysis we have found that it is not only the total number of atoms within the discrete doping region that matters, but the location of these atoms plays very important role.

Illustrated in Fig. 4.3, are the potential profiles and the current stream lines for two impurity distributions. Note that the significant current crowding in the upper panel near the critical source end of the channel leads to smaller drain current for $V_D = 50$ mV and a range of V_G values. This, in turn, results in larger threshold voltage for this device.

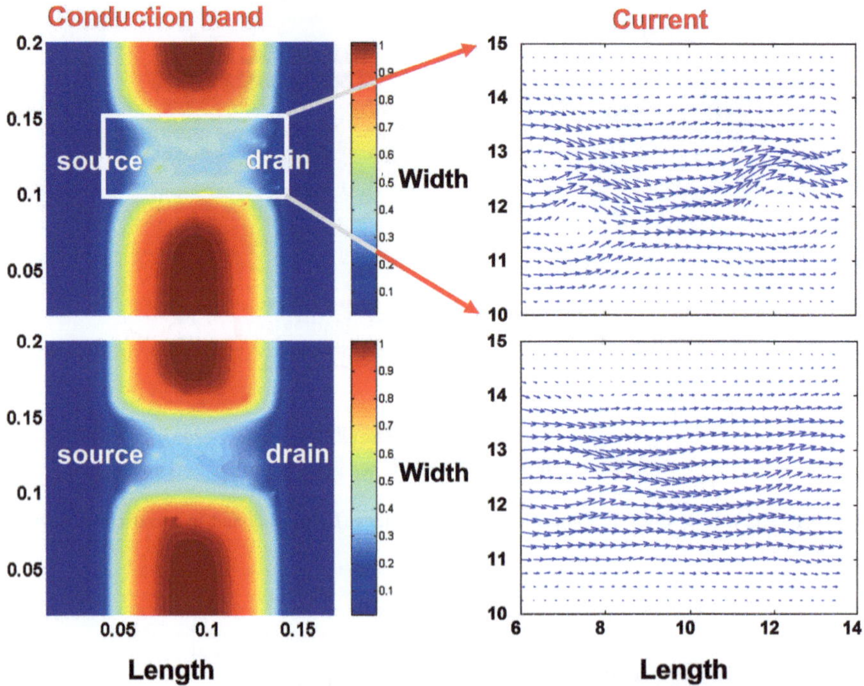

Fig. 4.3 Potential profile and current stream lines in a 50 nm gate-length MOSFET device. Shown here are two impurity distributions in the device active region

With the down-scaling of MOSFET devices, electrons in the conducting channel are in ever closer proximity to the high-density electron gases present in the source and drain regions—separated from each other by as little as tens of nanometers—and in the polycrystalline Si gate—separated from the channel by as little as 1.5 nm of SiO_2. As studied in [4], the role of these long-range Coulomb interactions is twofold: (1) The interaction between electrons in the channel and the high-density electron gases in the source and drain regions can be pictured classically as a reshaping of the electron distribution in the channel caused by the potential-fluctuations, associated with plasma oscillations in the source and drain regions, leaking into the low-density channel. Quantum-mechanically, this corresponds to emission and absorption of plasmons by the channel electrons. While these processes do not subtract directly momentum from the electron gas, their net effect is a thermalization of the hot electrons energy distribution in the channel, the resulting higher energy tail being affected by additional momentum-relaxation processes (phonons, ionized impurities). This causes, indirectly, a reduction of the effective electron velocity in the channel, and so, a depression of the transconductance as the channel length is reduced below about 4 nm. On the other hand, the interaction between channel-electrons and electrons in the gate (Coulomb drag across the very thin insulator) results in a direct loss of momentum of the electrons in the channel. Semi-classically, this interaction—also plasmon-mediated—has been studied by a group at IBM [5] predicting a significant depression of the electron velocity for SiO_2 layers thinner than about 2–3 nm. This behavior has also been observed experimentally [6], recent results being in quantitative agreement with early theoretical estimates.

2.3 Monte Carlo Device Simulations of Discrete Impurity Effects

In the past, the effect of discrete dopant random distribution in MOSFET channel has been assessed by analytical or drift-diffusion (DD) approaches. The first DD study consisted in using a stochastically fluctuating dopant distribution obeying Poisson statistics [7]. 3D atomistic simulators have also been developed for studying threshold voltage fluctuations [8, 9] Even though the DD/HD (hydrodynamic) methods are very useful because of their simplicity and fast computing times, it is not at all clear whether such macroscopic simulation schemes can be exploited into the atomistic regime. In fact, it is not at all clear how such discrete electrons and impurities are modeled in macroscopic device simulations due to the long-range nature of the Coulomb potential. 3D DD/HD macroscopic models may be accurate for modeling the threshold voltage fluctuations (since the device is in the off-state) but they are definitely not accurate when examining the on-state current fluctuations.

Three-dimensional Monte Carlo (MC) simulations should provide a more realistic transport description in ultra-short MOSFETs, in particular in the on-state. The MC procedure gives an exact solution of the Boltzmann transport equation. It, thus, correctly describes the non-stationary transport conditions. Even if microscopic simulations such as the MC method are concerned, the treatment of the electrons and impurities is not straightforward due to, again, the long-range nature of the Coulomb potential. The incorporation of the long-range Coulomb potential in the MC method has been a long-standing issue [10]. This problem is, in general, avoided by assuming that the electrons and the impurities are always screened by the other carriers so that the long-range part of the Coulomb interaction is effectively suppressed. The complexity of the MC simulation increases as one takes into account more complicated screening processes by using the dynamical and wave-vector dependent dielectric function obtained from, for example, the random phase approximation. Indeed, screening is a very complicated many-body matter [11].

A novel approach has been introduced by the ASU group [12], in which the MC method is supplemented by a *molecular dynamics* (MD) routine. In this approach, the mutual Coulomb interaction among electrons and impurities is treated in the drift part of the MC transport kernel. Indeed, the various aspects associated with the Coulomb interaction, such as dynamical screening and multiple scatterings, are automatically taken into account. Very recently, the MC/MD method has been extended for spatially inhomogeneous systems. Since a part of the Coulomb interaction is already taken into account by the solution of the Poisson equation, the MD treatment of the Coulomb interaction is restricted only to the limited area near the charged particles. It is claimed that the full incorporation of the Coulomb interaction is indispensable to reproduce the correct electron mobility in highly doped silicon samples.

Although real space treatments eliminate the problem of double counting of the force, a drawback is that the 3D Poisson equation must be solved repeatedly to properly describe the self-consistent fields which consume over 80 % of the total simulation time. To further speed up simulations, the ASU team has, for the first time, utilized a 3D Fast Multi-pole Method (FMM) [13–16] instead. The FMM allows calculation of the field and the potential in a system of n particles connected by a central force within $O(n)$ operations given certain prescribed accuracy. The FMM is based on the idea of condensing the information of the potential generated by point sources in truncated series expansions. After calculating suitable expansions, the long range part of the potential is obtained by evaluating the truncated series at the point in question and the short range part is calculated by direct summation. The field due to the applied boundary biases is obtained at the beginning of the simulation by solving the Laplace equation. Hence the total field acting on each electron is the sum of this constant field and the contribution from the electron–electron and electron–impurity interactions handled by the FMM calculations. The image charges, which arise because of the dielectric discontinuity, are handled by the method of images [17].

Next we present several success stories on the use of our 3D particle-based device simulation code that properly takes into account the short range and the long range Coulomb forces (Fig. 4.4). We begin with the example of energy relaxation of the

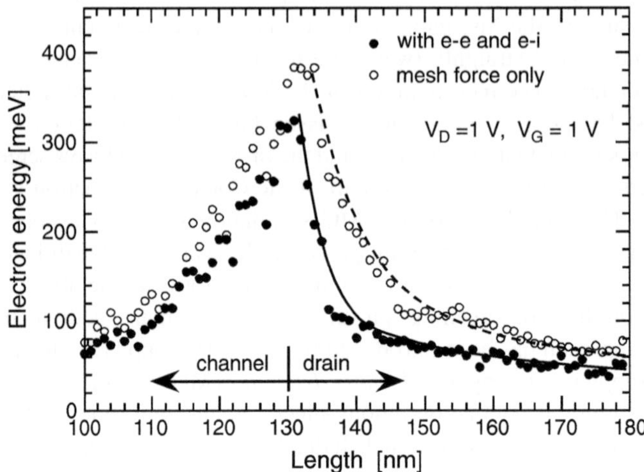

Fig. 4.4 Average energy of the electrons coming to the drain from the channel. The applied bias equals $V_D = V_G = 1$ V. *Filled* (*open*) *circles* correspond to the case when the short-range *e–e* and *e–i* interactions are included (omitted) in the simulations

carriers at the drain end of the MOSFET channel. The simulated device has channel length $L_G = 80$ nm, channel width $W_G = 80$ nm and oxide thickness $T_{ox} = 3$ nm. The lateral extension of the source and drain regions is 50 nm. The channel doping equals 3×10^{18} cm^{-3}. The applied bias is $V_G = V_D = 1$ V. Only those electrons that entered the channel region from the source side were "tagged" and their energy and position was monitored and used in the average energy calculation. From the average velocity simulation results, it follows that the short-range electron–electron (*e–e*) and electron–impurity (*e–i*) interaction terms damp the velocity overshoot effect, thus increasing the transit time of the carriers through the device; thus reducing its cut-off frequency. It is also quite clear that when we use the mesh force only, i.e. we skip the MD loop that allows us to correct for the short-range *e–e* and *e–i* interactions, those electrons that enter the drain end of the device from the channel never reach equilibrium. Their average energy is more than 60 meV far into the drain region. Also, the average energy peaks past the drain junction. The addition of the short-range Coulomb forces to the mesh force via the MD loop, leads to rapid thermalization of the carriers once they enter the drain region. The characteristic distance over which carriers thermalize is on the order of a few nm.

Second important example we have chosen in this paper is the quantitative prediction of the threshold voltage fluctuations versus device gate width, channel doping and oxide thickness, that are shown in Fig. 4.5. Also shown in this figure are the analytical model predictions derived from experimental studies by Mizuno, Okamura, and Toriumi [2], according to which

$$\sigma_{vt} = \left(\frac{\sqrt[4]{q^3 \varepsilon_s \phi_b}}{\sqrt{2}\varepsilon_{ox}} \right) \frac{T_{ox} \sqrt[4]{N}}{\sqrt{L_{eff} W_{eff}}}, \tag{4.1}$$

where N is the average channel doping density, ϕ_b is the built-in potential, T_{ox} is the oxide thickness, L_{eff} and W_{eff} are the effective channel length and width, and ε_s and ε_{ox} are the semiconductor and oxide permittivity, respectively. Stolk, Widdershoven and Klaassen [18] generalized the analytical result by Mizuno and his co-workers, by taking into account the finite thickness of the inversion layer, depth-distribution of charges in the depletion layer and the influence of the source and drain dopant distributions and depletion regions. For a uniform channel dopant distribution, the analytical expression for the threshold voltage standard deviation simplifies to

$$\sigma_{vt} = \left(\frac{\sqrt[4]{q^3 \varepsilon_s \phi_b}}{\sqrt{3}} \right) \left[\frac{k_b T}{q} \cdot \frac{1}{\sqrt{4\varepsilon_s \phi_b N_a}} + \frac{T_{ox}}{\varepsilon_{ox}} \right] \frac{\sqrt[4]{N}}{\sqrt{L_{eff} W_{eff}}}. \tag{4.2}$$

In Eq. (4.2), the first term in the square brackets represents the surface potential fluctuations whereas the second term represents the fluctuations in the electric field. The decrease of the threshold voltage fluctuations with increasing the width of the gate is due to the averaging effects, in agreement with the experimental findings by Horstmann et al. [19]. We want to point out that we still observed significant spread

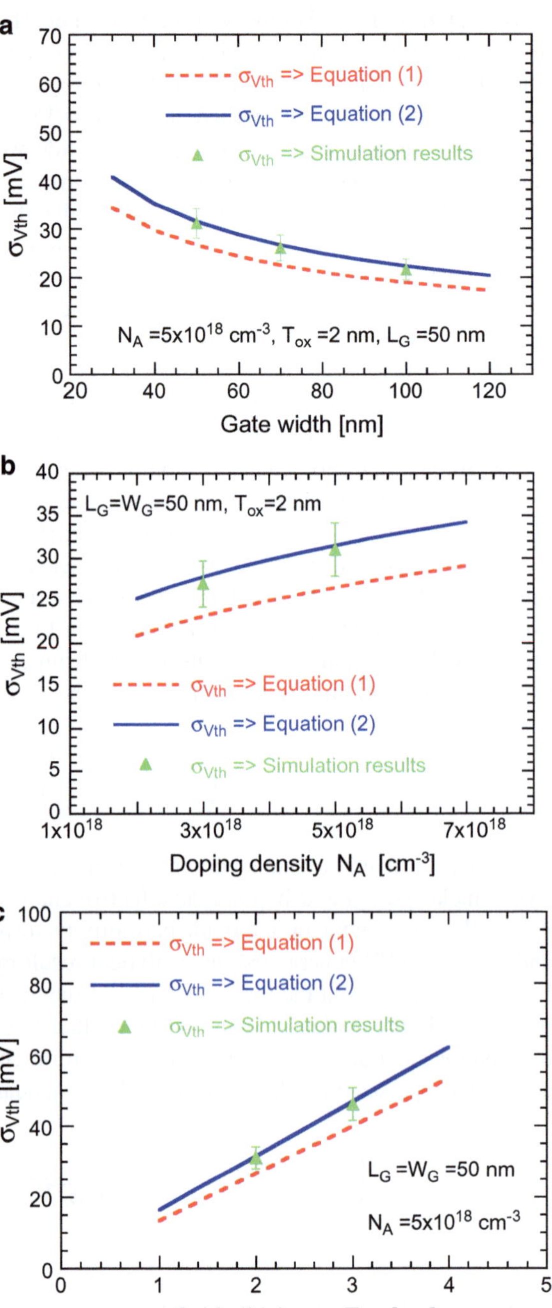

Fig. 4.5 Variation of the threshold voltage with (**a**) gate width, (**b**) channel doping, and (**c**) oxide thickness

of the device transfer characteristics along the gate voltage axis even for devices with $W_G = 100$ nm. This is due to the nonuniformity of the potential barrier, which allows for early turn-on of some parts of the channel. As expected, the increase in the channel doping leads to larger threshold voltage standard deviation $\sigma_{V_{TH}}$. These results also imply that the fluctuations in the threshold voltage can be even larger in devices in which counter ion implantation is used for threshold voltage adjustments. Similarly, the increase in the oxide thickness leads to linear increase in the threshold voltage standard deviation. The results shown in Fig. 4.5(a–c) also suggest that reconstruction of the established scaling laws is needed to reduce the fluctuations in the threshold voltage. In other words, within some new scaling methodology, T_{ox} should become much thinner, or N_A much lower that what the conventional scaling laws give.

To further elaborate on the role of Coulomb interactions, in Example 3, as suggested by the results presented in Example 2, we consider 50 nm channel length narrow-width SOI device with undoped channel. The results presented in Fig. 4.6 also suggest that there are fluctuations in the device threshold voltage for devices fabricated on the same chip due to unintentional doping and random positioning of the impurity atoms. This can also be deduced from the scatter of the experimental data from [20]. The simulation results of the transfer characteristics with a single impurity present in different regions in the channel of the device, shown in the top panel of Fig. 4.6, clearly demonstrate the origin of the threshold voltage shifts for devices with 10 and 5 nm channel width. The width dependence of the threshold voltage for the case of a uniform (undoped) and a discrete impurity model is shown in the bottom panel of Fig. 4.6. These results suggest that *both size-quantization effects and unintentional doping must be concurrently considered to explain threshold voltage variation in small devises.*

3 Random Telegraph Signal

3.1 Importance of Random Trap Fluctuations

The single most important interface in semiconductor technology is that between silicon and its thermally grown oxide. This interface with its propensity to surface micro-roughness after in-situ fabrication plays a crucial role in the performance of today's high speed MOSFET devices. The degree of perfection of the interface has been stipulated to be really exacting in terms of process integrity where a typical device-quality interface has defect densities on the interfacial plane of the order of 10^8–10^{10} cm^{-2} eV^{-1} resulting in defect densities of the order of 1–100 defects per square micron assuming the defects are located within 1 eV of energy distribution from the Fermi energy. As the device area is shrunk to aggressively scaled sub-μm^2 size with scaling-preserved process tolerances, the number of defect densities do show an upward trend and considering that 10^{11} cm^{-2} eV^{-1} values are

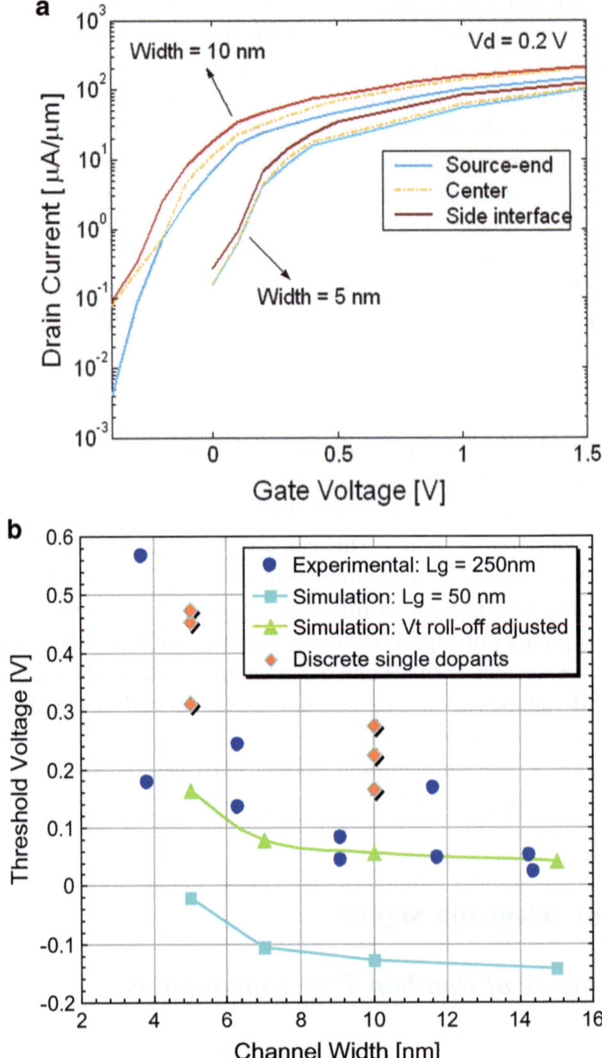

Fig. 4.6 *Top Panel*: Transfer characteristics of the device with 10 and 5 nm channel widths and different location of the impurity atoms. *Bottom Panel*: Width dependence of the threshold voltage for the case of a uniform and a discrete impurity model. Clearly seen in this figure are two trends: (**a**) Threshold voltage increase with decreasing channel width due to quantum-mechanical size quantization effects, and (**b**) Scatter in the threshold voltage data due to unintentional doping

at least readily encountered, the number of defects seem to reduce to less than 3 in number per square micron within an eV energy distribution for a device size of $W \times L = 50\ nm \times 50\ nm$. The reason reliability concern did not arise in wide gate area technology generation is because with a good number of traps lying within a few eV of Fermi energy, the spatial distribution of energy levels is more tighter making the energy barrier values ΔE_B a fraction of an eV. Hence the carriers trapped in traps easily reemit to inversion layers making RTS amplitude variation almost nonexistent. But in today's aggressively scaled device size, even though the trap numbers are countable and sparse, these traps can be located deep within the oxide with higher ΔE_B differential and a carrier once trapped in a trapped site, may stay there for a prolonged period of time and never get reemitted to inversion layer causing severe RTS amplitude drawbacks. The surface level trap lying close to the interface can block the carrier flow in the channel by causing local potential fluctuations where significant spread of RTS variation can be observed for even a single trap closer to source side to mid-channel zone impeding the carrier flow. Also one way carrier gets trapped and get detrapped is through tunneling from inversion layer to a trap location and in earlier technological generation with thicker gate oxide, tunneling was not as significant as it is today with the gate-oxide reaching nanometric thickness. The conclusion is with the channel electrons being random, presence of very few defects will suffice to cause notable RTS related device operation failure for current ongoing technological generation of MOSFETs,

Figure 4.7 (top panel) illustrates the measurement data from a 90 nm SRAM design [21]. The minimum supply voltage (V_{ccmin}), which is highly sensitive to device threshold voltage, exhibits a similar pattern in the time domain as that of RTF. The impact of RTF in this case is more than 200 mV, which is catastrophic to the yield and low-power design of SRAM. Therefore, accurate and physical models of RTF are essential to predict and optimize circuit performance during the design stage. Currently, such models are not available for circuit simulation. The compound between RTF and other sources of variation, such as RDF, further complicates the

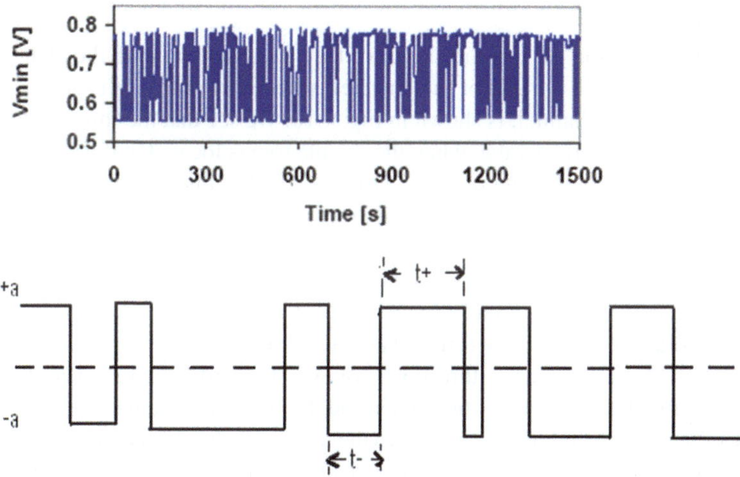

Fig. 4.7 *Top panel*: The fluctuation of SRAM V_{ccmin} due to RTF. *Bottom panel*: A random telegraph signal (namely x(t)) produced by a carrier trap

situation especially in extremely scaled CMOS design. Figure 4.7 (bottom panel) shows a typical two level RTS signal time domain characteristics.

3.2 Monte Carlo Device Simulations of Random Traps at the Semiconductor/Oxide Interface

In this research work the integration of random defects positioned across the channel at the Si/SiO$_2$ interface from source end to the drain end in the presence of different random channel and bulk dopant distributions are used to conduct Ensemble Monte-Carlo (EMC) based numerical simulation of key device performance metrics for 45 nm gate length MOSFET device. The two main performance parameters that affect RTF based reliability measurements are percentage change in relative drain current fluctuation, particularly in the saturation region where most digital circuits operate and percentage change in threshold voltage that affects the device transconductance and on-current drive.

The simulator described in the previous section is presently being used in the investigation of the random trap fluctuations in 45 nm technology node MOSFET device where, in addition to the randomness of the position and the actual number of the impurity atoms in the whole simulated domain of the device, a random double-charged trap is introduced in the middle section of the channel and moved from the source-end to the drain end of the channel. An example of a discrete impurity pattern and a double trap located at the source-end of the channel is shown in Fig. 4.8. The effective channel length of 45 nm technology node is taken to be 35 nm.

We consider ensemble of 20 devices with different random dopant distribution. The threshold voltage of each of these 20 devices without the presence of the trap is shown in Fig. 4.9.

The total variation of the threshold voltage as a function of the double trap position in the middle portion of the channel, when moved from the source end to the drain end of the channel, is shown in Fig. 4.10. We see that the threshold voltage increases from its average value when the double trap is located at the source end of the channel. This is due to the fact that carriers see additional large potential barrier due to the presence of the charged double trap and are reflected back in the source contact. The threshold voltage reduces when the double trap is moved away from the source injection barrier because when the electrons are injected in the channel, even though the electric field is small (due to small drain bias applied when measuring threshold voltage), they slowly drift towards the drain contact.

In Fig. 4.11 we depict the threshold voltage standard deviation variation as a function of the double trap position when the double trap is being moved from the middle of the source end of the channel to the middle of the drain end of the channel. An explanation of the results given in Fig. 4.11 is schematically shown in Fig. 4.12. At threshold voltage, the sheet electron density in the channel is small, therefore screening is not important. Traps near the source end of the channel have the largest

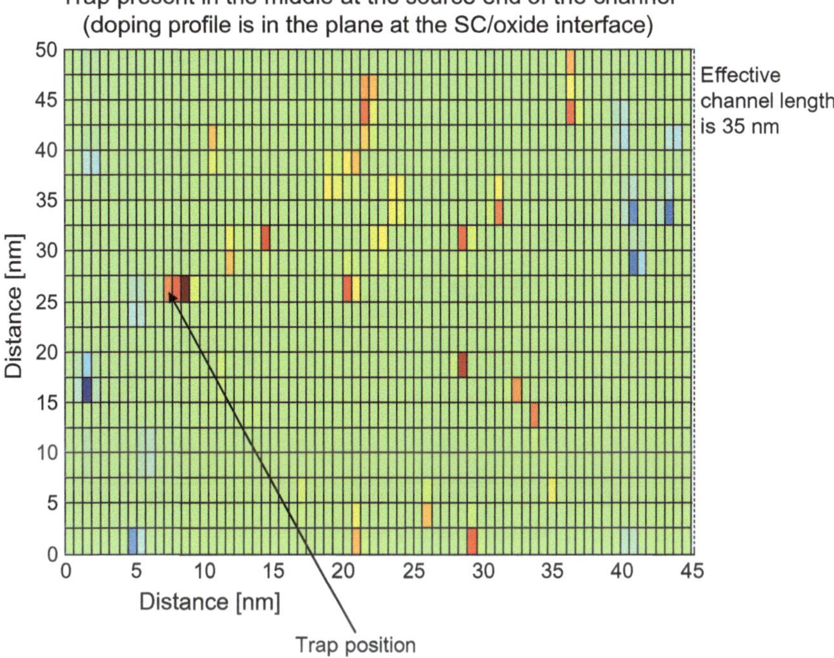

Fig. 4.8 Random dopant distribution and a double trap located in the middle of the source end of the channel

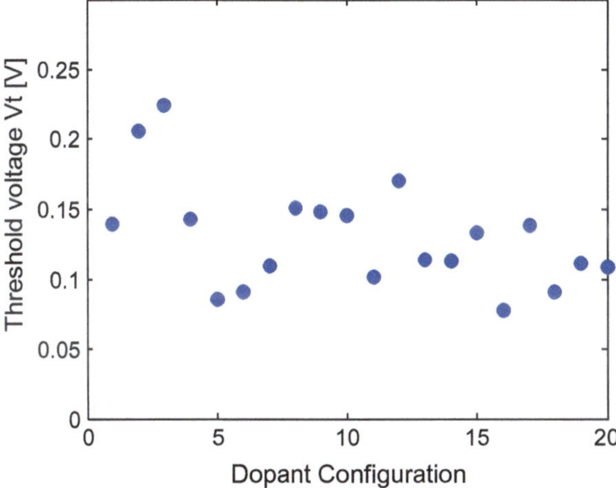

Fig. 4.9 Threshold voltage fluctuations due to random dopant fluctuations (without traps) for a statistical ensemble of 20 devices with different number and different distribution of the impurity atoms

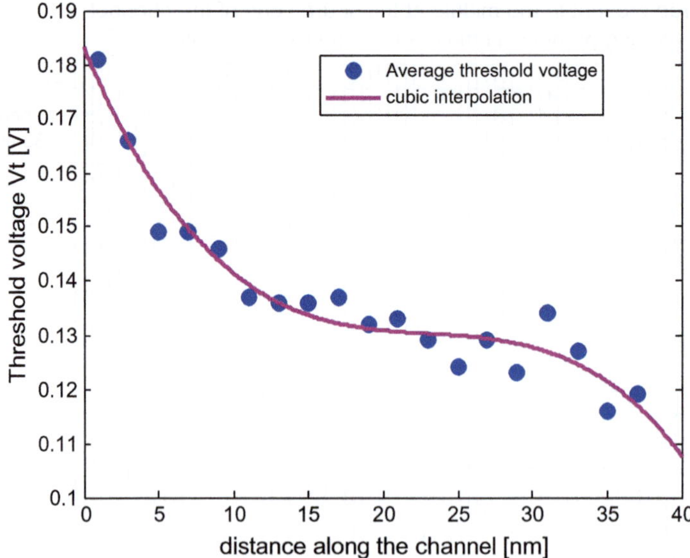

Fig. 4.10 Threshold voltage variation with trap position variation. These results are averages over the ensemble of 20 devices

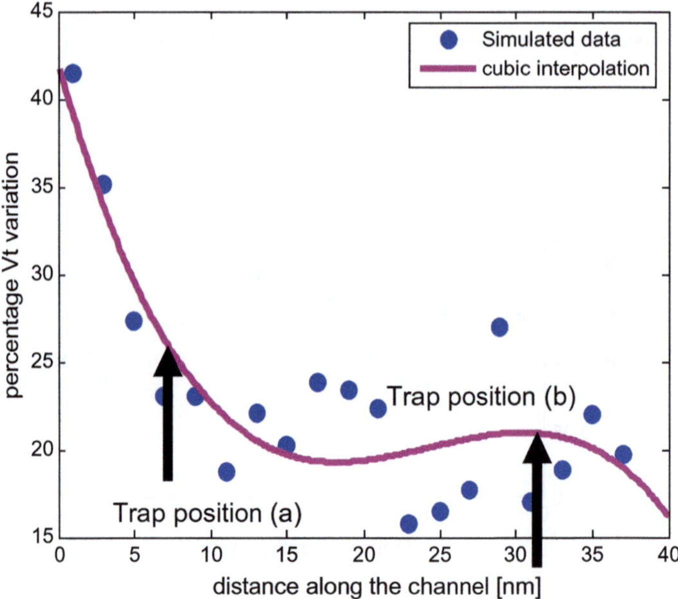

Fig. 4.11 Percentage threshold voltage fluctuation due to a double trap located at the semiconductor/oxide interface and different position along the middle section of the channel. Twenty devices with different random dopant distributions have been averaged out. Detailed description of the trend of the results shown in this figure is given in Fig. 4.12

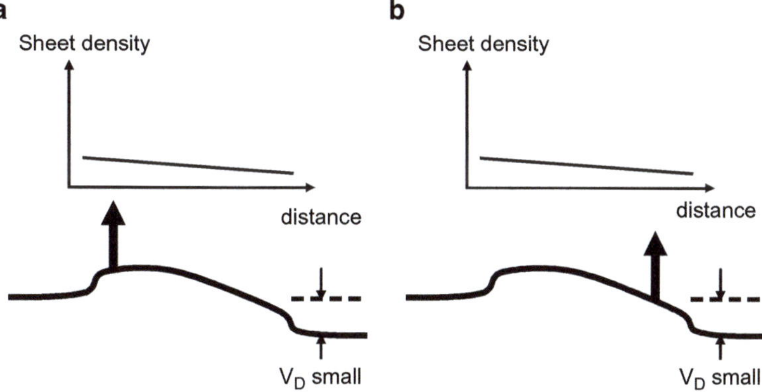

Fig. 4.12 Schematic explanation of the results from Fig. 4.11

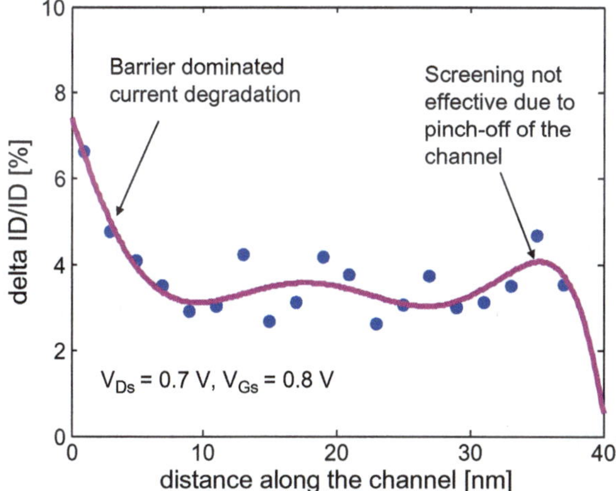

Fig. 4.13 ON-current degradation vs. trap position. The statistical ensemble used here consists of the first seven devices from Fig. 4.9 with different number and different distribution of the impurity atoms

influence since they are major obstacles to the electrons because of the large input barrier depicted in case (a) shown on the left panel of Fig. 4.12. Traps near the drain end of the channel have smaller influence since electrons are accelerated by the small electric field—case (b) shown on the right panel of Fig. 4.12.

Figure 4.13 shows the ON-current degradation as a function of the double trap position. As depicted on the figure, near the source end of the channel the current degradation due to the presence of a negatively charged double trap is large because the double trap introduces additional barrier for the current flow. When the double trap is in the middle section of the channel, the current degradation is smaller. Traps near the drain contact, where the electron density is pinched off for the bias

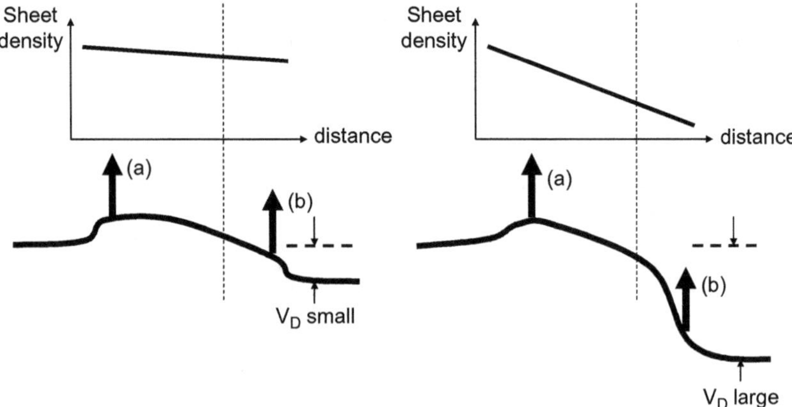

Fig. 4.14 Under small gate and drain bias, we have the situation depicted in Fig. 4.12. For large gate bias: (**a**) the sheet density increases which means that screening increases; (**b**) traps near the drain are surrounded with large number of electrons for small V_D, therefore screening of the Coulomb potential is large and there is smaller degradation of the current; (**c**) for large V_D traps near the drain are surrounded with smaller number of electrons, therefore screening is smaller and we have slightly larger I_D degradation

conditions used, are not effectively screened and a notable increase of the current degradation is observed. At the drain contact, the degradation drops practically to zero because there traps are effectively screened by the electrons. The expected current trends under small and large drain voltages are explained in Fig. 4.14.

4 Conclusions

Unlike most of the analytical revelation of published articles, where it has been reported that fluctuation in drain current amplitude variation and threshold voltage variation tend to diminish at strong inversion and saturation conditions imposing higher gate and drain bias owing to the screened out potential due to high inversion charge density at the surface with associated improvement of Coulombic-scattering related mobility, our simulations conducted at saturation bias conditions on a 45 nm MOSFET reveal that for different random dopant distributions in the channel and bulk, the fluctuation pattern exhibited by drain current amplitude variation and threshold voltage variation are truly statistical and random in nature, i.e., for some specific random dopant distributions, the fluctuation nature is well controlled whereas for some other random dopants, the fluctuation pattern shows significant transitions between local peaks and valleys. From our EMC simulation, it has been demonstrated that fluctuations in the drain current amplitude and the threshold voltage have been dependent on particular random dopant distribution type, i.e., its number within the channel area and its position, in addition to having strong correlation on strategically positioned interface traps along the channel from source to drain.

In order to truly represent the amplitude variation to show more dependence on spatial positioning of trap than specific random dopant type, the expectation value of the statistic (drain current or threshold voltage amplitude change) or the average term needs to be studied out of a significant number of possible random channel dopant distributions. These averaged-out values of drain current and threshold voltage fluctuations as extracted by the simulations can be analytically rationalized through the concept of device physics of scaled MOSFETs. We propose herein the following analytical physical explanations supporting the simulated results obtained for drain current fluctuations and threshold voltage fluctuations with associated fluctuations in its standard deviation induced by trap's occurrences in the channel close to the interface spatially located between source and drain.

4.1 Local Surface Potential Fluctuations

It has long been known that random dopant fluctuations cause fluctuations in the surface potential along the channel from source to drain and also from interface to some depth into the bulk. The fact that at strong inversion and saturation condition, the ready availability of inversion charge density smears out the surface potential variation, thus possible fluctuation in drain current and threshold voltage, needs a second look. With considerable scaling of gate length along with narrow width MOSFET devices, there are only few 100 atoms available per device area of scaled MOSFET and considering their random positioning along the channel and also along the depth from interface towards bulk, based on the fact that the depletion region available underneath the channel itself has a spatial random distribution owing to the random doping of the bulk, the inversion charge density formed by surface band bending is not continuous along the interface from source to drain, rather it is discrete and spatially random and nonuniform. Therefore the very possibility of screened out or smeared out surface potential at the interface is never a reality and depending upon the local fluctuation enhanced inhomogeneity of inversion charge density even at high gate bias, there will be sizeable spatial variation of surface potential at the interface from source towards drain. In addition to experiencing fluctuation due to random discreteness of channel dopant position and its number, the local surface potential fluctuation is also enhanced by the strategic location of trap or a number of traps interspersed between one another and distributed along the channel from source to drain. A repulsive trap center originating from the presence of acceptor like traps will enhance the spike in local potential amplitude at the trap's position and if this trap happens to be situated right at the source side or even a few nm away from the source side, the spike in potential barrier is going to significantly impede the carrier flow and thereby reduce the number of carriers that reach the drain side and contribute to current value. A trap's positioning at the source side being repulsive type consequently causes significant negative shift in current value and significant rise in fluctuation percentage value of relative drain current amplitude. This phenomenon is not uncommon in modern process controlled MOSFET fabrication as most RTF related phenomenon are caused by acceptor like traps being repulsive in nature. As the drain bias increases ensuring saturation conditions, the spatial conduction energy band gradually decreases towards the drain, enhancing drift related carrier transport where the carriers either enter the condition of velocity saturation or velocity overshoot near the drain zone and easily surmount any local barrier constituted surface potential fluctuation. Hence if the traps are near the drain side, the fluctuation pattern reduces to a steady and stable value. This physical aspect is corroborated by all the simulation outputs of present research work regarding drain current fluctuation. When the double-trap is located at the pinch-off region there is slight increase in the current degradation.

For the threshold voltage fluctuations and fluctuation in standard deviation of threshold voltage as extracted by our simulation results, the analytical explanation holds and since a trap's presence at the source side of the channel is causing maximum resistance to carrier flow by showing a peak of local surface potential at trap's position, therefore this will affect the inversion charge density available at threshold and shift the threshold voltage by altering the gate bias conditions at threshold. Therefore relative threshold voltage amplitude reaches a high peak near source side trap positions and due to presence of high enough energetic carriers along the drain region, a trap's positioning there would not alter the inversion condition to the extent that a considerable shift in threshold voltage amplitude will be expected. Hence, the threshold voltage shift for drain side-positioned traps will be minimal. This physical aspect is also vindicated by the available data statistics that have been shown in the plots shown previously for threshold voltage variation with its fluctuation in standard deviation of threshold voltage. The explanation presented in this subsection from device physics standpoint is attributed to the condition that a large number of channel random dopant distributions are considered and have been averaged to arrive at the values of final drain current fluctuation statistic and threshold voltage fluctuation statistic. Expectedly, the randomness of the distribution of channel and bulk dopants will induce comprehensive randomness in local surface potential fluctuations making the after effect on the drain current fluctuations and threshold voltage fluctuations to be purely statistical. In essence, the plots shown with different random dopant distributions related drain current amplitude fluctuation and threshold voltage variation reveal a very important observation that strategically devised random channel dopant distributions through ion implantation can more successfully suppress the variability in drain current and threshold voltage and excel against RTF related device failure.

4.2 Fluctuation in Carrier Mobility

Each trap acts as a potential scattering center and impacts the scattering related transport efficiency. When the traps are closely apart and randomly positioned along the channel from source to drain, trap's interaction with carrier electron in the inversion channel and fixed ions in the underlying depletion region modifies the short range trap-to-electron–electron and trap-to-electron–ion Coulombic potential giving rise to significant Coulomb scattering related mobility reduction. Since our EMC transport kernel is equipped with properly accounting for this Coulomb effect, the results of the simulations perfectly reveal the fact that due to the random reduction of transport mobility of carrier due to trap's position and interactions with nearby electrons and ions, drain current amplitude reduction and hence larger fluctuation of drain current amplitude is expected which will be spatially nonuniform. The Coulomb related short range and long range e–e and e–ion interaction with traps will also result in significant threshold voltage shift at inversion condition on the gate, hence is responsible for trap related threshold voltage fluctuation. The modification of Coulomb potential seen by a trap surrounding its vicinity where a significant or no carrier may reside and significant or no ion may reside stemming from their mutual random and discrete nature, is also severely impacted by a trap's proximity to a nearby trap. Since for all our simulations, the traps were very closely positioned to one another within 1 nm separation, this will induce more Coulomb potential change and changes in local short range trap-electron and trap-ion interaction affecting carrier mobility. The rapid screening of Coulomb potential only occurs at the drain side with high bias on drain and hence at those locations (except the pinch-off region), mobility reduction is minimal and the fluctuation pattern for both relative drain current and threshold voltage is minimized. In addition, depending on the discrete positioning and number density of channel dopants, the particular trap-to-electron and trap-to-ion interaction for both short and long range Coulomb forces will be fully distinctive and random causing randomness in mobility reduction spreading over different random channel and bulk dopant distributions. Larger spikes and surges in local surface potential values expected near the source side, will contribute to associated variation in short range and long range Coulomb force related scattering probability of carrier transport for a trap positioned there.

4.3 Interface Conditions

From the EMC simulation results of our present research work on RTF related drain current and threshold voltage fluctuation, a close observation reveals the important viewpoint that for interface trap occurrences, only traps within a thousandth fraction of a nm from the interface in the channel are considered for study. It has been shown by the exhaustive work of Professor Dragica Vasileska and her co-researchers that both the drain current and threshold voltage are strongly correlated to within a few nm from the interface to the vertical channel depth. Since the traps designed in our simulation are almost within a considerable fraction of a nm from the interface, it is vindicated from previous simulation study with present EMC code that both drain current fluctuation pattern and threshold voltage fluctuation pattern will show strong positive correlation with trap's proximity to the interface. This feature has been found to be another potent factor for reasoning out wide fluctuation patterns observed for both drain current amplitude and threshold voltage variation with its standard deviation. In addition, the interface roughness and chemical imperfections due to trap's positioning as scattering centers at random locations from source to drain are also a significant source for interface roughness scattering related mobility reduction which from theory is found to be one of most dominant causes for mobility reduction at high gate and drain bias, i.e. saturation bias condition of the MOSFET which we employed in our simulation for on current fluctuations study. The interface roughness will reduce the overall effective mobility with the traps acting as scattering sites randomly distributed within the inversion channel, being in conjunction with Coulomb-force related mobility reduction and will impact spatial random and statistical variation in both relative drain current amplitude.

References

1. Keyes R. W., The effect of randomness in the distribution of impurity atoms on FET thresholds, *Appl. Phys.* vol. **8**, **251**–259 (1975).
2. Mizuno T., Okumtura J. and Toriumi A., Experimental study of threshold voltage fluctuation due to statistical variation of channel dopant number in MOSFET's, *IEEE Trans. Electron Devices* 41, 2216–2221 (1994).
3. Vasileska D., Gross W. J., Kafedziski V. and Ferry D. K., Convergence properties of the Bi-CGSTAB method for the solution of the 3D Poisson and 3D electron current continuity equations for scaled Si MOSFETs, *VLSI Design* 8, Nos. 1-4, 301–305, (1998).
4. Fischetti M. V., Towards Fully Quantum Mechanical 3D Device Simulations, *Journal of Computational Electronics* 1, 81–85 (2002).
5. Fischetti M. V., Effect of the electron–plasmon interaction on the electron mo-bility in silicon, *Phys. Rev. B* 44, 5527–5534 (1991).
6. Lilly M. P., Eisenstein J. P., Pfeiffer L. N., and West K. W., Coulomb Drag in the Extreme Quantum Limit, *Phys. Rev Lett.* 80, pp. 1714–1717, (1998).
7. Wong H. S. and Taur Y., Three-dimensional "atomistic" simulation of 50 nm FETs, in *Proc. IEDM*, 29.2.1 (1993).
8. Gross W. J., Vasileska D. and Ferry D. K., 3D Simulations of Ultra-Small MOSFETs with Real-Space Treatment of the Electron–Electron and Electron–Ion Interactions, *VLSI Design*, Vol. 10, 437–452 (2000).
9. Asenov A., Random Dopant Induced Threshold Voltage Lowering and Fluctuations in Sub-0.1 um MOSFETs: A 3-D Atomistic Simulation Study, *IEEE Trans. Electron Devices* 45, 2505–2513 (1998).
10. Sano N., Matsuzawa K., Mukai M., and Nakayama N., On discrete random dopant modeling in drift-diffusion simulations: physical meaning of 'atom istic dopants, *Microelectronics Reliability*, Volume 42, Issue 2, 189–199 Feb. (2002).

11. Ferry D. K., Kriman A. M., Kann M. J., and Joshi R. P., Molecular dynamics extensions of Monte Carlo simulation in semiconductor modeling, *Computer Physics Comm.*, vol. 67, 119–134 (1991).
12. Gross W. J., Vasileska D. and Ferry D. K., A Novel Approach for Introducing the Electron–Electron and Electron–Impurity Interactions in Particle-Based Simulations, *IEEE Electron Device Lett.* 20, No. 9, 463–465 (1999).
13. Greengard L. and Rokhlin V., A fast algorithm for particle simulations, *J. Comput. Phys.* 135, 280–292 (1997).
14. Beatson R. and L. Greengard, A short course on fast multipole methods, in Wavelets, Multilevel Methods and Elliptic PDEs (Leicester, 1996), ser. *Nu mer.Math. Sci. Comput.* New York: Oxford Univ. Press, 1–37 (1997).
15. Cheng H., Greengard L., and Rokhlin V., A Fast Adaptive Multipole Algo-rithm in Three Dimensions, *J. Comput. Phys.* 155, 468–498 (1999).
16. FMMPART3D user's guide, version 1.0 ed., *MadMax Optics*, Hamden, CT, USA.
17. Khan H. R.,Vasileska D., Ahmed S. S., Ringhofer C. and Heitzinger C., Mod-eling of FinFETs: 3D MC Simulation Using FMM and Unintentional Doping Effects on Device Operation, *Journal of Computational Electronics*, Vol. 3, Nos. 3-4, 337–340 (2005).
18. Stolk P. A., Widdershoven F. P. and Klaassen D. B. M., *IEEE Trans. Electron Devices* **45**, 1960–1971 (1998).
19. Horstmann J. T., Hilleringmann U. and Goser K. F., *IEEE Trans. Electron Devices* **45**, 299–306 (1998).
20. Majima H., Ishikuro H., and Hiramoto T., *IEEE Electron Dev. Lett.* 21, 396–398 (2000).
21. Agostinelli T., Arca M., Caironi M., Ferrero V., Natali D. and Sampietro M., Trapping effects on the frequency response of dithiolane-based planar photodetectors, *ICSM0 International Conference on Science and Technology of Synthetic Metals.*

Chapter 5
On-Chip Characterization of Statistical Device Degradation

Takashi Sato and Hiromitsu Awano

Abstract Bias temperature instability (BTI) is one of the most critical degradation mechanisms that occur in modern semiconductor devices. The degradation due to BTI is transient, and known to be greatly influenced by bias voltages and temperature, making it very difficult to detect possible BTI-related failures during manufacturing test. Characterization and modeling of BTI is hence extremely important to protect a chip from BTI-related failures. In this chapter, an array structure that accelerates the statistical characterization of BTI is described. By overlapping the stress-application period for each device, measurements on hundreds or thousands of devices can be conducted concurrently. Test chip measurement results that provides a statistical insight on the parameters of BTI-related degradation process are also presented.

1 Introduction

The quality of the semiconductor–insulator interface in silicon devices has been a major topic of research. Defects in the interface between silicon (Si) and silicon dioxide (SiO_2) not only lower device performance of the metal–oxide–semiconductor (MOS) transistor but may also lead to severe reliability degradation. The lifetime of a device decreases when the quality of the interface film is insufficient. Owing to the appropriate treatments of the interface, such as optimizing the annealing process using hydrogen (H) gas, which passivates the Si-dangling bond by H, modern silicon devices have become considerably more stable and are thus durable. Enhanced device reliability has allowed semiconductor circuits to be adapted in a broad range of applications.

Despite the significant quality improvement, unsaturated bonds are still considered to exist. The interface states and the associated change in the surface potential is again becoming an increasing concern for circuit designers who utilize devices fabricated using nanometer-scale technology. As device dimensions become even smaller, a single charge that is trapped in the interface region can greatly distort

T. Sato (✉) • H. Awano
Kyoto University, 36-1 Yoshida-Hommachi, Sakyo-ku, Kyoto 606-8501, Japan
e-mail: takashi@i.kyoto-u.ac.jp

R. Reis et al. (eds.), *Circuit Design for Reliability*,
DOI 10.1007/978-1-4614-4078-9_5

channel formation. The temporal change in the charges in the channel region will cause a drift in circuit performance or, in the worst case, will result in a malfunction of the circuit.

As semiconductor circuits are utilized in various applications, diverse levels of reliability are demanded. Certain applications impose very stringent reliability requirements. Examples include transportation systems such as automated car-driving systems or medical applications. On the other hand, in another application, cost lowering may be pursued as the highest priority, requiring a certain level of reliability to guarantee equal or longer than product lifespans. It has now become a common practice to design for the reliability of a circuit in addition to its performance.

A device model that incorporates a transient change in the device parameters is a key enabler to facilitate the design of device and circuit reliability. As a solid basis for building such a model, it is important to establish an accurate measurement methodology that will contribute to understanding the mechanisms behind the device degradation. In particular, the phenomena of bias temperature instability (BTI) are drawing increasing attention.

In this chapter, we survey characterization methods for the purpose of device modeling that includes the effects of temporal parameter degradation. We begin by quickly reviewing the physical mechanisms of temporal threshold voltage (Vth) changes: random telegraph noise (RTN), negative bias temperature instability (NBTI), and positive bias temperature instability (PBTI), which are considered to be interrelated and important in miniaturized devices. Then, we review the work to accurately capture the RTN, NBTI, and PBTI phenomena. Particular focus is placed on the approaches that utilize on-chip circuits to enhance the accuracy and fidelity of the measurements. Finally, the characterization of statistical degradation will be reviewed with examples.

1.1 Transient Change in Device Parameters

BTI is considered to be one of the major limiting factors for long-term performance and reliability of semiconductor devices [25, 30].

NBTI and PBTI degradations are observed in pMOS and nMOS transistors, respectively. These phenomena are observed as a gradual increase in threshold voltage during circuit operation. The threshold voltage increase is accelerated under a strong vertical electric field and under an elevated temperature. It is also known that the increased threshold voltage will be partially recovered once the stress condition of the electric field or temperature is alleviated. Hence, their models naturally become dependent on history. The prediction of a device parameter at a certain time in the future has been found to be very difficult.

Physical mechanisms of threshold voltage shifts due to NBTI have long been under debate. As device dimensions decreases, there is an urgent need for developing a model that is physically correct and thus predictable for both degradation and recovery phenomena. One promising hypothesis is the formation of a positive oxide charge due to interface traps. The other explanation is the hydrogen-ion diffusion generated by broken Si-H bonds.

Another developing concern is PBTI, which is the degradation in nMOS transistors. PBTI is attributed to the application of new materials such as high-k gate insulators and metal gates. In addition to the interface state degradation, the capture of carriers in the intrinsic defects in high-k insulator stacks causes a threshold voltage shift. For example, the void formulated in an HfO_2 dielectric gives a rise to the threshold voltage [32].

The degradation processes in pMOS and nMOS transistors originate in discrete charges. The effects of BTI are expected to become more discrete and larger in devices fabricated by using scaled technologies [17]. Hence, the characterization of device-dependent degradation has gathered keen interest.

1.2 Measurement Requirements

It is essential to incorporate BTI effects into device models. Good experimental data is critically important to gain a full understanding of the BTI phenomena and to improve the reliability of a circuit.

Variation in the NBTI parameter values has not been fully investigated despite its importance. One major difficulty involved in the measurement of BTI phenomena is the inclusion of very slow components in the threshold voltage change. Even if the degradation is accelerated under high-temperature and high-stress-voltage conditions, hours or even days are required to observe threshold voltage change because of slow components. This has made it almost impossible to collect statistical data on a large number of devices in a practical time frame.

It is also known for the BTI phenomena that the threshold voltage recovery incorporates a very rapid component as well. In the rapid recovery of small devices, stair-like changes in threshold voltage have been observed. This suggests the release of a charge from the interface trap [9]. Such a wide-spread time response of the threshold voltage recovery makes measurement even more difficult.

In most of the existing measurement structures, large-channel-area devices have been used. Even when individual small-area devices are used, the equivalent device area is considered to be large. An example is a ring oscillator structure whose frequency change reflects the device degradation [12]. Although the oscillation frequency indicates an average degradation of the devices, it is more important to separate the degradation of a single transistor to create an accurate model.

Device size, i.e., the channel area of a device, has a strong impact on the temporal threshold voltage changes. In particular, in the context of trap-based models [3, 8, 9, 20], device size is a critical parameter because it is closely related to the expected number of defects in a device. The amplitude of threshold voltage change caused by the capture and release of a charge in an interface trap is also severely affected. Understanding the variability in NBTI parameters for devices that are approximately the sizes of logic LSIs is thus critically important. Additionally, note that as device size decreases, the measurements become more difficult because small parasitic elements in the measurement system may significantly alter the measurement results.

2 Circuit Structures for the Measurement of Device Degradation

Reflecting the importance and difficulty of characterizing device degradation on the basis of the underlying physics, various circuit structures have been proposed [5, 7, 11, 12, 14].

2.1 *Measurements Using Off-Chip Equipment*

In this section, we review the measurement techniques for the characterization of BTI of a single transistor using off-chip equipment. The off-chip equipment refers to parametric analyzers with source-measurement units (SMUs).

The most straightforward and widely used method is to measure I–V curves through direct probing because BTI is observed as a change in threshold voltage. Once the full I–V curve is obtained before and after stress application, most of the device parameters, such as threshold voltage or carrier mobility, can be extracted to analyze their temporal changes. The drawback of this measurement is a relatively long interruption of the bias voltages. The bias conditions during voltage sweeping for I–V curve measurement are, in most cases, different from the stress or recovery bias in the BTI measurement.

The threshold voltage measurement using the *constant current method* defines the threshold voltage to be the gate-to-source voltage at which a particular current, e.g., 10^{-7} A, in which the transistor is in the linear region, is observed [16]. Because this measurement focuses on observing the threshold voltage only, this measurement evades full I–V curve measurement. However, the constant current method still requires a device to depart from a strong stress or recovery bias condition. Though the interruption is short, partial recovery is still unavoidable during the measurement. This may lead to under-estimation of the degradation. In [19], the use of an operational-amplifier-based on-chip current generator is proposed. This type of circuit can potentially shorten the settling time of the output node voltage as compared to what can be achieved using off-chip parametric analyzers.

On-the-fly (OTF) measurement [4] attempts to eliminate the bias interruptions associated with the threshold voltage measurement. In this method, instead of measuring I-V curves, the change in the drain current is measured while the stress voltage is continuously applied to the device. Hence, the degradation of a device can be observed without recovery associated with the stress interruption. An ultra-fast on-the-fly (UFOTF) method improved timing resolution to 1 μs [13]. This method utilizes a current-to-voltage converter to isolate the device of interest from an oscilloscope probe. However, these methods require the transconductance of the device in order to convert the measured current to the corresponding threshold voltage. Because transconductance may also be subject to variation between devices, error associated with the conversion may be involved. In addition, the sensitivity of the threshold voltage to the drain current tends to be very weak in the saturation region, in which a strong stress bias condition for the device is achieved. In this case, it is difficult to obtain high accuracy.

In [14], the drain current measurement of a transistor in the weak inversion region, instead of the saturation region, is proposed to utilize the exponential sensitivity. In order to overcome a slow settling time due to the extremely small charging and discharging current of a device in the weak-inversion region, the device of interest has to be large, and an assist circuit is required. In the literature, it is reported that a measurement delay as fast as 400 ns has been achieved. Because this method also relies on the drain current measurement, a conversion is also required to obtain the threshold voltage by using separately acquired transconductance.

2.2 *On-Chip Measurements*

A ring oscillator is a widely used circuit structure for device characterization because it is relatively easy to implement and measure. Ring oscillators are also applied for BTI measurements. In [12], Kim et al. proposed a ring-oscillator-based degradation sensor, which is known as a silicon odometer. It accurately measures device degradation by monitoring the beat frequencies of two ring oscillators, one that is fresh and another that has experienced stress periods. This sensor is suitable for monitoring dynamic degradation behavior, such as on-chip path-delay degradation. The advantages of utilizing ring-oscillators are twofold: they do not require additional off-chip measurement equipment, and they can be placed in environments where the circuits are in operational condition. The drawback is that this technique cannot measure the degradation of a single transistor because all transistors that compose the ring oscillator are involved in the degradation of the oscillation frequency.

In an attempt to enhance the sensitivity of several selected transistors in the ring-oscillator sensor, an inhomogeneous ring oscillator sensor was proposed [6]. In this type of sensor, an inverter stage is designed to have significantly larger delay than the other stages by connecting a pass gate transistor—pMOSFET or nMOSFET—between the output of the preceding inverter and the inverter stage of interest. The device type is selected such that the gate voltage of the inverter is limited by the threshold voltage, making its delay very sensitive to the threshold voltage of the pass gate transistor. To keep track of the degradation of SRAM cells, a ring oscillator composed of embedded SRAM was proposed in [26].

3 BTIarray for Statistical Characterization of Device Degradation

As device dimensions continue to shrink, the variability of device parameters is becoming an increasing concern. Similarly, the variability of the parameter degradation may differ between devices. Even if two identical devices have experienced the same BTI stress, their degradation behaviour differs because of inherent variability. In order to accurately account for the impact of degradation during circuit designs, the modeling of degradation variability, in addition to the average degradation trend, is very important [28, 29]. For this purpose, measurements for a large number of devices are vital in characterizing the statistical properties of BTI phenomena. However, the methods that were previously reviewed target the accurate capture of the degradation of a single transistor. Considering the long measurement period, they are difficult to apply for characterizing a statistically significant number of devices.

One promising approach is to construct an array circuit on a chip to facilitate overlap of a long biasing time when collecting degradation data on a large number of devices. In the following subsections, the circuit design and experimental results of an array-based on-chip approach will be reviewed.

3.1 Stress and Recovery Time Overlapping

When measuring BTI phenomena, bias voltages have to be applied for a long time to observe a detectable degradation in the threshold voltage. The largest obstacle for statistical BTI characterization is the very long time required for the measurement of multiple devices.

Upon close look examination of the BTI measurements, it is observed that each threshold voltage measurement takes up only a very short time as compared to the stress or recovery period. Most of the measurement period is used for applying the bias voltages to the device. In such a condition, which is typical in BTI characterizations, the total period for the measurement of multiple devices will be significantly shortened by the overlapping stress periods of different devices.

A straightforward approach is to conduct I–V curve measurements for multiple devices using a high-pin-count probe. In that case, all electrodes of the devices need to be equipped with individual pad electrodes for probing. The use of a probe card with a very high pin-count correspondingly requires a large number of SMUs, which is impractical. Signal synchronizations become more complicated and costly

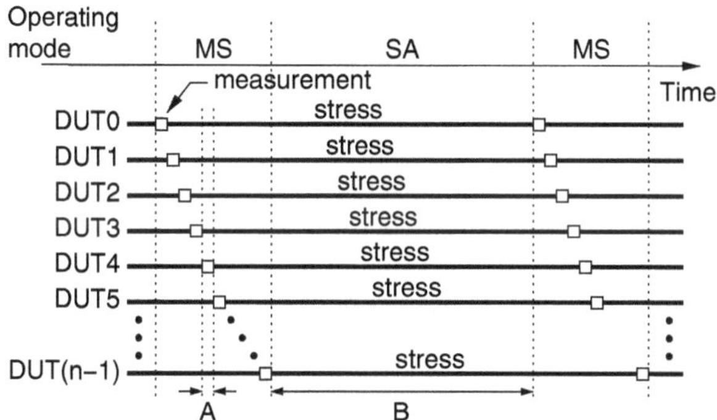

Fig. 5.1 Stress period overlapping on n devices. Stress periods with equal durations can be achieved for all devices

as a higher number of parametric analyzers are used. Another drawback of this type of approach is area overhead. A set of four probing pads is necessary for the measurement of a device. Because a probe needle has to physically contact a probing pad, the pad area cannot be arbitrarily small. A probing pad occupies a significantly larger area than the device of interest does, which significantly reduces the area efficiency. This area gap between the device and probing pad will increase as transistors are miniaturized.

To resolve the above inefficiency, an array-based approach that simultaneously measures the degradation of many devices has been proposed, which is called BTIarray [22]. The concept of the bias overlapping in the BTIarray is illustrated in Fig. 5.1. In [22], the measurement results of 128 transistors are also presented, which validates the stress overlapping concept. A similar idea has been proposed in [24] with only simulation results.

The bold lines in Fig. 5.1 represent the stress periods in which *devices under test* (DUTs) are in a stress condition. The small box represents a period in which the threshold voltage of a selected DUT is measured. A single voltage source is shared, and the voltage is distributed for all DUTs in the array to apply an equal stress voltage for an exactly equal period, while the threshold voltage measurements are performed individually. With the aid of on-chip circuitry, probing pads and hence the SMUs are shared by the all DUTs in the array.

The application of stress or recovery bias voltages is controlled by an on-chip circuit that is implemented with the DUT. In contrast to the parallel probing, a single supply voltage is shared among all the DUTs on the chip. The circuit structure and measurement process will be reviewed in detail in the next section. It is also crucially important to completely automate the measurement process in the statistical device measurements. The measurement environment of the BTIarray will also be described in the following section.

3.2 Circuit Structure of BTIarray

BTIarray provides several modes to realize a series of measurement procedures of many DUTs. Figure 5.1 depicts example modes and procedures. In time slot A, a single DUT is selected for threshold voltage measurement, and all other DUTs are in a stress condition. During time slot B, a stress bias voltage is applied for all DUTs. When observed from a DUT, the DUT is always in one of three conditions—stress, recovery, or measurement. Considering the typical measurement procedure of BTI degradation, the BTIarray should provide at least four operating modes:

- stress all (SA): apply stress bias for all DUTs simultaneously,
- recovery all (RA): apply recovery bias for all DUTs simultaneously,
- measurement stress (MS): measure threshold voltage of a selected DUT, and simultaneously apply stress bias for other DUTs,
- measurement recovery (MR): measure threshold voltage of a selected DUT, and simultaneously apply recovery bias for other DUTs.

In [22], to realize all of the above operational modes, the unit structure of the BTIarray is defined as shown in Fig. 5.2a. This circuit is called a DUT unit. The DUT unit consists of a DUT cell and a control logic circuit. The DUT unit is designed so that it can be arranged adjacent to the other DUT units in the horizontal and vertical directions to form a device array similar to [23]. The array structure facilitates sharing of the supply voltage and control signal networks among many DUTs. Forming an array also maximizes DUT density by sharing probing pad electrodes, realizing a compact test structure. Control logic generates switch control signals according to the combination of the control and address-select signals: MEAS, STRS, and MSEL. The internal signal MSEL becomes logical '1' when the DUT unit is selected by the address signal.

A simplified schematic of a DUT cell that implements a pMOS-transistor as the DUT is shown in Fig. 5.2b. In the DUT cell, three terminals of the DUT are connected to different voltage sources through pass-gate switches to realize a specified operating mode. Voltages of V_{SS}, V_{STR}, V_{REC}, and V_{DD} are shared among all DUTs in the circuit, each of which is connected to a constant voltage source. V_{DD} and V_{SS} are nominal supply and ground voltages, respectively. V_{CC} is the terminal to connect a current source for the threshold voltage measurement using the constant current method. All switches are realized by CMOS pass-gate transistors.

An example construction of a BTIarray is shown in Fig. 5.2c [22]. In this example, an array of 128 DUTs is formed. In order to select a DUT in the array, 7-bit address signal is decoded into 8-bit and 16-bit select signals, using X- and Y-decoders, respectively.

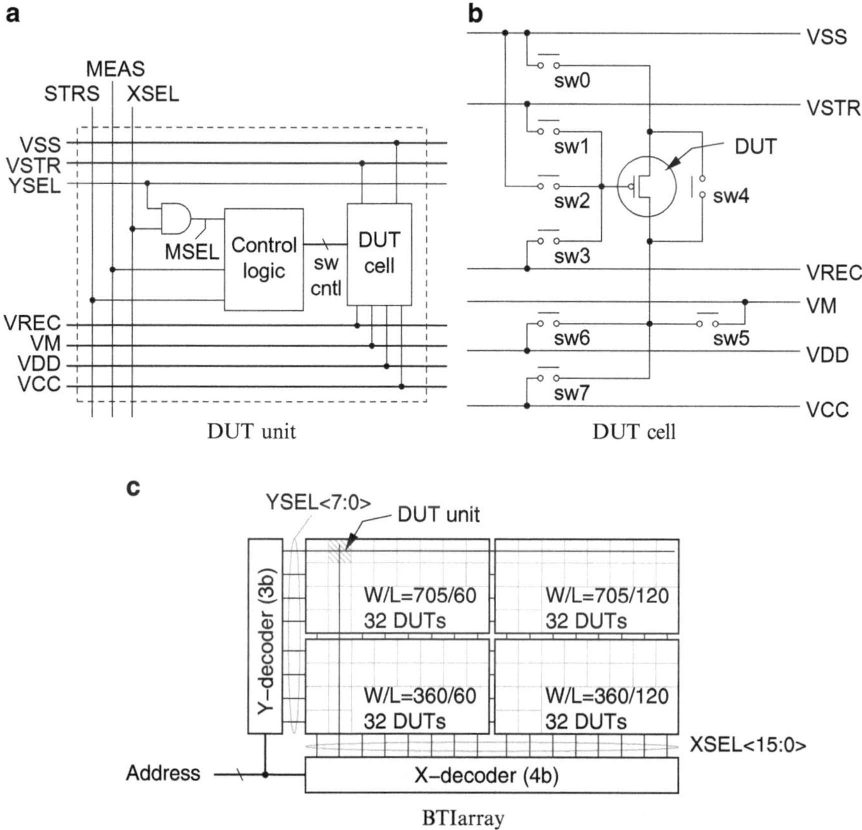

Fig. 5.2 Circuit structure of BTIarray for a pMOS DUT. (**a**) The DUT unit consists of "control logic" and a "DUT cell." All signals pass through the DUT unit, allowing an array circuit to be easily formed. Control logic generates switch control signals for the DUT cell. (**b**) Schematic diagram of a DUT cell. Switches are realized by pass-gate transistors. (**c**) Placement of DUT units in an implementation of a BTIarray [22]. An array of 128 DUTs is formed. Power supply and control signals are omitted for clarity except for XSEL and YSEL

Switch Control

In the SA mode, stress voltage is applied for a DUT in the DUT cell by closing switches sw1, sw4, and sw6. Other switches are controlled to be open. The stress voltage is V_{STR} minus V_{DD}, where V_{STR} is more negative than V_{DD}. In the RA mode, switches sw0, sw3, and sw6 will be closed, and others are left open. V_{REC} is a recovery voltage and nominally equals V_{DD}, which turns devices off.

In the two measurement modes, MS and MR, a measurement is carried out for a DUT that is selected by an address signal. Other DUTs that are not selected are all in either a stress or a recovery condition. DUT selection is specified by XSEL and YSEL signals, both of which are generated as one-hot signals by the decoders in

Table 5.1 An example truth table of control logic

MEAS	STRS	MSEL	Mode
0	0	*	Recovery
0	1	*	Stress
1	0	0	Recovery (in MR)
1	0	1	Measurement (in MR)
1	1	0	Stress (in MS)
1	1	1	Measurement (in MS)

An asterisk (*) in the table indicates "*don't care*"

Fig. 5.2c. Threshold voltage is measured by the constant-current method [16]. The voltage that appears at node V_M is captured during a constant current is applied to Vcc. Hence, the switches sw0, sw2, sw5, and sw7 are closed in the selected DUT cell. The switches of other DUT cells that are not selected are equal to those of the SA mode or RA mode, depending on whether the current mode is MS or MR, respectively.

Control Logic Design

The control logic circuit in the DUT unit in Fig. 5.2a can be realized according to the truth table in Table 5.1. In the BTIarray realization in [22], mode switching is achieved by the combination of three control signals: "MEAS," "STRS," and "MSEL." A DUT is selected for the measurement only when both MEAS and MSEL are "1," i.e., the array is in one of the measurement modes, and the address of the DUT is selected. Otherwise, the constant bias voltages that correspond to either the stress or recovery mode is applied to all the DUTs in parallel.

Advantages of BTIarray

BTIarray achieves substantial reduction during the measurement period. Because the stress period is usually several or many orders of magnitude longer than the time required for threshold voltage measurement, the total measurement period becomes very close to that of a single DUT, regardless of the number of DUTs in the array. In addition, with the array structure, the time for the mode transitions can be made short and equal for all DUTs. This is a preferable characteristic when collecting statistical data over many DUTs. During the entire measurement period, all voltage sources and a current source are maintained at constant values so that the change in bias voltages is realized by the on-chip pass-gate switches. Quick current and voltage steering is achieve by this construction because parasitic capacitance existing in the array side of the switches is significantly smaller than that of an off-chip parameter analyzer. Hence, fast transitions between the operational modes are achieved, which contributes to precise equalization of the stress and recovery periods.

Table 5.2 Chip specifications [1]

	pMOS array	nMOS array
Technology	65-nm, 11-metal-layer CMOS	
Circuit area	225.2 μm× 62.3 μm (DUT array), 489.2 μm × 332.8 μm (incl. pads)	
Implemented DUTs	128 DUTs, 32 each size	
width(nm)/length(nm)	705/60, 705/120, 360/60, 360/120	360/60, 360/120, 180/60, 180/120
Precision at 10 ms/acq.	0.07 mV rms	0.05 mV rms

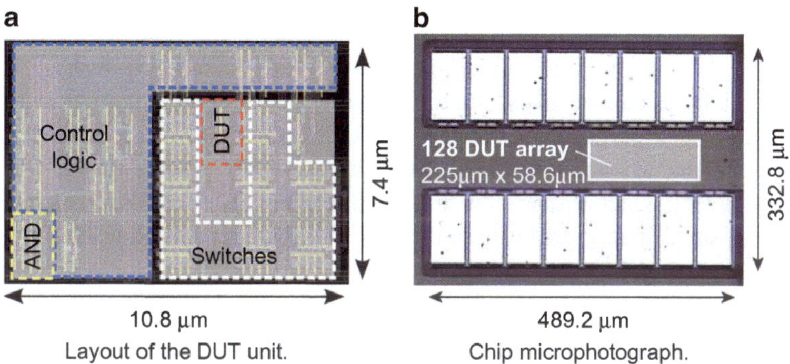

Fig. 5.3 An example implementation of the BTIarray. (**a**) Layout of the DUT unit, showing selected layers up to the second level wiring. (**b**) A total of 128 DUTs are implemented sharing only 16 probing pads [22]

3.3 Implementation Example

The implementation and measurement results of the BTIarray for pMOS transistors were first presented in [22]. Then, in [1], measurement results of BTIarrays for both pMOS and nMOS transistors were presented. The nMOS array was designed as a separate array from the pMOS array, but the circuit structure of the two arrays is similar. The only difference between the two arrays is the switched polarities of the stress and recovery voltages, and the connection of the DUT.

The specifications of the array design using a 65-nm process are summarized in Table 5.2. In both arrays, four transistor sizes were implemented as DUTs in combination with two channel lengths and two channel widths.

The layout diagram of the pMOS DUT unit including the pass-gate switches and control logic circuit is shown in Fig. 5.3a. The layout size of a DUT unit is 10.8 μm × 7.4 μm. Maximum layout regularities have been pursued by sharing the single layout skeleton in Fig. 5.3a for four different DUT sizes. As a result, the layout designs of the switches and the control logic were completely identical except for the sizes of transistors that were located at the center of the DUT unit. This regularity minimized layout-dependent device parameter variation. Additionally, it greatly eased the layout design of the arrays.

As observed Fig. 5.2c, the number of DUT units with an equally sized DUT is 32. In these designs, the DUT units are regularly placed to form a 4×8 array. Four 4×8 arrays are again arranged to construct an entire array; thus, 128 DUTs are implemented in total. A microphotograph of a test chip is presented in Fig. 5.3b. The layout size of the array including decoders and periphery circuit is 225.2 μm $\times$ 62.3 μm, and the total test circuit area including the probing pads is 489.2 μm $\times$ 332.8 μm.

Considering the circuit structure of the DUT unit and formation of the entire array structure, the number of DUTs in an array can be easily scaled. More recently, it has been reported that 3,996 DUTs have been successfully implemented in an array circuit by extending the concept of BTIarray without compromising measurement accuracy [2]. A slight circuit modification was made to suppress leakage current caused by a large number of off-state pass-gate switches to maintain measurement accuracy. For that purpose, the concept of a dummy supply voltage [23] is utilized to eliminate the leakage current. If the dummy supply voltage is not used, the constant current, which should flow through a selected DUT to develop its threshold voltage, becomes inaccurate because the current will be reduced significantly by the leakages.

3.4 Measurement Automation Through Scripting

When measuring of device degradation for many transistors, it is important to precisely control stress and recovery periods. The signal timing is particularly critical because a BTI-induced threshold voltage shift has components that quickly respond to the change in the bias voltages. Even a short stress interruption can lead to a partial recovery of the threshold voltages. Other conditions such as equalizing the temperature of the transistors are also important when statistical degradation results for a large number of DUTs are to be collected.

When measuring statistical degradation, it is also important to make the experiments completely reproducible. Because of the long measurement period, complicated commands have to be sent to the circuit many times in order to observe the response to the changing bias voltages. The ability to repeat exactly the same measurement procedure for multiple arrays is also required to further increase the number of samples.

In order to fulfill all these requirements, a scripting language interface for the BTIarray measurement is proposed [21]. The automation of setting measurement parameters and of issuing commands to the instruments is realized by the instrumentation as depicted in Fig. 5.4. A scripting language, *BTIscript*, has been developed in order to specify signal switch timing, voltage levels, chuck temperature, and other parameters, at a higher abstraction level. The BTIscript was realized on the basis of the Python scripting language. When writing the measurement scenario using the BTIscript, all types of powerful features equipped

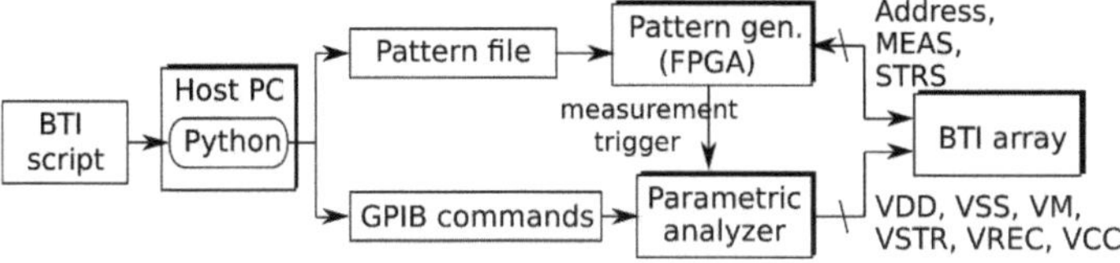

Fig. 5.4 Instrumentation for the measurement of BTIarray. Stress, recovery, and measurements can be entirely written in a macro language, BTIscript

Table 5.3 Subset of macro commands realized as functions in Python used in BTIscript

Macro	Description
stress_all(*duration*)	All DUTs are biased to stress for *duration* seconds.
recovery_all(*duration*)	All DUTs are biased to recover for *duration* seconds.
measure_stress_all(*burst*)	Measure all DUTs in ascending order. Measurement of a DUT is conducted for *burst* times while all other DUTs are biased to stress.
measure_stress_one(*adrs, burst*)	Measure single DUT specified by address *adrs*. Measurement of a DUT is conducted for *burst* times while all other DUTs are biased to stress.
measure_recovery_all(*burst*)	Measure all DUTs in ascending order. Measurement of a DUT is conducted by *burst* times while all other DUTs are biased to recover.
measure_recovery_one(*adrs, burst*)	Measure single DUT specified by address *adrs*. Measurement of a DUT is conducted by *burst* times while all other DUTs are biased to recover.

Parameters in parentheses may have a default value as in a regular Python function. The addition of new commands that are dedicated to a specific measurement is easy [21]

with the Python language, such as predefined and user-defined classes, subroutine calls, and flow controls, can be fully utilized.

The scripting language is useful to efficiently develop and debug measurement scripts because of its user-friendliness. Without such an environment based on a programming language, lengthy, complicated, and thus error-prone commands would have to be hand-written. In particular, iterations and procedure calls with parameters are very useful to represent measurement procedures both formally and compactly. Macro commands covering typical measurement procedures for BTI degradation are listed in Table 5.3.

To understand the versatility of the script, let us proceed through an example script in Fig. 5.8 by comparing the commands to the measurement results, such as in Fig. 5.9. Other measurement examples can also be found in [21, 27, 29].

Figure 5.5 shows a block diagram of the measurement system. A hand-written BTIscript is first converted into a pattern file and GPIB commands. The pattern file defines the transition timings of the control signals, and the commands from GPIB and triggers from the FPGA-based pattern generator control the parametric analyzer.

The generated pattern file is transferred to the FPGA via a USB interface. Sequences of bit patterns and output timings are transferred from the host PC via

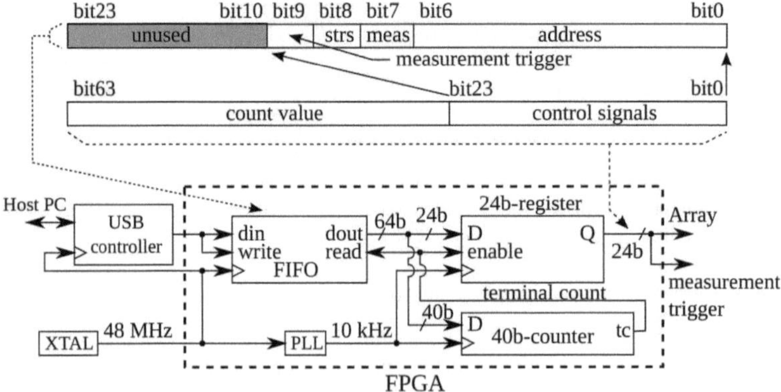

Fig. 5.5 Block diagram of the FPGA-based pattern generator. A 24-bit command output is generated to control the BTIarray and a parametric analyzer

USB. Each atomic command is formatted in a 64 bit command word. The upper 40 bits and lower 24 bits correspond to the count value and the state of the control signals, respectively. The count value determines the duration for which the control signals of a mode are applied. It is used as an initial value of the 40-bit countdown counter. When the counter reaches zero, the next command word is fetched from the FIFO queue to reload the counter and the output register.

The GPIB commands setup output voltages, output current, and the measurement parameters such as integration time. Once the measurement is started, threshold voltage measurement is conducted by the parametric analyzer according to the trigger signal generated in the FPGA.

3.5 Measurement Accuracy

The timing accuracy of the FPGA-based measurement environment was evaluated using an oscilloscope, referencing the trigger output that is generated by the parametric analyzer at the end of each measurement operation. When the target interval is 10 ms, the standard deviation of the transition timing was 19.0 μs in the FPGA-based pulse generator [1]. That of a PC-based command control [21, 22] using a GPIB interface was 452 μs.

In [1], the precision of the voltage measurements was also evaluated, which was approximately 50 μV rms or lower. It is reported that the voltage precision was evaluated using a DUT that had very small fluctuations because small stair-like fluctuations in the threshold voltage due to RTN were observed in almost all DUTs. The RTN of a DUT can be easily measured by repeating threshold voltage measurements in the MR mode. An example RTN waveform captured using the BTIarray fabricated using a 65-nm process is shown in Fig. 5.6. In the figure, multilevel RTN is clearly captured.

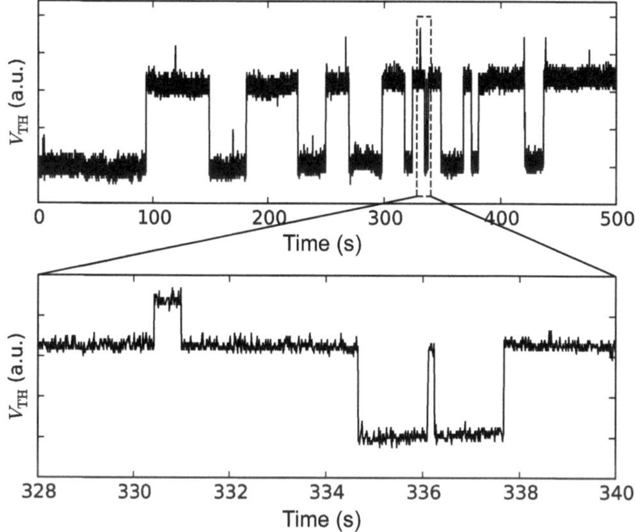

Fig. 5.6 Example threshold voltage measurement of a pMOS device, showing multilevel random telegraph noise (DUT size: W/L = 360/60 nm/nm)

Ideally, the currents and voltages of the measurement paths should be switched instantly by the on-chip pass-gate switches. However, in reality, there is a lower limit on the transition time because of parasitic capacitances in the circuit and the measurement system. Hence, the output voltage node requires a finite time to settle to a final value. The settling time of the output voltage of the BTIarray has also been evaluated in [1] using an oscilloscope. The measured settling time was distributed between 10 and 60 μs for almost all DUTs in the pMOS and nMOS arrays. The minimum interval of the measurements can be reduced to 100 μs or shorter in the 128-DUT array.

4 Characterization Example of Statistical Degradations

4.1 Measurement Scenario

Figure 5.7 depicts the scenario for a BTI degradation measurement. The BTIscript that defines the scenario is also shown in Fig. 5.8. First, the output values of the voltage sources and current source are set (lines 1–8). Next, measurement timings are defined for later use (lines 10–12). Then, the threshold voltages of all DUTs in recovery bias mode are measured to obtain initial threshold voltages for the DUTs before applying stress (line 21). The measurement interval is set to 10 ms, which is determined by the minimum interval of the parametric analyzer. This step

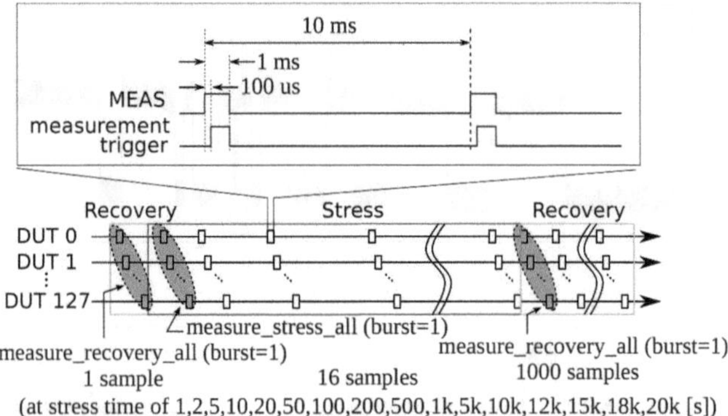

Fig. 5.7 Scenario for BTI degradation measurement

```
# constant supply voltage settings
vdd = 1.8
vss = 0.0
vrec = 1.8
vstr = 0.0      # for nMOS array: vstr = 1.8

# constant supply current setting for pMOS array
icc = 600e-9    # for nMOS array: icc = -600e-9

# measurement timings
tlist = [0, 1, 2, 5, 10, 20, 50, 100, 200, 500, 1000, \
         2000, 5000, 10000, 12000, 15000, 18000, 20000]

# measurement object generation
obj = btiarray(vdd, vss, vrec, vstr, icc)

# stress period calculation
time_interval = [tlist[i+1]-tlist[i] for i in range(len(tlist)-1)]

# measurements
obj.measure_recovery_all(burst=1)
for t in time_interval:
    obj.stress(t)
    obj.measure_stress_all(burst=1)
for _ in range(1000):
    obj.measure_recovery_all(burst=1)
```

Fig. 5.8 Script that realizes measurement scenario of Fig. 5.7. Stress, recovery, and measurement processes are written using macro commands. After voltage sources and current source are initialized and measurement interval is defined, threshold voltages of DUTs in recovery bias mode are measured 1,000 times. Then, all DUTs undergo stress for 20,000 s during which time threshold voltage measurements are conducted intermittently to reduce effect of interrupting stress. Finally, all DUTs returned to recovery mode, and threshold voltages are measured 1,000 times requires $1 \times 128 \times 10\,\text{ms} = 1.28\,\text{s}$ in total. Note that, during these measurements, the DUTs that are not selected for measurement are kept in recovery bias mode to avoid degradation. After that, all DUTs enter stress bias mode (lines 22–24). Stress is continuously applied for 20,000 s. Measurements are intermittently conducted during the stress period for all DUTs in series at 1, 2, 5, 10, 20, 50, 100, 500, 1,000, 5,000, 10,000, 12,000, 15,000, 18,000 and 20,000 s. The macro command "*measure_stress_all*" is repeatedly issued by the "for" statement according to the list of measurement intervals. Then, all DUTs return to recovery bias mode, and measurements are conducted for all DUTs 1,000 times (lines 25–26). Repetitions of 1,000 are realized by issuing the macro command "*measure_recovery_all*" in a for loop.

The upper part of Fig. 5.7 shows the detailed timing of the control signals used for acquiring a threshold voltage of the selected DUT in the BTIarray. First, a "MEAS" signal is asserted. Then, the measurement trigger is issued after 100 μs have elapsed. This timing was determined according to the settling time of the output node voltage. This ensures that the parametric analyzer starts an integration of V_M after the node voltage has been stabilized.

The threshold voltage measurement is based on the constant current method [16] in which the gate voltage of the DUT is switched from V_{STR} to V_{SS} when the DUT is under measurement. The source voltage becomes the threshold voltage of the selected DUT, which reduces the stress on the DUT under measurement. Because the change in the threshold voltages contains a very fast component, partial recovery may occur during the measurement. To minimize this recovery, the "MEAS" signal is negated immediately after the sampling time required by the parametric analyzer has elapsed.

4.2 *Threshold Voltage Shifts for Small DUTs*

Examples of the measured threshold voltage shift are shown in Fig. 5.9. The results for two DUT sizes are presented: W/L = 360 nm/60 nm and 705 nm/120 nm. Ten traces were randomly selected from each size. In the 20,000 s stress period, the threshold voltage increased for all DUTs, but the traces of the increases differed among DUTs. The variation in the threshold voltage increases was larger for DUTs with a smaller channel area. Because the variation is expected to increase as the transistor size decreases, statistical representations of the degradation parameters are required.

In addition, a nonmonotonic increase in the threshold voltage was observed for almost all of the DUTs. Although its cause is still not understood, there are two possible hypotheses. The first hypothesis is recovery during measurement. As described above, the measured DUTs experience less stress voltage than the other fully stressed DUTs. This may lead to device-dependent partial recovery during measurement.

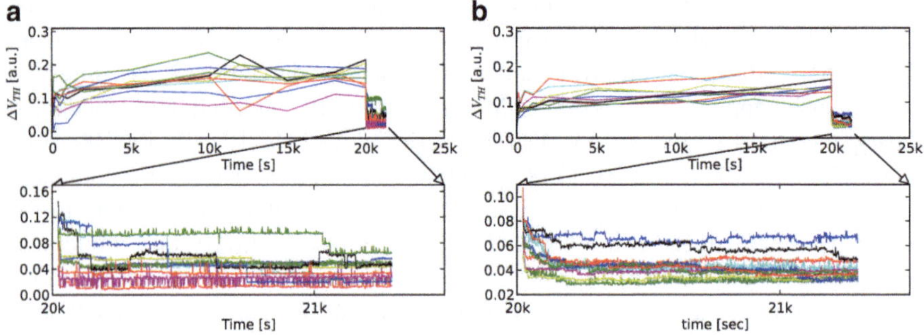

Fig. 5.9 Temporal changes in the threshold voltage (pMOS array) for ten randomly selected DUTs. Variation in threshold voltage increases was larger for smaller-area DUTs. In the recovery period, discrete changes in threshold voltage increases were observed for most of the DUTs. The discrete changes are more distinct for smaller-area DUTs. (**a**) DUT size: W/L=360 nm/60 nm. (**b**) DUT size: W/L=705 nm/120 nm

The other hypothesis is the interaction between carrier traps and emissions during the stress period. The nonmonotonic increase may have been caused by the carrier trap and emission process at interface defects, which is considered to be the major mechanism of BTI degradation [10, 31]. This is a stochastic process; thus, the fluctuation in the threshold voltage is observed at any time.

The time courses of threshold voltage recovery are magnified to the right of the figures. The discrete recovery of the threshold voltages was more prominent for the DUTs with a small channel area. Smoother recovery was again observed for the DUTs with a larger channel area. This trend held true for the reminder of the DUTs. The origin can also be explained by the averaging effect of carrier trap and emission process. Under an assumption of uniform defect density over all areas of the measured chip, a larger number of interface traps should be found in the DUTs of a larger channel area than in those of a smaller channel area. The trapping and emission of carriers occur more often in the larger DUTs, and the threshold voltage shift caused by the trap of a single carrier is smaller than that of the smaller DUTs.

To investigate the stair-like changes in more detail, the response to the stress and recovery bias was repeatedly measured for the DUT with a small channel area. A stress with a duration of 1,000 s was first applied, and then, recovery mode was applied for 1,280 s, during which 1,000 measurements were conducted for each DUT. This set of stress-recovery measurements was repeated 100 times to characterize the stochastic recovery process. The upper figure in Fig. 5.10 shows the recovery period of the threshold voltage shift. The results of 30 trials, which are again randomly selected from 100 trials, are presented.

In the recovery period, the threshold voltage shifts of an equal magnitude were found throughout different repetitions. The magnitude of the threshold voltage shift was determined by the defect location in the channel. Hence, the defects could be identified by the magnitude. This characterization method is known as *time dependent defect spectroscopy* (TDDS) [9]. The associated TDDS plot is also

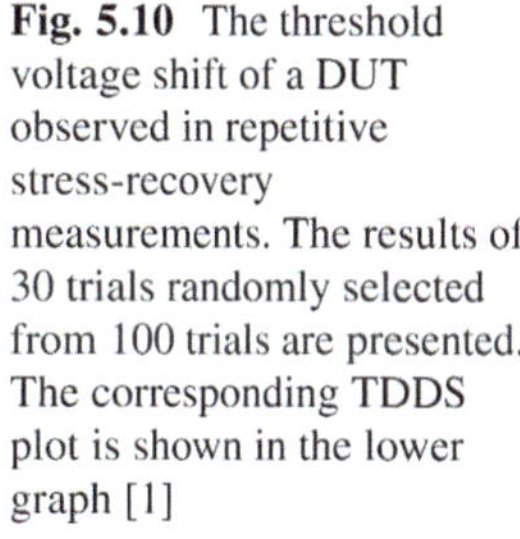

Fig. 5.10 The threshold voltage shift of a DUT observed in repetitive stress-recovery measurements. The results of 30 trials randomly selected from 100 trials are presented. The corresponding TDDS plot is shown in the lower graph [1]

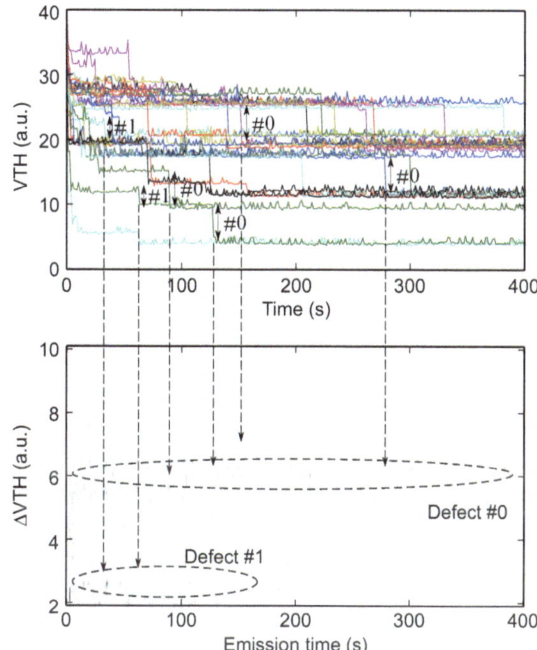

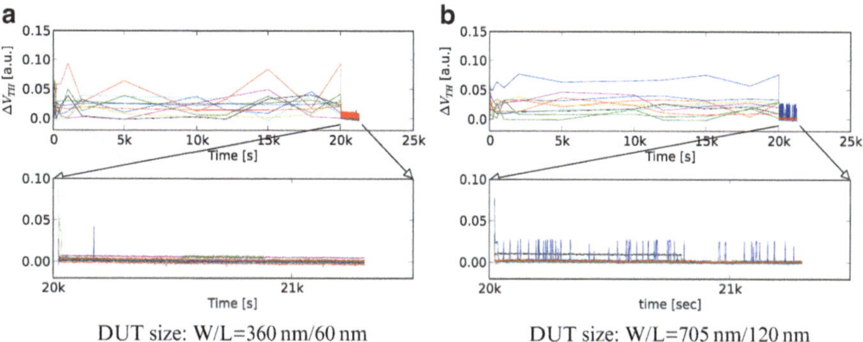

DUT size: W/L=360 nm/60 nm

DUT size: W/L=705 nm/120 nm

Fig. 5.11 Temporal changes in threshold voltage for ten randomly selected DUTs, measured for an nMOS array. Both the degradation and recovery occurred in a very short time

shown in Fig. 5.10. There are two major clusters of defects. The emission time constant of Defect #0 assumes a value of approximately 10 to 50 s, whereas that of Defect #1 distributes wider in the range of 100 to 400 s. Characterizing the range and distribution of the emission time is important in device lifetime extension methods, such as a circuit that alternately uses one of two identical circuits [15].

Similarly, Fig. 5.11 shows the results of the nMOS array obtained using the same measurement scenario as the pMOS array. The DUT set is again randomly selected. By comparing the graphs with those in [1], readers can find the similar trend of the temporal threshold voltage changes for nMOS DUTs. Although smaller than that of pMOS DUTs, the threshold voltage of an nMOS DUT is increased. The threshold voltage shift seemingly reached a plateau relatively quickly within 5,000 s, and remained almost constant for the reminder of the stress time. In this device, nMOS DUTs seemingly lack "slow trap" components [18]. In addition, regardless of the device size, the threshold voltages of most of the DUTs recovered almost completely once the stress was removed, which is different from the case of the pMOS DUTs.

4.3 Statistical Model Parameter Extraction for NBTI

As we have observed in Fig. 5.9, there was a large difference in the threshold voltage shift even for DUTs of equal channel-area size. The threshold voltage increase during a stress period has been commonly modeled using a power law model in which the stress period and threshold voltage increase are related by $\Delta V_{\rm th} \propto t^n$, where t is the stress period, and n is the power law exponent that varies between transistors.

The distributions of the power law exponent for 720 DUTs (180 DUTs for each size) measured using five pMOS arrays at 125 °C are depicted in Fig. 5.12. The

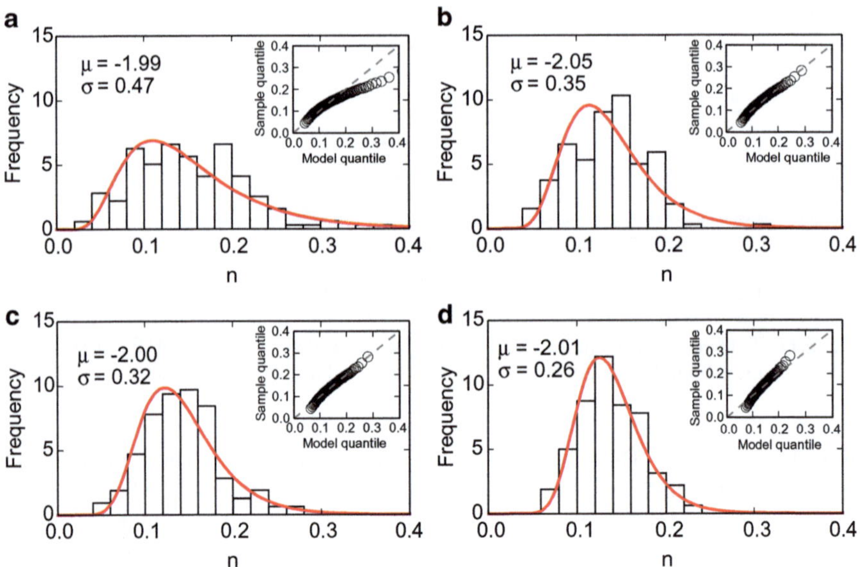

Fig. 5.12 Measured distributions of power law exponent and their log-normal approximations for four sizes of DUTs [1]. The channel-area dependency of the variation is evident. (**a**) W/L = 360 nm/60 nm. (**b**) W/L = 360 nm/120 nm. (**c**) W/L = 705 nm/60 nm. (**d**) W/L = 705 nm/120 nm

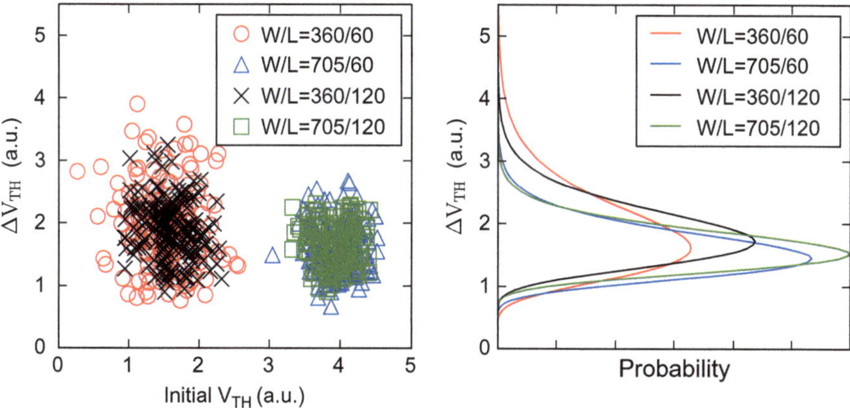

Fig. 5.13 Correlations between threshold voltages of fresh DUTs and the threshold voltage shift after a stress with a duration of 20,000 s [1]. The approximated distributions of the threshold voltage shift are also plotted

variance in the exponent decreased as the DUT size increased. All distributions are approximated by a log-normal distribution, which is indicated by a solid line. The two shape parameters of each distribution, μ and σ, are also provided. An average of the log-normal distribution μ, which represents the logarithmic mean of the power-law exponent, was approximately -2 regardless of the DUT size.[1] The inset graph shows quantile-quantile plot for the log-normal approximation.

The standard deviation of the power-law exponent σ increases as the channel area of the DUTs decreases. The standard deviation of the power-law exponent is inversely proportional to the channel area of a device. The inverse dependency between the device channel area and exponent variance has also been reported in [2]. The exponent variation increases at a faster rate than the threshold voltage, whose variation is inversely proportional to the square root of the channel area.

The accurate consideration of correlations between device parameters is an effective way to avoid overdesign. Figure 5.13 shows a scatter plot between threshold voltages of fresh DUTs and threshold voltage shifts after a stress period of 20,000 s. The results of the four DUT sizes are shown. For the transistors of these sizes, initial threshold voltages and the threshold voltage shifts fit normal and log-normal distributions, respectively. As expected, no correlation was found between the initial threshold voltage and its shift because the variation sources are different for these variations.

[1] $\mu = -2$ approximately corresponds to an exponent $n = 0.15$.

4.4 Measurement Efficiency

The time to complete the scenario in Fig. 5.7 was approximately 5 h and 55 min whereas that for a single DUT was approximately 5 h and 33 min. If a single DUT measurement is serially conducted for 128 DUTs, the measurement would require approximately 710 h, which is 29.6 days, even when the time for moving the probe needles to the next DUT is ignored. Considering that 128 DUTs are integrated in an array, BTIarray has thus achieved an increase in speed of a factor of 120 with this measurement scenario, enabling month-long measurements to be completed within a day. An ideal increase in speed of a factor of 128 could not be achieved because the time required for threshold voltage measurements is finite. The parametric analyzer requires 10 ms for the voltage measurement. Therefore, the acquisitions of the threshold voltages for 128 DUTs for a total of 1,018 measurements require approximately 22 min. The use of a faster SMU shall further shorten the total measurement time.

5 Chapter Summary

Characterization techniques for the transient changes in device parameters have been reviewed. Aiming to efficiently characterize degradation statistics, an on-chip circuit structure called BTIarray has been developed. In the BTIarray, transistors are placed in an array such that the stress bias voltages will be applied for all the devices in parallel by sharing a single source voltage. By applying bias voltages in parallel, the aggregate stress time required for a large number of devices is substantially shortened. The array also offers a function to measure the threshold voltage of a selected device while maintaining the stress voltages to other devices, i.e., the measurement of a device can be carried out in parallel while other devices are in stress or recovery conditions. With the array circuit structure, measurement conditions, including timing, voltage, temperature, and other parameters, will be equalized. In order to facilitate the description of a scenario in which trigger timings for stress, recovery, and measurements are given, a scripting language environment has been provided. Array-based circuit design as well as a development and debugging environment will become even more useful to achieve accurate and reproducible measurements.

For the proof of concept, the experimental results of BTIarray, fabricated using a 65-nm CMOS process, are also presented. Using an array consisting of 128 devices, transient changes in the threshold voltages were measured. With the array structure, a month-long measurement time has been shortened to less than 6 h, while maintaining sufficient voltage and timing precision required in statistical characterizations. The measurement results have shown the following statistical observations. (1) Degradation and recovery slopes vary between devices even for devices with equal channel areas. (2) The variance of the time exponent is inversely proportional to the channel area. (3) Most of the measurement results involve a stair-like recovery trace, which suggests the interface trap charge is playing a key role in the threshold voltage shift.

Acknowledgements Part of this work has been supported by MEXT/JSPS KAKENHI Grant Numbers 22360143 and 20431992.

References

1. Awano, H., Hiromoto, M., Sato, T.: BTIarray: A time-overlapping transistor array for efficient statistical characterization of bias temperature instability. IEEE Trans. Dev. & Mat. Reliab. (2014)
2. Awano, H., Hiromoto, M., Sato, T.: Variability in device degradations: Statistical observation of NBTI for 3996 transistors. In: Proc. European Solid-State Dev. Res. Conf. (ESSDERC) (2014)
3. Campbell, J., Lenahan, P., Krishnan, A., Krishnan, S.: Observations of NBTI-induced atomic-scale defects. IEEE Trans. Dev. & Mat. Reliab. **6**(2), 117–122 (2006)
4. Denais, M., Parthasarathy, C., Ribes, G., Rey-Tauriac, Y., Revil, N., Bravaix, A., Huard, V., Perrier, F.: On-the-fly characterization of NBTI in ultra-thin gate oxide PMOSFET's. In: IEDM Tech. Dig., pp. 109–112 (2004)
5. Du, G., Ang, D., Teo, Z., Hu, Y.: Ultrafast measurement on NBTI. IEEE Trans. Electron Device Lett. **30**(3), 275–277 (2009)
6. Fujimoto, S., Islam, A.M., Matsumoto, T., Onodera, H.: Inhomogeneous ring oscillator for within-die variability and RTN characterization. IEEE Trans. Semicond. Manufact. **26**(3), 296–305 (2013)
7. Ghosh, A., Rao, R.M., Brown, R.B., Chuang, C.T.: On-chip negative bias temperature instability sensor using slew rate monitoring circuitry. In: ACM IEEE Intl. Symposium on Low Power Electronics and Design (2009)
8. Grasser, T., Kaczer, B., Goes, W.: An energy-level perspective of bias temperature instability. In: Proc. Intl. Reliab. Phys. Symp. (IRPS), pp. 28–38 (2008)
9. Grasser, T., Reisinger, H., Wagner, P., Schanovsky, F., Goes, W., Kaczer, B.: The time dependent defect spectroscopy (TDDS) for the characterization of the bias temperature instability. In: Proc. Intl. Reliab. Phys. Symp. (IRPS), pp. 16–25 (2010)
10. Huard, V., Denais, M.: Hole trapping effect on methodology for DC and AC negative bias temperature instability measurements in pMOS transistors. In: Proc. Intl. Reliab. Phys. Symp. (IRPS), pp. 40–45 (2004)
11. Kim, J.J., Rao, R., Schaub, J., Ghosh, A., Bansal, A., Zhao, K., Linder, B., Stathis, J.: PBTI/NBTI monitoring ring oscillator circuits with on-chip Vt characterization and high frequency AC stress capability. In: Proc. VLSI Technology symposium, pp. 224–225 (2011)
12. Kim, T., Persaud, R., Kim, C.: Silicon odometer: An on-chip reliability monitor for measuring frequency degradation of digital circuits. IEEE J. Solid-State Circ. **43**(4), 874–880 (2008)
13. Kumar, E., Maheta, V., Purawat, S., Islam, A., Olsen, C., Ahmed, K., Alam, M., Mahapatra, S.: Material dependence of NBTI physical mechanism in silicon oxynitride (SiON) p-MOSFETs: A comprehensive study by ultra-fast on-the-fly (UF-OTF) i_{DLIN} technique. In: IEDM Tech. Dig., pp. 809–812. IEEE (2007)
14. Matsumoto, T., Makino, H., Kobayashi, K., Onodera, H.: A 65 nm complementary metal-oxide-semiconductor 400 ns measurement delay negative-bias-temperature-instability recovery sensor with minimum assist circuit. Japan. J. Appl. Phys. **50**(4), 04DE06 (2011)
15. Matsumoto, T., Makino, H., Kobayashi, K., Onodera, H.: Multicore large-scale integration lifetime extension by negative bias temperature instability recovery-based self-healing. Japan. J. Appl. Phys. **51**, 04DE02 (2012)

16. Ortis-Conde, A., García Sánchez, F.J., Liou, J.J., Cerdeira, A., Estrada, M., Yue, Y.: A review of recent MOSFET threshold voltage extraction methods. Microelec. Reliability **42**, 583–596 (2002)
17. Rauch, S.E.: Review and reexamination of reliability effects related to NBTI-induced statistical variations. IEEE Trans. Dev. & Mat. Reliab. **7**(4), 524–529 (2007)
18. Reisinger, H., Blank, O., Heinrigs, W., Gustin, W., Schlunder, C.: A comparison of very fast to very slow components in degradation and recovery due to NBTI and bulk hole trapping to existing physical models. IEEE Trans. Dev. & Mat. Reliab. **7**(1), 119–129 (2007)
19. Reisinger, H., Blank, O., Heinrigs, W., Muhlhoff, A., Gustin, W., Schlunder, C.: Analysis of NBTI degradation- and recovery-behavior based on ultra fast VT-measurements. In: Proc. Intl. Reliab. Phys. Symp. (IRPS), pp. 448–453 (2006)
20. Reisinger, H., Grasser, T., Gustin, W., Schlunder, C.: The statistical analysis of individual defects constituting NBTI and its implications for modeling DC-and AC-stress. In: Proc. Intl. Reliab. Phys. Symp. (IRPS), pp. 7–15 (2010)
21. Sato, T., Awano, H., Shimizu, H., Tsutsui, H., Ochi, H.: Statistical observations of NBTI-induced threshold voltage shifts on small channel-area devices. In: Proc. Intl. Symp. Quality Electronic Design (ISQED), pp. 306–311. IEEE (2012)
22. Sato, T., Kozaki, T., Uezono, T., Tsutsui, H., Ochi, H.: A device array for efficient bias-temperature instability measurements. In: Proc. European Solid-State Dev. Res. Conf. (ESSDERC), pp. 143–146 (2011)
23. Sato, T., Ueyama, H., Nakayama, N., Masu, K.: Accurate array-based measurement for subthreshold-current of MOS transistors. IEEE J. Solid-State Circ. **44**(11), 2977–2986 (2009)
24. da Silva, M., Kaczer, B., Van der Plas, G., Wirth, G., Groeseneken, G.: On-chip circuit for massively parallel BTI characterization. In: IEEE Intl. Integrated Reliability Workshop Final Report (IRW), pp. 90–93 (2011)
25. Stathis, J.H., Zafar, S.: The negative bias temperature instability in MOS devices: A review. Microelec. Reliability **46**(2–4), 270–286 (2006)
26. Tsai, M.C., Lin, Y.W., Yang, H.I., Tu, M.H., Shih, W.C., Lien, N.C., Lee, K.D., Jou, S.J., Chuang, C.T., Hwang, W.: Embedded SRAM ring oscillator for in-situ measurement of NBTI and PBTI degradation in CMOS 6T SRAM array. In: VLSI Design, Automation, and Test (VLSI-DAT), 2012 International Symposium on, pp. 1–4. IEEE (2012)
27. Velamala, J., Sutaria, K., Shimuzu, H., Awano, H., Sato, T., Wirth, G., Cao, Y.: Logarithmic modeling of BTI under dynamic circuit operation: Static, dynamic and long-term prediction. In: Proc. Intl. Reliab. Phys. Symp. (IRPS), pp. CM.3.1–CM.3.5 (2013)
28. Velamala, J.B., Sutaria, K.B., Sato, T., Cao, Y.: Aging statistics based on trapping/detrapping: Silicon evidence, modeling and long-term prediction. In: Proc. Intl. Reliab. Phys. Symp. (IRPS), pp. 2F2.1–2F2.5 (2012)
29. Velamala, J.B., Sutaria, K.B., Sato, T., Cao, Y.: Physics matters: statistical aging prediction under trapping/detrapping. In: Proc. ACM/IEEE Design Automation Conf. (DAC), pp. 139–144 (2012)
30. Wang, W., Yang, S., Bhardwaj, S., Vrudhula, S., Liu, F., Cao, Y.: The impact of NBTI effect on combinational circuit: Modeling, simulation, and analysis. Trans. VLSI **18**(2), 173–183 (2010)
31. Wirth, G., da Silva, R., Kaczer, B.: Statistical model for MOSFET bias temperature instability component due to charge trapping. IEEE Trans. Electron Dev. **58**(8), 2743–2751 (2011)
32. Zafar, S., Callegari, A., Gusev, E., Fischetti, M.V.: Charge trapping related threshold voltage instabilities in high permittivity gate dielectric stacks. J. Appl. Phys. **93**(11), 9298–9303 (2003)

Chapter 6
Compact Modeling of BTI for Circuit Reliability Analysis

Ketul B. Sutaria, Jyothi B. Velamala, Athul Ramkumar, and Yu Cao

Abstract The aging process due to Bias Temperature Instability (BTI) is a key limiting factor of circuit lifetime in contemporary CMOS design. Threshold voltage shift induced by BTI is a strong function of stress voltage and temperature. Furthermore, BTI consists of both stress and recovery phases, depending on the dynamic stress conditions. This behavior poses a unique challenge for long-term aging prediction for a wide range of stress patterns encountered in today's circuits. Traditional approaches usually resort to an average, constant stress waveform to simplify the lifetime prediction. They are efficient, but fail to capture the reality of circuit operation, especially under Dynamic Voltage Scaling (DVS) or in analog/mixed signal designs where the stress waveform is much more random. In this chapter, we present a suite of modeling solutions that enable *aging simulation under all dynamic stress conditions*. The key innovation of this chapter is to develop compact models of BTI when the stress voltage is varying. The results cover the underlying physics of two leading mechanisms, Reaction–Diffusion (R–D) and Trapping/Detrapping (T–D). Moreover, silicon validation of these models is performed at 45 and 65 nm technology nodes, at both device and circuit levels. Leveraging the newly developed BTI models under DVS and random input waveforms, efficient aging simulation is demonstrated in representative digital and analog circuits. Our proposed work provides a general and comprehensive solution to circuit aging analysis under random stress patterns.

1 Introduction

Aggressive scaling of CMOS technology has been driving the evolution of electronics and the increase in the total number of transistors per chip. Moore's law predicts that the total number of transistors placed on a chip will be approximately doubled every 18 months. Benefiting from the scaling of CMOS technology, system-on-chips (SoCs) further improve the system capability by integrating both analog

K.B. Sutaria • J.B. Velamala • A. Ramkumar • Y. Cao (✉)
School of Electrical, Computer and Energy Engineering,
Arizona State University, Tempe, Arizona 85287, USA
e-mail: ycao@asu.edu

R. Reis et al. (eds.), *Circuit Design for Reliability*,
DOI 10.1007/978-1-4614-4078-9_6

and digital circuits on a single platform. Present SoCs pack close to 800 million transistors incorporating analog and mixed signal units such as data converters (A/D, D/A), PLLs, etc., providing higher performance at a lower cost.

On the other side, aggressive CMOS scaling results in serious reliability threats such as Bias Temperature Instability (BTI), Channel Hot Carrier (CHC) and Time Dependent Dielectric Breakdown (TDDB) [1–10]. These reliability effects become more pronounced partially due to process scaling techniques that improve device and circuit performance. Moreover, scaling of supply voltage and threshold voltage does not go hand in hand with scaling of device feature size. This greatly increases vertical and lateral electrical field causing the exacerbation of reliability issues. As an aftermath of these issues, circuit performance degrades over time and eventually results in functionality error, which is called circuit aging.

Threshold voltage shift (ΔV_{th}) due to negative BTI (NBTI) in PMOS devices is the dominant aging mechanism in scaled digital and analog circuits. Accurate characterization of the aging rate is essential to lifetime evaluation and guard banding [11–24]. One major challenge in reliability prediction of NBTI comes from the recovery phase, which causes the dependence on the stress pattern and its history. The situation is more complex under Dynamic Voltage Scaling (DVS) in low power digital design. Analog/mixed signal circuits face similar challenges due to random input pattern. Modulation of voltage levels due to such operations results in partial recovery of V_{th} and thus, the average or the worst case pattern does not provide an accurate prediction of the lifetime. Figure 6.1 presents the operation pattern in a 65 nm dual core Intel processor and ΔV_{th} under such an operation.

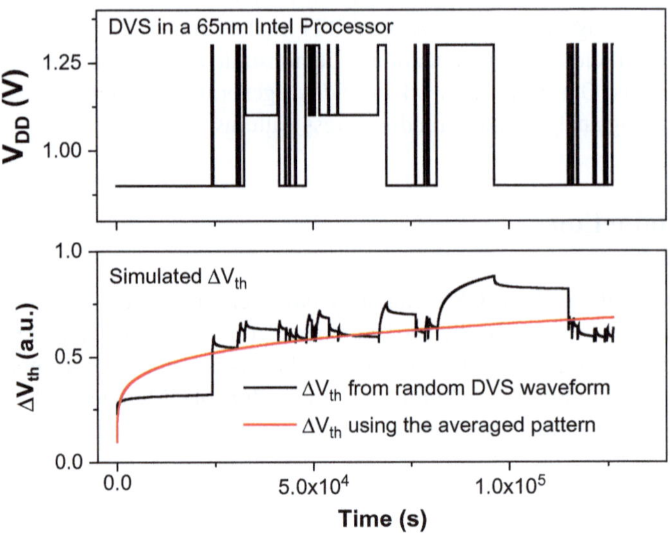

Fig. 6.1 ΔV_{th} predicted from the averaged voltage significantly deviates from that simulated by dynamic aging models

When V_{DD} fluctuates, the degradation is strongly non-linear and varies according to the stress condition. The average value of the stress voltage predicts a lower or higher V_{th} shifts, as compared to that by the real-time model. Therefore, real-time prediction of V_{th} shift allows a realistic and accurate assessment of circuit aging, avoiding over-design to compensate the degradation. Overall, it is critical to understand NBTI with recovery effects, simulate its effect under dynamic stress voltages, and ensure reliable circuit operation for the desired lifetime.

1.1 BTI Aging Models: Challenges and Needs

To date, research work on aging mechanisms has mainly focused on device and reliability physics [24–31]. Proprietary efforts also exist in leading industrial companies to develop their own reliability models and tools [32–35]. Such knowledge needs to be propagated into circuit design and CAD tools to assess the impact of device degradation on various circuit performance metrics. As BTI is the dominant aging mechanism, the accuracy and efficiency of BTI models directly affect the quality of reliability analysis.

In a contemporary SoC, a wide variety of circuit operation patterns co-exists. The diversity of circuit operation presents a unique challenge to predict the lifetime under the BTI effect: A device operating under constant stress voltage needs a static aging model; different from static stress, devices under AMS operation may experience totally random stress during their lifetime. For digital circuits, they employ DVS extensively to balance the workload and power consumption. Under such operation, a static model is insufficient to accurately determine circuit aging. Therefore, a model which can analyze aging under any random input stress pattern is necessary.

Different from AMS circuits, large-scale logic designs encounter periodic inputs with different duty cycles and frequency. Although a random stress model can handle such situations, long-term aging simulation is not efficient and requires expensive simulation time. To improve the efficiency, a long-term BTI model is preferred which predicts an upper bound of threshold voltage shift for a periodic input. In summary, a set of new BTI models are needed to improve circuit reliability prediction in both VLSI and AMS design under dynamic operations.

In reliability physics, NBTI was first assumed to be induced by the Reaction–Diffusion (R–D) mechanism, where holes initiate the breaking of Si–H bonds (reaction phase) at the silicon substrate/oxide interface [4, 28, 31]. The broken hydrogen species diffuse away from the interface into gate oxide and the poly-silicon gate (diffusion phase) and thus, interface charges are generated. This mechanism leads to the classical power-law model of V_{th} shift. More recent research suggests a role of hole Trapping/Detrapping (T–D) for V_{th} shift of a device under stress [36–51]. According to the T–D theory, if the trap gains sufficient energy it may capture a charge carrier, thus reducing the number of available carriers in the channel and modulating the threshold voltage of a device. A logarithmic time dependence of V_{th} shift is observed under T–D as opposed to the power-law behavior. There are research works which have suggested a combination of R–D and T–D. Whether T–D or R–D dominates the V_{th} shift behavior depends on the fabrication technology. For reliable circuit design, it is important to have an access to both R–D and T–D based aging models in order to accurately determine the guard-band of a design. In this chapter, we present a set of BTI models based on R–D and T–D theory which can predict threshold voltage shift under static, random and long-term stress input. Such a set of aging models provides an accurate and efficient solution for reliable design.

2 Reliability Physics: Device-Level Modeling of BTI

The primary impact of BTI at the device level is the gradual increase in transistor V_{th}, whereas the degradation of other device parameters are less pronounced. Since the threshold voltage directly affects the delay of a digital gate, operating frequency of a logic path decreases temporally. Similarly, in AMS circuits, increase in V_{th} degrades the gain, trans-conductance and other performance metrics. To estimate circuit aging rate, the fundamental step is to model device V_{th} shift under given stress voltage and temperature. In this section, compact models based on both R–D and T–D mechanisms are presented.

Two prevalent theories, Reaction–Diffusion and Trapping/Detrapping, are proposed to explain BTI. Figure 6.2 shows the cross section of a PMOS device which shows the difference between these mechanisms. R–D is a two-step process namely: Reaction and Diffusion. According to the R–D theory, the stress voltage causes covalent bonds (Si–H) at interface to break, which is *Reaction*. In the *Diffusion* step the broken hydrogen atoms combine to form H_2, which diffuses towards gate. For current thin-oxide devices, diffusion in poly-gate dominates the incremental behavior of V_{th} shift. The interface states left at Si–SiO_2 due to H_2 diffusion increase the threshold voltage. This leads to a power law relation (t^n) with time exponent (n) 1/6, which should be independent of process parameters. ΔV_{th} exponentially depends on the stress voltage and temperature [28, 31].

According to the T–D theory shown in Fig. 6.2b, there exist a number of defect states with different energy levels, and capture and emission time constants. Threshold voltage increases when a trap captures a charge carrier from the channel

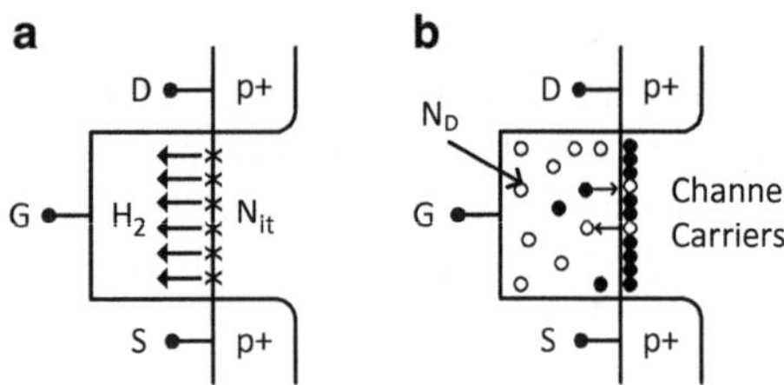

Fig. 6.2 Two mechanisms for BTI: (**a**) Reaction Diffusion (*RD*), and (**b**) Trapping/Detrapping (*TD*)

of a MOS device. Reduced number of channel carriers also causes drain current to decrease. The probability of trapping and detrapping is a function of capture and emission time constants, respectively. The gradual change in the number of occupied traps results in a logarithmic time evolution of V_{th} shift, different from the power law behavior in R–D. The stress voltage and temperature has an exponential impact on the degradation rate, similar as that in the R–D theory [47].

2.1 Static BTI Models

2.1.1 Reaction: Diffusion Based Static BTI Model

There are two critical steps based on the Reaction–Diffusion theory. *Reaction*: Si–H or Si–O bonds at the substrate/gate oxide interface are broken under the electrical stress [16–18]. The interface charges are induced in gate oxide and further in poly-gate, which cause the increase of V_{th}. Given the initial concentration of the Si–H bonds (N_0) at the interface and the concentration of inversion carriers (P), the generation rate of the interface traps is given by [28]:

$$\frac{dN_{IT}}{dt} = k_f\left(N_0 - N_{IT}\right)P - k_r N_H N_{IT} \tag{6.1}$$

where, k_f and k_r are forward and reverse reaction rates. The generation rate is exponentially dependent on the stress voltage and temperature. During the initial phase of stress period, the trap generation is slow with respect to time. Thus dN_{IT}/dt ~ 0 and $N_{IT} << N_0$, reducing Eq. (6.1):

$$N_H N_{IT} = \frac{k_f}{k_r} N_0 P \tag{6.2}$$

The hydrogen atoms combine to form H_2 molecules. The relationship between H and H_2 molecule is given as $N_{H2} = k_h {N_H}^2$, where k_h is the rate constant.

Diffusion: Generated hydrogen species diffuse away from Si–SiO_2 interface towards the gate. This diffusion is driven by the gradient of the density. This process is governed by the following equation [28]:

$$\frac{dN_{H_2}}{dt} = C\frac{d^2 N_{H_2}}{dx^2} \tag{6.3}$$

C is the diffusion constant for hydrogen molecules in poly–Si, which depends on the activation energy. To solve differential Eqs. (6.1) and (6.3) to derive a compact model, an approximate profile of H_2 concentration in gate oxide and poly–Si is assumed, as shown in Fig. 6.3. N_{H2} is the concentration of hydrogen molecules at distance 'x' from Si–SiO_2 interface at time (t), where x is given by $\sqrt{Ct}$. Diffusion of H_2 molecules in oxide is much faster than in poly–Si. The total number of interface traps generated for a given stress time is solved as [31]:

$$N_{IT} = 2\int_{0}^{x(t)} N_{H_2}(x)dx \tag{6.4}$$

H_2 diffusion is divided in silicon oxide and poly-gate. Fast diffusion of H_2 in oxide and small thickness of dielectric lead to a very small difference in H_2 concentration between Si–SiO$_2$ interface and SiO$_2$–Poly interface. Figure 6.3 shows the approximate profile of hydrogen concentration where a fitting parameter δ is introduced to account for the fraction drop of H_2 concentration. Using this approximation, Eq. (6.4) can be written as [31],

$$N_{IT} = 2\int_{0}^{t_{ox}} N_{H_2}(x)dx + \int_{t_{ox}}^{\sqrt{Ct}+t_{ox}} N_{H_2}(x)dx \tag{6.5}$$

$$\approx 2\left(\frac{1}{2}(1+\delta)\cdot N_{H_2}(0)\cdot t_{ox} + \frac{1}{2}N_{H_2}(0)\cdot\sqrt{Ct}\right) \tag{6.6}$$

$N_{H2}(0)$ is H_2 concentration at Si–SiO$_2$ interface while δ $N_{H2}(0)$ is the density at Si–Poly interface. Finally, using $N_{H2} = k_h {N_H}^2$, N_{IT} can be represented as

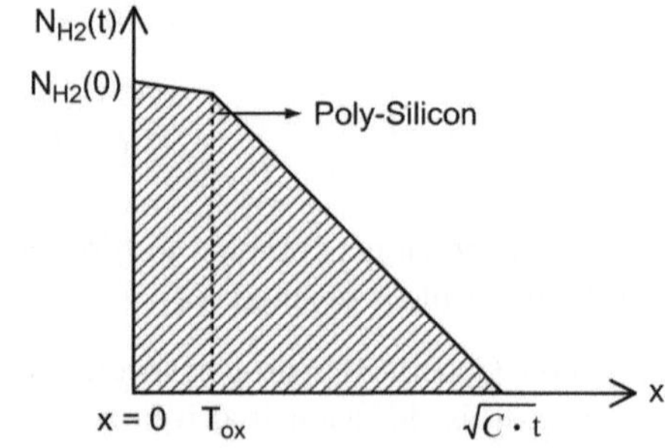

Fig. 6.3 Approximate diffusion profile of hydrogen atoms under constant stress

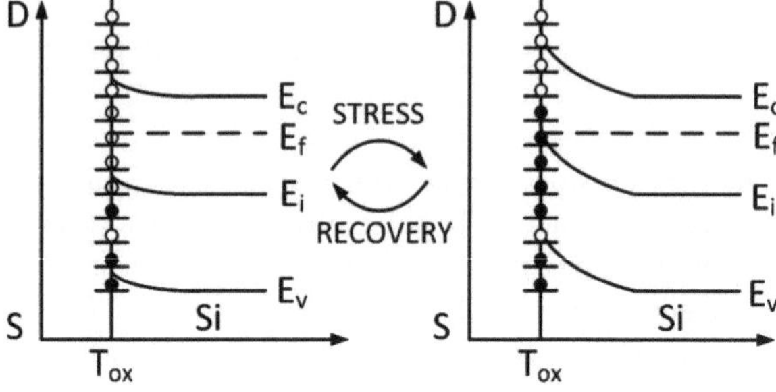

Fig. 6.4 Illustration of statistical trapping/ de-trapping process in gate oxide for a PMOS device, leading to threshold voltage shift

$$N_{IT} = \left(\frac{\sqrt{k_h} k_f N_0 P}{k_r}\right)^{\frac{2}{3}} \left((1+\delta)\, t_{ox} + \sqrt{Ct}\right)^{\frac{1}{3}} \tag{6.7}$$

where, inversion hole density $P = C_{ox}\ (V_{gs} - V_{th})$. Based on the interface charges, threshold voltage shift can be derived as: $\Delta V_{th} = qN_{IT}/C_{ox}$. This R–D based model predicts the V_{th} shift under any given constant stress voltage, temperature and time. The final form of the degradation is given as:

$$\begin{aligned} \Delta V_{th} &= \tfrac{qN_{IT}}{C_{ox}} = A \cdot \left((1+\delta)\, t_{ox} + \sqrt{Ct}\right)^{2n} \\ A &= \left(\tfrac{qt_{ox}}{\varepsilon_{ox}}\right) \sqrt[1/2n]{K_2 C_{ox} \left(V_{gs} - V_{th}\right) \exp\left(\tfrac{2E_{ox}}{E_0}\right)} \end{aligned} \tag{6.8}$$

C is the diffusion constant which incorporates the temperature dependence. Time exponent n is 1/6 when the diffusion species is H_2. The static model equation is verified by 45 nm silicon data along with the T–D model (Fig. 6.5).

2.1.2 Trapping/De-trapping Based Static BTI Model

Several works have presented the evidence of the T–D mechanism, such as the discrete V_{th} shift, especially during the recovery phase in NBTI observed with the fast measurement techniques [52, 53]. Figure 6.4 illustrates the physical picture of T–D: when a negative bias voltage is applied to the gate of a PMOS device, the trap energy (relative to the Fermi energy level) is modulated. If the trap gains sufficient energy, it may capture a charge carrier, thus reducing the number of available carriers in the channel [38]. The charged trap state modulates the local V_{th} and acts as a scattering source, reducing the effective mobility [38]. Faster traps (with shorter time constants) having a higher probability of capturing carriers; the occupation probability increases with voltage and temperature. Trapping and de-trapping events are stochastic in nature and hence a compact model is based on the statistics of trap properties.

The basic modeling assumptions based on T-D theory are the same as the ones used in modeling of low-frequency noise, since the charge trapping dynamics (capture and emission time statistics) that contribute to the degradation of device performance over time is similar [40–44]. The main assumptions of the trap properties are:

- The number of traps follows a Poisson distribution, which is common for a discrete process.
- Capture and emission time constants are uniformly distributed on the logarithmic scale. This microscopic assumption is critical to derive the logarithmic time evolution at the macro scale.

- The distribution of trap energy is approximated as a U-shape, which is verified by silicon measurement and key to the voltage and temperature dependence of the aging effect.

Based on the T–D theory, the V_{th} shift at a given stress time is the result of number of traps ($n(t)$) occupied by the channel carriers. The probability of a particular trap, initially empty (0), to be occupied (1) after an elapsed time t is given by $P_{01}(t)$. This occupation probability can be calculated by observing that

$$P_{01}\,(t+dt) = P_{01}(t)p_{11}(dt) + P_{00}(t)p_{01}(dt) \tag{6.9}$$

where $p_{01}(dt) = 1/\tau_c$ and $p_{11}(dt) = 1 - p_{10}(dt) = 1/\tau_e$. Integrating it from t_0 to t:

$$P_{01}\,(t+t_0) = \frac{\tau_{eq}}{\tau_c}\left(1 - e^{-t/\tau_{eq}}\right) + P_{01}\,(t_0)\,e^{-t/\tau_{eq}} \tag{6.10}$$

where $1/\tau_{eq} = 1/\tau_c + 1/\tau_e$. $\tau_{c,}$ τ_e are random in nature, representing capture and emission time constants respectively, and dependent on bias point and temperature. The values are determined by [38]:

$$\tau_c = 10^p\,(1 + e^{-q}) \tag{6.11}$$

$$\tau_e = 10^p\left(1 + e^{+q}\right) \tag{6.12}$$

where $p \in [p_{min}, p_{max}]$. p_{min} and p_{max} define the time constants for fastest and slowest traps respectively (p_{min} ~1 and $p_{max} > 10$). This assumption of the existence of defects with wide distribution of time constants is in line with recent NBTI data [38]. Since p is assumed to be uniformly distributed, the characteristic time constants are uniformly distributed on logarithmic scale. The parameter q is given by $(E_T - E_F)/kT$, where E_T is the trap energy and E_F is the Fermi energy level. The trap energy (relative to Fermi energy) is a function of applied electric field. Consequently, τ_c and τ_e are dependent on voltage and temperature.

The occupation probability of the trap at time t, assuming that it is under constant stress from time $t_0 = 0$ is obtained by substituting $P_{01}(0) = 1 - P_{01}(0) = 0$ in Eq. (6.10), integrating P_{01} and multiplying with the number of available traps, the average number of occupied traps obtained by substituting the logarithmic distribution of time constants, and the U shaped distribution of trap energies:

$$n(t) = \frac{N}{\ln 10\,(p_{\max} - p_{\min})}\int_0^{E_{T\max}} \frac{g\,(E_T)\,dE_T}{1 + \exp\left(-\frac{E_T - E_F}{kT}\right)} \cdot \int_{10^{-p\min}t}^{10^{-p\max}t} \frac{e^{-u} - 1}{u}du \tag{6.13}$$

where $g(E_T)$ is the trap energy distribution, and p_{min} and p_{max} represent fast and slow traps, respectively. The trap energy, E_T changes as a function of electric field (E_{ox}). Assuming p_{min} ~1 and $p_{max} > 10$, and E_T ~$1/E_{ox}$,

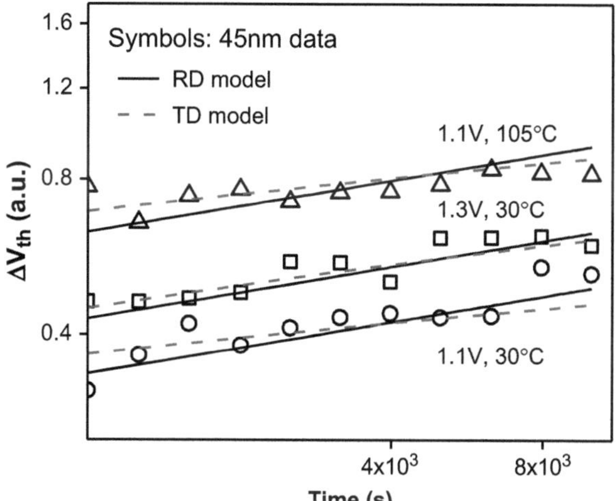

Fig. 6.5 Both R–D and T–D based compact static model matches 45 nm silicon data under constant stress for different voltage and temperature

$$n(t) = \frac{N}{\ln 10\,(p_{\max} - p_{\min})} \exp\left(\frac{\beta V_g}{t_{ox} kT}\right) \exp\left(\frac{-E_0}{kT}\right) \left[A + B \log 10^{-p\max} t\right] \tag{6.14}$$

Equation (6.14) describes the aging under a constant stress voltage and temperature. Similar as previous R–D model, it is an exponential function of the stress voltage, temperature and T_{ox}. Furthermore, it has a statistical nature with N, an index for the number of traps per device. For the simplicity, Eq. (6.14) is written as:

$$\Delta V_{th}(t) = \phi\left[A + B \log\left(1 + Ct\right)\right] \tag{6.15}$$

Equations (6.8) and (6.15) represent V_{th} shift based on R–D and T–D theory respectively. Aging prediction of these models is compared with 45 nm silicon data in Fig. 6.5. Within the range of data fluctuations, both models match well with measurement under the static stress. The exponential dependence on stress voltage and temperature are correctly captured by both aging models.

2.2 *BTI Models Under Random Stress Patterns*

Today's circuits typically have a reduced activity factor (or duty cycle) and dynamic voltage scaling (DVS), to reduce power consumption. Therefore, a significant portion of the operation is under lower supply voltage, resulting in large recovery. Since the degradation is highly sensitive to the stress voltage, DVS leads to different amounts of circuit aging. For a random stress pattern, it is necessary to derive voltage dependent recovery to accurately predict V_{th} shift.

2.2.1 Reaction: Diffusion Based Aging Model for Random Input

The hydrogen atoms that are generated during the stress phase will recover if the stress voltage is removed completely. Atoms close to Si–SiO_2 interface anneal the broken Si–H bond while atoms deep in Poly-gate continue to diffuse away, leading to incomplete recovery of V_{th}. Approximate profile of hydrogen species is shown in Fig. 6.6. Hydrogen atoms diffuse quickly in oxide and anneal some of Si–H bonds in a short time. If a stress voltage is removed after time (t_1), the interface charges generated at the given time are $N_{IT}(t_1)$. The total number of charges to be annealed is given by $N^A{}_{IT}(t)$. Interface charges at a given time t is given by

$$N_{IT}(t) = N_{IT}(t_1) - N_{IT}^A(t) \tag{6.16}$$

as shown in Fig. 6.6; the number of annealed traps can be divided in to two parts: (1) recombination of H_2 in oxide, and (2) back diffusion of H_2 in poly-gate [31]. Thus we have

$$N_{IT}^A(t) = 2\left(\xi_1 t_e + \frac{1}{2}\sqrt{\xi_2 C\,(t - t_1)}\right) N_{H_2}(0) \tag{6.17}$$

In Eq. (6.17) above, ξ_1 and ξ_2 are the back diffusion constants. From Eqs. (6.7) and (6.17), we get

$$N_{IT}^A(t) = N_{IT}(t)\left(\frac{2\xi_1 t_e + \sqrt{\xi_2 C\,(t - t_1)}}{(1+\delta)\,t_{ox} + \sqrt{Ct}}\right) \tag{6.18}$$

Substituting Eq. (6.18) in Eq. (6.16), simplifying equations and using $\Delta V_{th} = qN_{IT}/C_{ox}$, we get the recovery model as:

$$\Delta V_{th}(t) = \Delta V_{th}(t_1)\left(1 - \frac{2\xi_1 t_e + \sqrt{\xi_2 C\,(t - t_1)}}{(1+\delta)\,t_{ox} + \sqrt{Ct}}\right) \tag{6.19}$$

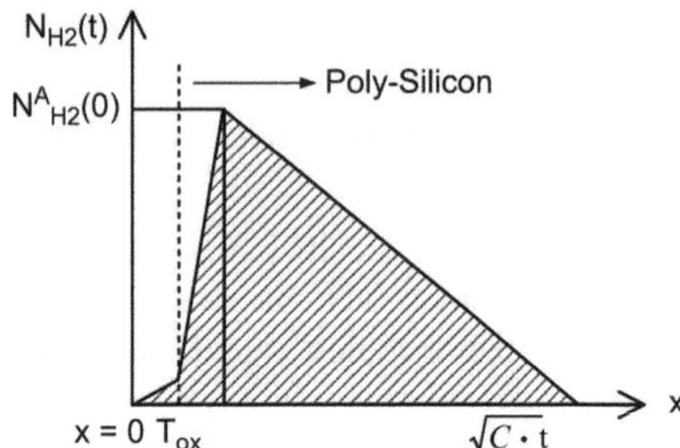

Fig. 6.6 Approximate diffusion profile of hydrogen atoms under recovery

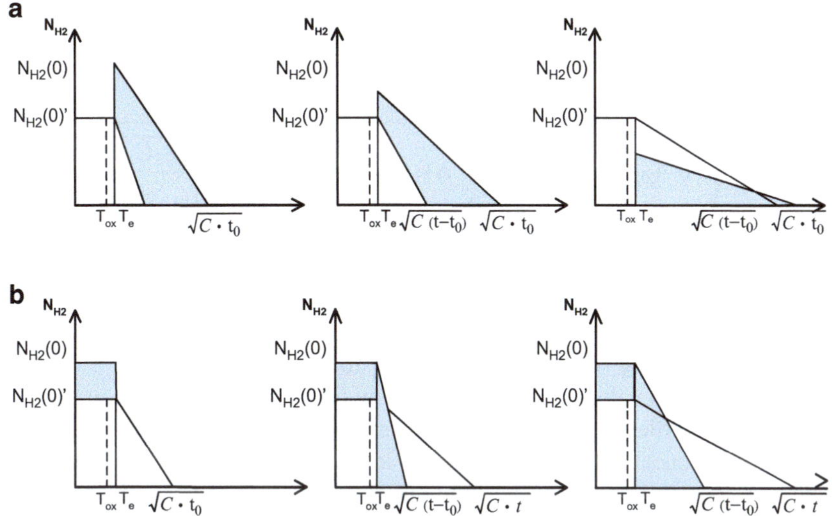

Fig. 6.7 Approximate diffusion profile of hydrogen atoms in PMOS under dynamic stress voltage: (**a**) from high to low and (**b**) from low to high

The equation above represents V_{th} recovery when stress voltage is completely removed. This model is not applicable to the V_{th} behavior when the stress voltage is lowered or increased. Under DVS operation, the stress voltage is changed up and down. Diffusion profile of hydrogen atoms when the stress voltage changes from high to low is different from that when the voltage changes from low to high [54]. Time-dependent diffusion profiles are shown in Fig. 6.7 for a voltage change from high (V_1) to low (V_2) at time (t_0). The resulting V_{th} shift is divided in to two components. The first component is the diffusion due to the lower voltage (V_2) which is the un-shaded region in the figure. The shaded part is the recovery component that gradually decreases with time. Eventually, the ΔV_{th} degradation is driven by the lower voltage. To derive a closed form solution, we start with ΔV_{th} under the lower voltage:

$$\Delta V_{th}(t) = A_2\left((1+\delta)\, t_{ox} + \sqrt{C\,(t-t_0)} + s(t)\right)^{2n} \tag{6.20}$$

where A_2 is the function of the lower voltage (V_2). s(t) is time-dependent initial distance of hydrogen diffused, shown as the shaded area in first graph of Fig. 6.7a. As time progresses, these hydrogen atoms recover, given by the equation:

$$\Delta V_{thr}(t) = \Delta V_{th}(T)\left(1 - \frac{2\xi_1 T_e + \sqrt{\xi_2 C\,(t-t_0)}}{(1+\delta)\, t_{ox} + \sqrt{Ct}}\right) \tag{6.21}$$

By substituting $t = t_0$ in Eq. (6.20) and equating the results with Eq. (6.21), we get a compact model which predicts V_{th} shift when stress voltage changes from a high to low voltage:

$$\Delta V_{th}(t) = \left(\sqrt[2n]{A_2}\sqrt{C\,(t - t_0)} + \sqrt[2n]{\Delta V_{th}\,(t_0)\left(1 - \frac{2\xi_1 T_e + \sqrt{\xi_2 C\,(t - t_0)}}{(1+\delta)\,t_{ox} + \sqrt{Ct}}\right)} \right)^{2n} \tag{6.22}$$

ξ_1 and ξ_2 are back diffusion constants, same as those in the recovery model. This equation is capable of predicting the non-monotonic behavior of initial recovery and eventually converging with rising V_{th} shift due to the lower voltage.

When the stress voltage changes from a low (V_2) to high (V_1) value at time (t_0), the diffusion profile approximation is shown in Fig. 6.7b. In this case, there is no recovery component. The diffusion due to low voltage continues to diffuse at the same rate. However, the diffusion front due to higher voltage is dominant, and V_{th} shift eventually converges. We start from Eq. (6.20) in the derivation. In this case, A_2 is the function of the higher voltage (V_1). The diffusion profile of hydrogen under V_1 follows the shaded region in Fig. 6.7b. V_{th} shift under the lower voltage (V_2) at time (t_0) is given by:

$$\Delta V_{th}\,(t_0) = A_1\left((1+\delta)\,t_{ox} + \sqrt{C t_0}\right)^{2n} \tag{6.23}$$

Substituting $t = t_0$ in Eq. (6.20) and equating with Eq. (6.23), we arrive at the model which describes the behavior when voltage transits from a low to a high value:

$$\Delta V_{th}(t) = \left[\sqrt[2n]{A_2}\left((1+\delta)\,t_{ox} + \sqrt{C\,(t - t_0)}\right) + \sqrt[2n]{A_1}\left((1+\delta)\,t_{ox} + \sqrt{Ct}\right)\right]^{2n} \tag{6.24}$$

Table 6.1 summarizes the models under random stress, and static models based on the R–D theory. Figure 6.8 validated the model prediction with 45 nm silicon data under random stress patterns. The stress and recovery behaviors of threshold voltage are accurately captured when device is stressed under arbitrary voltage (Fig. 6.8a). Fig. 6.8b emphasizes the model prediction of the recovery component. In this figure, a device is stress under 1.3 V and after 1,000 s, the stress voltage is lowered to 1.2 V. As a result of excessive stress until 1,000 s, the device initially recovers and then V_{th} shift increases under the lower voltage. This non-monotonic behavior is well predicted by the dynamic model [54].

Table 6.1 Summary of R–D based BTI models

Constant stress	$\Delta V_{th}(t) = A \cdot t^n$
Random Input Stress	*Low to high voltage transition*: $\Delta V_{th}(t) =$ $\left[\sqrt[2n]{A_2}\left((1+\delta)\,t_{ox}+\sqrt{C\,(t-t_0)}\right)+\sqrt[2n]{A_1}\left((1+\delta)\,t_{ox}+\sqrt{Ct}\right)\right]^{2n}$
	High to low voltage transition: $\Delta V_{th}\,(t+t_0) =$ $\left(\sqrt[2n]{A_2}\sqrt{C(t)}+\sqrt[2n]{\Delta V_{th}\,(t_0)\left(1-\frac{2\xi_1 t_e+\sqrt{\xi_2 C(t)}}{(1+\delta)t_{ox}+\sqrt{C(t+t_0)}}\right)}\right)^{2n}$

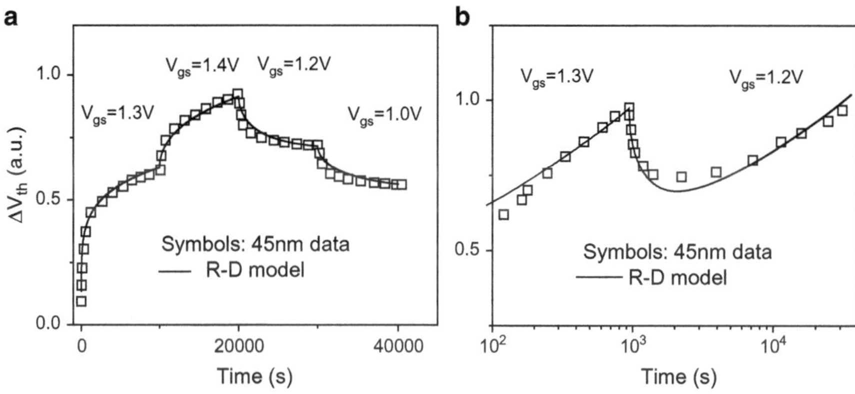

Fig. 6.8 Random input model validation (**a**) arbitrary stress pattern and (**b**) non-monotonic behavior of V_{th} shift under a lower voltage

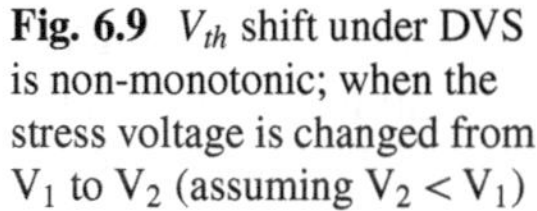

Fig. 6.9 V_{th} shift under DVS is non-monotonic; when the stress voltage is changed from V_1 to V_2 (assuming $V_2 < V_1$)

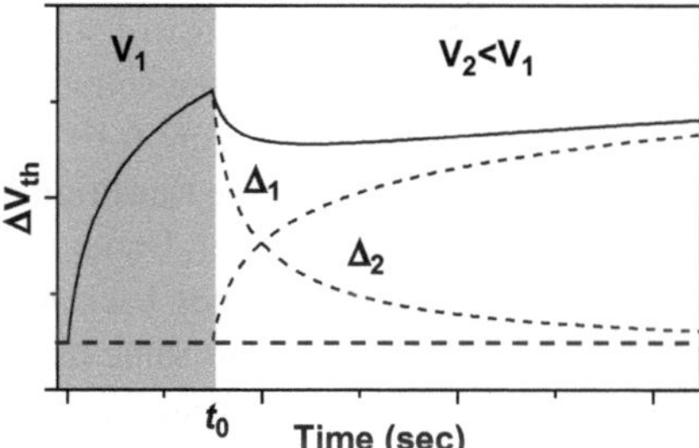

2.2.2 Trapping/De-Trapping Based Random Input Aging Model

In this section, trapping/de-trapping based models are presented which can handle random stress waveform. Since the degradation is highly sensitive to the voltage (Eq. (6.15)), dynamic voltage scaling leads to different amounts of circuit aging. The voltage tuning is categorized into two cases. In case 1, stress voltage is changed from lower voltage V_1 to a higher voltage V_2 and in case 2, V_2 is lower than V_1 at time t_0. To handle such a voltage transition, using Eq. (6.10) and a non-zero time, t_0, to calculate the occupation probability at time t (time elapsed after t_0), as shown in Fig. 6.9 [48].

To handle such a voltage transition, using a non-zero time, t_0 to calculate the occupation probability at time t (time elapsed after t_0) using Eq. (6.10):

$$P_{01}\left(t+t_0\right)=\frac{\tau_{eq2}}{\tau_{c2}}\left(1-e^{-t/\tau_{eq2}}\right)+P_{01}\left(t_0\right)e^{-t/\tau_{eq2}} \tag{6.25}$$

where τ_{eq2}, τ_{c2} represent the time constants under voltage V_2. Using Eqs. (6.5) and (6.6), $\tau_{eq1}=\tau_{eq2}$, since τ_{eq} depends only on parameter p, which is independent of the voltage. Substituting this property in Eq. (6.25):

$$P_{01}\left(t+t_0\right)=\frac{\tau_{eq}}{\tau_{c1}}\left(1-e^{-t/\tau_{eq}}\right)-\frac{\tau_{eq}}{\tau_{c2}}\left(e^{-t/\tau_{eq}}-e^{-(t+t_0)/\tau_{eq}}\right) \tag{6.26}$$

where τ_{c1} and τ_{c2} correspond to voltages V_1 and V_2. Following similar steps as in static model derivation, we arrive at a closed form solution:

$$\Delta V_{th}(t)=\phi_2\left[A+B\log\left(1+Ct\right)\right]+\phi_1\cdot B\left[\log\left(\frac{1+C\left(t+t_0\right)}{1+Ct}\right)\right] \tag{6.27}$$

where ϕ_1 corresponds to the voltage V_1 and ϕ_2 corresponds to V_2. The degradation in Eq. (6.27) is physically interpreted as a sum of two components, Δ_1 and Δ_2 which are proportional to ϕ_1 and ϕ_2 respectively. When the voltage is changed to a lower voltage, traps emit some of the charge carriers, and the number of occupied traps reaches a new equilibrium. Δ_2 dominates initially, which contributes to the recovery. If the operation under V_2 continues for a longer time, Δ_1 eventually takes over and ΔV_{th} increases. Such a non-monotonic behavior is correctly predicted. Table 6.2 presents the summary of static and dynamic models based on trapping/de-trapping theory [48]. When the voltage is increased, the degradation rate rises at the point of voltage change. Figure 6.10 validates the dynamic model. Non monotonic behavior when stress voltage transition to a lower value is correctly predicted by T–D based random input model. Under this condition, the device experience recovery period, before the stress goes back to the equilibrium condition. Eventually, the degradation rate goes to the same as the constant stress under the lower voltage. This behavior is predicted from Eq. (6.27), where the second component dominates initially, resulting in the recovery; after $t >> 200$ s, the second component decays down and the first component takes over, leading to the stress behavior under the second voltage. Experimental results from the test chip well validate these non-monotonic behaviors, as shown in Fig. 6.10, supporting further study on aging

Table 6.2 Summary of T–D based BTI models

Constant stress	$\Delta V_{th}(t)=\phi\cdot[A+B\log(1+Ct)]$	
Random Input Stress	$\Delta V_{th}(t+t_0)=\Delta_1+\Delta_2$	$\Delta_1=\varphi\left(A+B\log\left(1+Ct\right)\right)$, $\Delta_2=\Delta V_{th}\left(t_0\right)\left(1-\frac{k+\log(1+Ct)}{k+\log(1+C(t+t_0))}\right)$

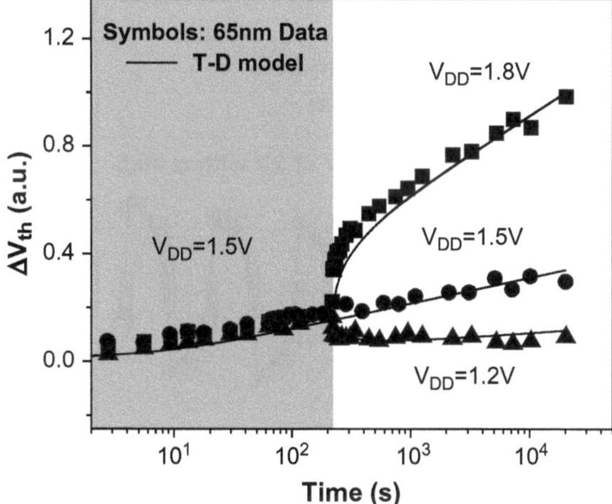

Fig. 6.10 The T–D model predicts the dynamic behavior under voltage tuning, such as the transition period, and the convergence to the constant stress condition

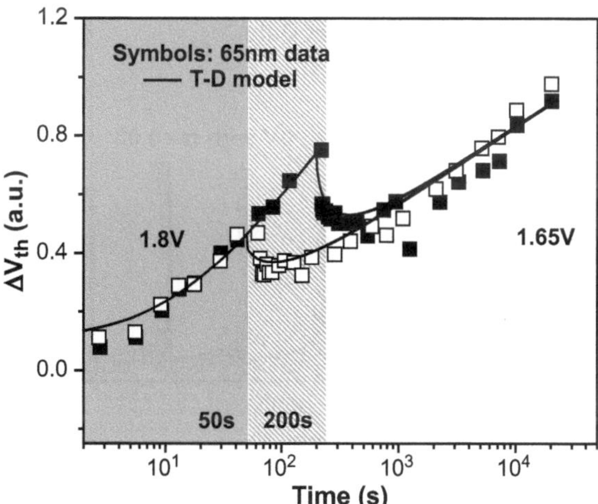

Fig. 6.11 V_{th} shift under voltage tuning from 1.8 to 1.65 V. The shift eventually converges to the final voltage, weakly dependent on previous stress history

prediction under DVS. The two components in Eq. (6.27) play an important role in long-term prediction under multiple cycles.

Higher recovery is seen when the device is stressed at 1.8 V compared to 1.65 V as more traps are captured under higher stress voltage. The T–D model captures such behavior. As stress voltage continues to be 1.2 V after 200 s, the temporary recovery is overwhelmed by the capture of traps under lower stress voltage. Increasing stress time causes the degradation to converge to this constant stress voltage at 1.2 V. This behavior is predicted from Eq. (6.27), where the second component dominates initially, resulting in the recovery, while the second component decays down and the first component takes over, leading to the stress behavior under the second voltage. Figure 6.11 evaluates the model prediction, with different periods under the same voltage. In this study, the device is initially stressed under 1.8 V, for 50 or 200 s;

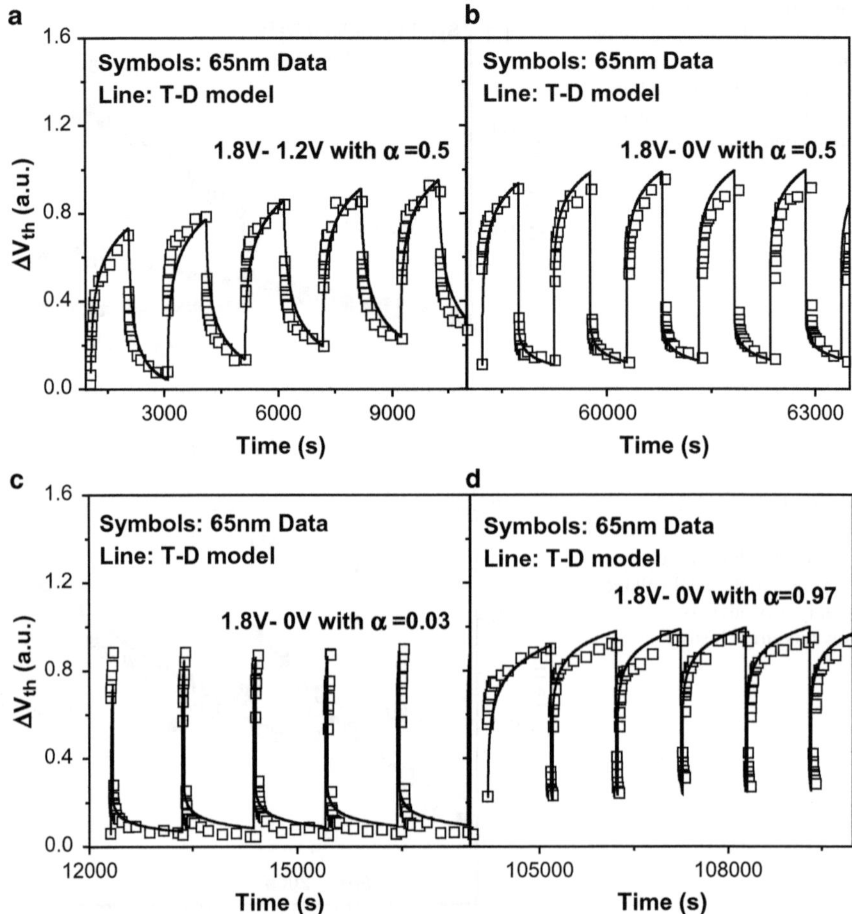

Fig. 6.12 Validation of the dynamic stress model under different voltages and duty cycles

then the voltage is switched to 1.65 V. As the stress voltage is lowered, a temporary recovery behavior is observed due to the emission of excessive amount of trapped charges. The T–D model captures such behavior in both cases. In the case where the device is first stressed under 1.8 V for 200 s, the stress time is higher as compared to the case where the initial stress is 50 s causing higher V_{th} shift. However, since in both cases, the device is later stressed for a much longer time (10 ks) at 1.65 V, the degradation converges to the constant stress condition at 1.65 V. This validation helps predict the aging under various switching activities (α) much needed for aging evaluation of digital circuits.

Figure 6.12 presents the measurement under different patterns of voltages and duty cycles. Figure 6.12a and b show the multiple cycle prediction stressed at 1.8, 1.2 V and 1.8, 0 V at $\alpha = 0.5$. When the voltage is lowered from 1.8 V, a recovery is observed from silicon data, which is well predicted from the model. Furthermore, if the voltage is lowered down to 0 V, the recovery is more pronounced (Fig. 6.12b). The cycle-to-cycle model in Eq. (6.27b) is scalable with different duty cycles. Figure 6.12c and d show the validation of our model with silicon data at very low ($\alpha = 0.03$) and very high ($\alpha = 0.97$) duty cycles, respectively. A voltage transition even for a short duration results in the sudden shift in threshold voltage, due to the fast trapping/de-trapping (Fig. 6.12c and d).

Random input models derived in this section are thoroughly validated with 45 and 65 nm data. These models provide an accurate and efficient way to identify aging under arbitrary stress patterns, allowing designers to adequately guard-band their design. Such models can be more beneficial to AMS designs where the number of transistors are less than that in a digital design and the input is highly random. However for large scale digital circuits, using random stress models is insufficient to predict the lifetime of a high-speed design. Thus, there is a clear need for long-term aging models based on both R–D and T–D theories, which can estimate a tight upper bound of aging, without tracking the behavior cycle by cycle. The following section presents the solutions.

2.3 *Long-Term BTI Model Under DVS*

Under the extensive usage of DVS, accurate V_{th} shift can be predicted using the random input stress models derived in previous section. However for a large scale logic design, applying such random stress models will not be efficient in circuit simulation. Most digital circuits operate at a regular frequency periodically with a representative duty cycle. A long-term model will be desirable for efficient aging prediction, which directly estimates aging at a given operation time without tracking the stress-recovery over many cycles. This long-term model predicts a tight upper bound after multi-cycle operations under DVS. Based on the multi-cycle model in previous section, stress ($\Delta V_{ths,m}$) and recovery ($\Delta V_{ths,m+1}$) in Fig. 6.13 can be connected to derive a long-term model based on both R–D and T–D theories.

2.3.1 Long-Term BTI Model Based on Reaction–Diffusion

It is possible to obtain a closed form solution to predict the upper bound of ΔV_{th} for different clock cycle (T_{clk}), duty cycle (α) and stress time for a circuit oscillating between two voltages. To derive a closed form solution, $\Delta V_{ths,m}$ and $\Delta V_{ths,m+1}$ are connected iteratively using random stress models as:

$$\Delta V_{ths,m+1} = \begin{bmatrix} \sqrt[2n]{A_2} X \left(1 + \beta_m^{1/2n} + \beta_m^{1/2n} \beta_{m-1}^{1/2n} + \ldots\right) + \\ \sqrt[2n]{A_1} \left(Y + \sqrt{C\alpha T_{clk}}\right) \left(1 + \beta_m^{1/2n} + \beta_m^{1/2n} \beta_{m-1}^{1/2n} + \ldots\right) \end{bmatrix}^{2n} \tag{6.28}$$

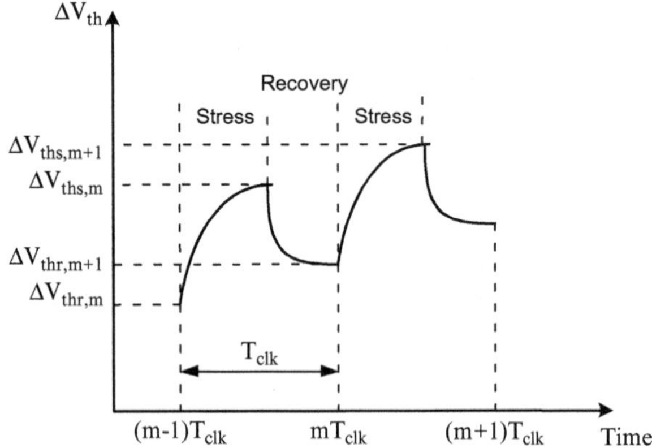

Fig. 6.13 V_{th} shift during the stress and recovery cycles, presenting the important parameters used in the multi-cycle model

where,

$$X = (1+\delta)\, t_{ox} + \sqrt{C\alpha T_{clk}}, \qquad Y = \sqrt{C\,(1-\alpha)\, T_{clk}}$$

$$\beta = 1 - \frac{2\xi_1 t_e + \sqrt{\xi_2 C\,(1-\alpha)\, T_{clk}}}{(1+\delta)\, t_{ox} + \sqrt{C m T_{clk}}}$$

Using $\beta_1 < \beta_2 < \cdots \beta_{m\text{-}1} < \beta_m$ and geometric series approximation, the upper bound of degradation is derived as:

$$\Delta V_{ths,m+1} = \left[\sqrt[2n]{A_2} X \left(\frac{1}{1-\beta^{1/2n}} \right) + \sqrt[2n]{A_1} \left(Y + \sqrt{C\alpha T_{clk}} \right) \left(\frac{1}{1-\beta^{1/2n}} \right) \right]^{2n} \tag{6.29}$$

The new long-term model is capable of predicting the upper bound of ΔV_{th} for cycle of two non-zero voltages [54].

Figure 6.14 confirms that for multiple cycles, the long-term model predicts the same result as the model under random input, given a constant duty cycle. Such model predicts a tight upper bound under the periodic stress.

2.3.2 Long-Term BTI Model Based on Trapping/De-Trapping

Similar to the derivation of the R–D model, a long-term model based on the T–D theory is obtained. Based on the multi-cycle model in the previous sub-section, $\Delta V_{ths,m}$ and $\Delta V_{ths,m+1}$ are connected by:

$$\begin{aligned} \Delta V_{ths,m+1} = {} & \phi_1 \left[A + B \log (1 + C\alpha T_{Clk}) \right] + \phi_2 \left[A + B \log (1 + C\,(1-\alpha)\, T_{Clk}) \right] \beta_{1,m} \\ & + \Delta_{Vths,m} \left(1 - \beta_{1,m} \right) \left(1 - \beta_{2,m} \right) \end{aligned} \tag{6.30}$$

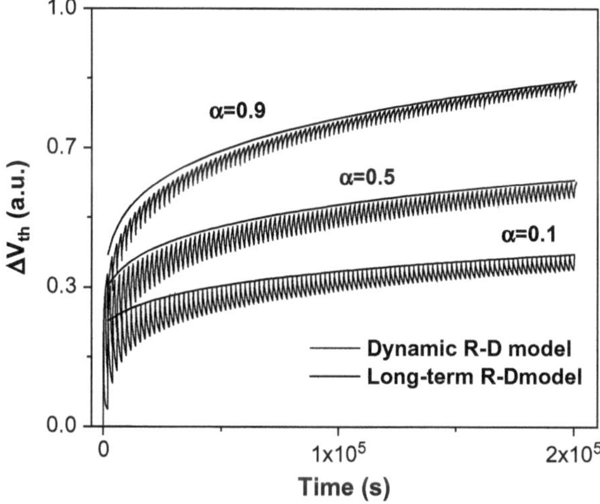

Fig. 6.14 Under constant duty cycles, the long-term model is consistent with the model for random input

Using Eq. (6.30) and repeatedly replacing the $\Delta V_{ths,m+1}$ by $\Delta V_{ths,i}$ for $i = m, \ldots, 1$, we get:

$$\Delta V_{ths,m} = \phi_1 \left[A + B \log\left(1 + C\alpha T_{Clk}\right)\right] \left(1 + \sum_{i=1}^{m} \prod_{j=m-i+1}^{m} \beta_{1,.j} \cdot \beta_{2,.j}\right) + \phi_2 \left[A + B \log\left(1 + C\left(1-\alpha\right) T_{Clk}\right)\right] \beta_{1,m} \left(1 + \sum_{i=1}^{m} \prod_{j=m-i+1}^{m} \beta_{1,.j-1} \cdot \beta_{2,.j}\right) \quad (6.31)$$

Since obtaining a closed-form solution for Eq. (6.31) is not straight-forward, we use the property $\beta_{1,m\text{-}1} < \beta_{1,m}$ and $\beta_{2,m\text{-}1} < \beta_{2,m}$,

$$\Delta V_{ths,m+1} \le \phi_1 \left[A + B \log\left(1 + C\alpha T_{Clk}\right)\right] \left(1 + \beta_{1,.m} \cdot \beta_{2,.m} + \left(\beta_{1,.m} \cdot \beta_{2,.m}\right)^2 + \ldots\right) + \phi_2 \beta_{1,m} \left[A + B \log\left(1 + C\left(1-\alpha\right) T_{Clk}\right)\right] \left(1 + \beta_{1,.m} \cdot \beta_{2,.m} + \left(\beta_{1,.m} \cdot \beta_{2,.m}\right)^2 + \ldots\right) \quad (6.32)$$

Equation (6.32) is a geometric series and the upper bound of degradation is:

$$\Delta V_{ths,m} = \phi_1 \left[A + B \log\left(1 + C\alpha T_{Clk}\right)\right] \frac{1}{1 - \beta_{1,.m} \cdot \beta_{2,.m}} + \phi_2 \left[A + B \log\left(1 + C\left(1-\alpha\right) T_{Clk}\right)\right] \frac{\beta_{1,.m}}{1 - \beta_{1,.m} \cdot \beta_{2,.m}} \quad (6.33)$$

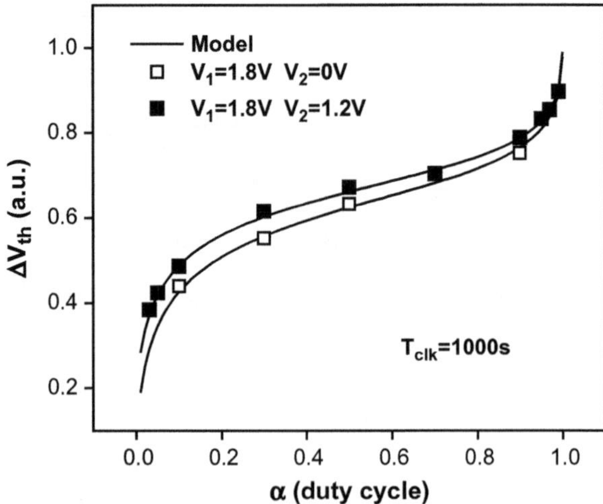

Fig. 6.15 Long-term model prediction for different duty cycle

Equation (6.33) is sensitive to the duty cycle, α (ratio of time under V_1 to time under V_2), time period (sum of operation times under V_1 and V_2 for a single cycle), and the stress voltage. The long term model captures the tight upper bound of cycle-to-cycle prediction. Furthermore, Fig. 6.15 presents the aging behavior under a wide range of duty cycles as validated with 65 nm silicon data. The degradation rate changes rapidly when α approaches 0 or 1, but gradual for intermediate duty cycles. This behavior is due to the sudden change in degradation at the beginning of the stress and recovery phase, due to the voltage dependent time constants. It is well predicted by a single equation in the long-term model [48].

3 Circuit Aging Simulation

New models are derived from basic physics and are comprehensively validated with discrete device data in previous sections. In this section, these models are further integrated with design tools at the circuit level. Different types of circuits have different requirements on reliability tools, due to the diversity in circuit operation patterns. Digital circuits usually operate at two distinct voltage levels (V_{DD} and GND) as opposed to continuous voltages in analog circuits. Thus long-term models are more appropriate for digital circuits and random input stress models are better for AMS designs. Aging models required to predict circuit lifetime under any given stress conditions are presented in previous sections. Based on them, we present the aging analysis of digital and AMS designs by implementing aging models in SPICE for real time simulation under any possible stress conditions, supporting design practice for reliability. The model prediction is validated with 45 nm ring oscillator data.

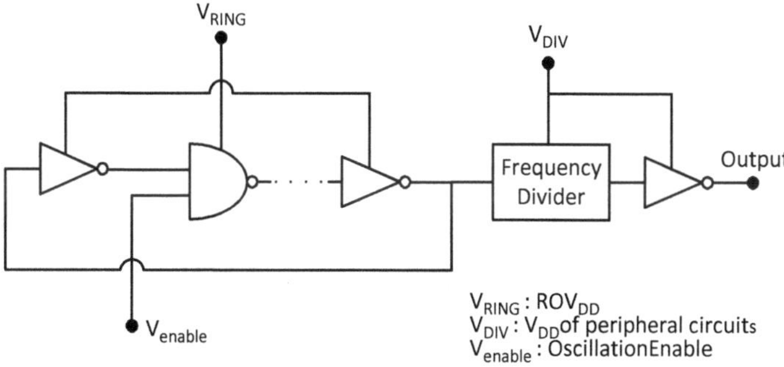

Fig. 6.16 Test circuit at 45 nm used for the validation

In order to estimate the impact of aging, a sub-circuit module is implemented in SPICE. Model equations (R–D or T–D based) described in previous sections are coded in Verilog-A, which take the stress voltage, temperature and time as inputs [31]. A voltage controlled voltage source then subtracts the V_{th} shift calculated by the model from V_g, emulating aging of a device and resulting in decrease in drain current. To improve the simulation speed, input voltages are quantized into discrete levels for random input models. Thus, there exists a trade-off between simulation speed and step size of the voltage. Stress time is the simulation time from the SPICE environment allowing designer to track circuit performance at real time and to closely monitor the degradation behavior.

3.1 Aging Analysis: Digital Circuits

To validate the new models at circuit level, silicon data is measured from 45 nm ring oscillators (ROs). The frequency change (ΔF/F) of RO is measured as a direct index of the degradation, which is proportional to PMOS transistor threshold voltage change under NBTI. Figure 6.16 presents the test circuit of ROs used for aging characterization. Frequency change in 11 stage ring oscillator is monitored during the test. The ring oscillator is activated by the enable (V_{enable}) pin. Different from traditional RO based aging test, the supply voltage of this circuit, V_{RING}, is switched between V_{DD} (stress mode) and 0 (recovery mode) at different duration, in order to emulate dynamic voltage scaling encountered in logic circuits. The data is collected at regular intervals at multiple supply voltages and temperature. The pin, V_{DIV}, is implemented to control the frequency divider and the output buffer. It also helps to eliminate the impact on frequency shift of ROs due to aging of peripheral circuits, thus giving a clean data [54].

ΔV_{th} of a device in RO has a strong dependence on the dynamic scaling of supply voltages due to the recovery effect. Figure 6.17a shows the basic test pattern in a ring oscillator with alternate active and sleep modes, each for 5,000 s duration. The

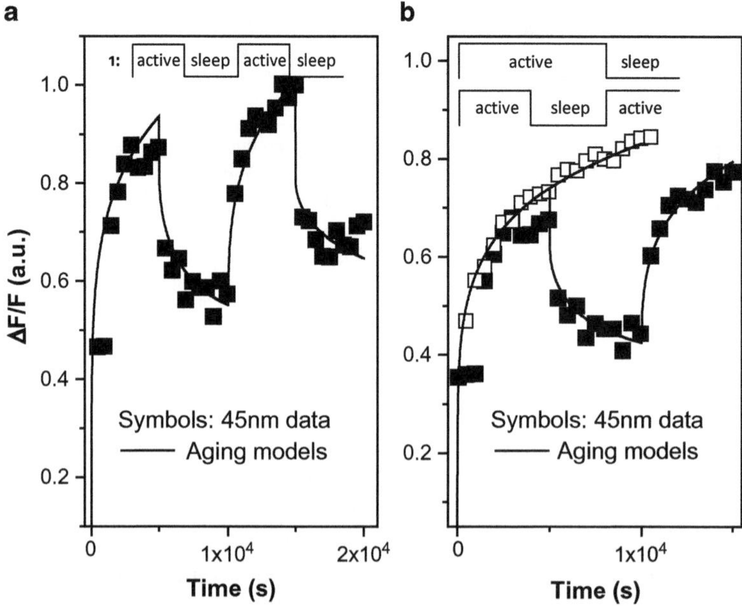

Fig. 6.17 (**a**) Model validation with circuit-level DVS data. (**b**) Aging under different stress patterns well captured by the model

increase of frequency degradation in the active mode and its decrease in the sleep mode are well predicted by the new random stress models based on R–D or T–D theory. Figure 6.17b shows the impact of the recovery on the frequency shift of RO. Although the stress times are same for both patterns, the recovery in PMOS device causes over all degradation to be less. These effects are successfully captured by the aging models described previously.

3.2 Aging Analysis: AMS Circuits

Analog/mixed signal circuits encounter more complex stress patterns. Furthermore, device degradation changes the DC biasing conditions, which in turn change circuit performance. Figure 6.18 shows a case study of the impact of NBTI on a PMOS input differential amplifier. In this case, we demonstrate that the amplifier is being used to amplify the output of a mixer circuit. The stress pattern of a device is a combination of two different frequencies signal ω_1 and ω_2. Conventional prediction with an average pattern fails to capture amplitude distortion (Fig. 6.19b). Using new random input stress models, an accurate aging analysis on the differential amplifier can be conducted. Note that the accuracy of aging prediction has a trade-off with simulation time in SPICE. Real-time stress and recovery analysis requires SPICE to

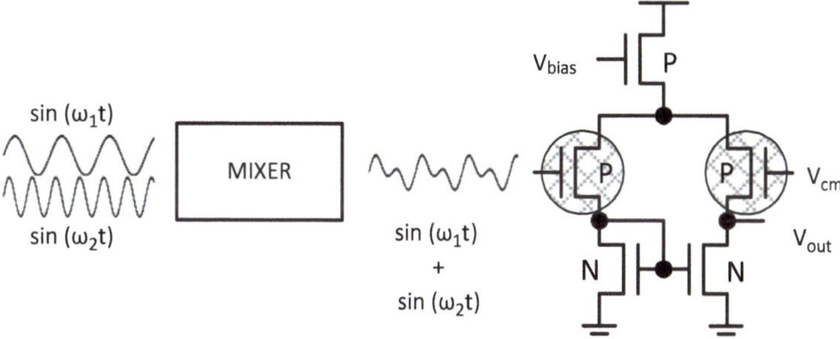

Fig. 6.18 A differential amplifier in aging analysis

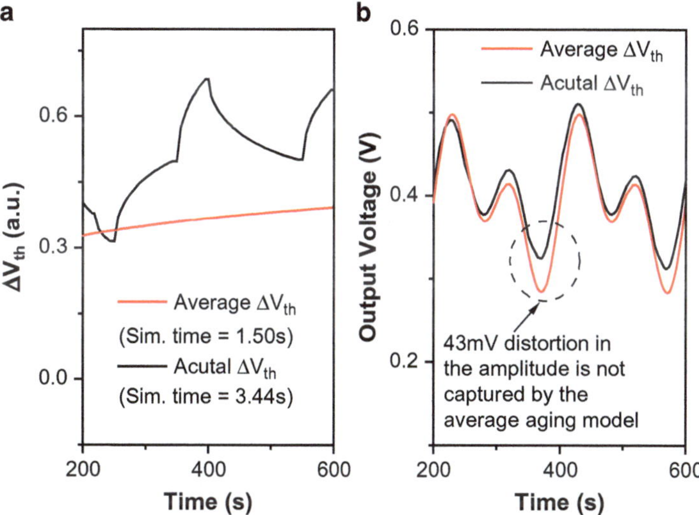

Fig. 6.19 Aging analysis with the average value fails to capture the distortion. (**a**) V_{th} shift prediction (**b**) output voltage of the amplifier

increase the internal time steps which take longer simulation time compared to the average aging stress pattern in Fig. 6.19.

4 Summary

Aging issue has become increasingly important as the technology scales. BTI, especially NBTI in a PMOS device, has emerged as a dominant mechanism in digital as well as AMS designs, impacting circuit performance over time. The challenges are further compounded with Dynamic Voltage Scaling (DVS) and random inputs that are encountered in a realistic operation. In this chapter, both R–D and T–D theories are comprehended in detail. Compact BTI models for both R–D and

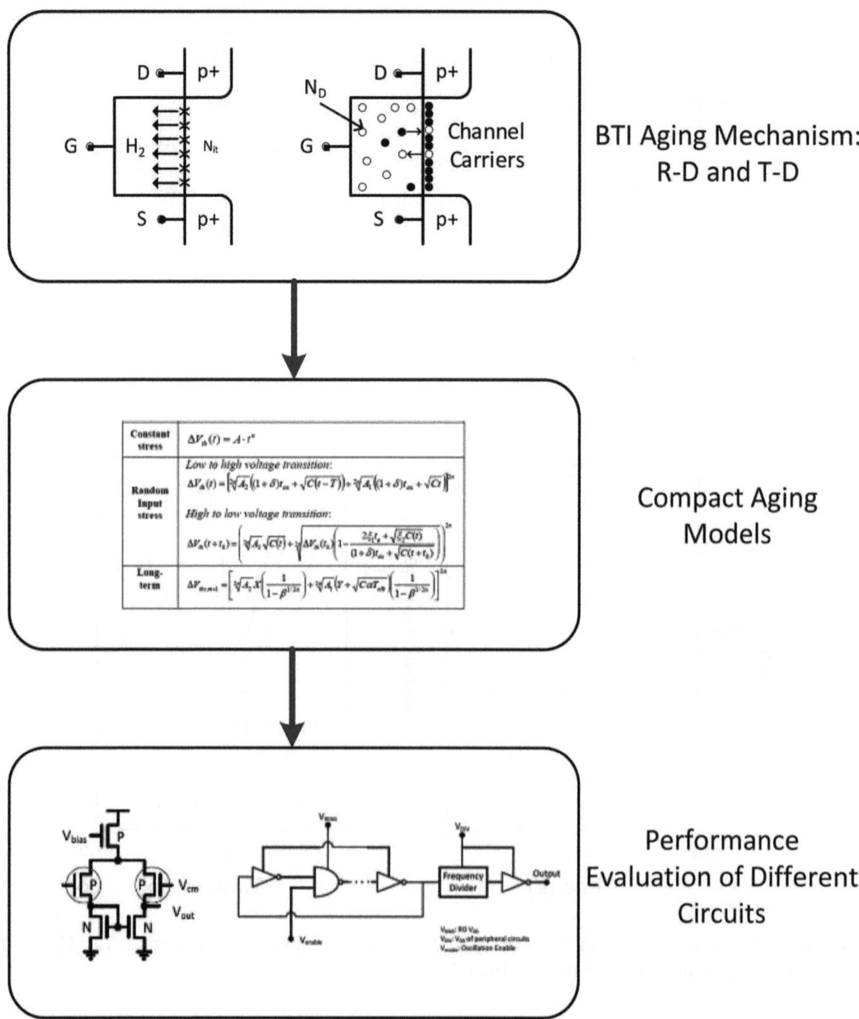

Fig. 6.20 Hierarchical development of BTI models for circuit aging analysis

T–D theories are then derived and validated using 45 and 65 nm silicon data. Figure 6.20 presents distinct levels of reliability modeling. Three different types of models are developed for diverse design needs: Static, Random stress and Long-term models. The models are implemented in the SPICE environment, through Verilog-A. Depending on the operation pattern, appropriate models can be applied to circuit-level aging analysis and help improve circuit reliability.

References

1. C. Constantinescu, "Trends and challenges in vlsi circuit reliability," *IEEE Computer Society*, pp. 14–19, 2003.
2. Semiconductor Industry Associate, *International Technology Roadmap for Semiconductors*, 2010. (Available at http://public.itrs.net).
3. D. K. Schroder and J. A. Babcock, "Negative bias temperature instability: Road to cross in deep submicron silicon semiconductor manufacturing," *Journal of Applied Physics*, vol. 94, no. 1, pp. 1-18, 2003.
4. K. O. Jeppson and C. M. Svensson, "Negative bias stress of MOS devices at high electric fields and degradation of MNOS devices," *J. Applied Physics*, vol. 48, no. 5, pp. 2004-2014, 1977.
5. K. Bernstein, D. J. Frank, A. E. Gattiker, W. Haensch, B. L. Ji, S. R. Nassif, E. J. Nowak, D. J. Pearson, N. J. Rohrer, "High-performance CMOS variability in the 65-nm regime and beyond," *IBM Journal of Research and Development*, vol. 50, no. 4/5, pp. 433-449, July 2006.
6. S. Borkar, T. Karnik, S. Narendra, J. Tschanz, A. Keshavarzi, V. De, "Parameter variations and impact on circuits and microarchitecture," *Design Automation Conference*, pp. 338-342, 2003.
7. F. Arnaud, L. Pinzelli, C. Gallon, M. Rafik, P. Mora, F. Boeuf, "Challenges and opportunity in performance, variability and reliability in sub-45 nm CMOS technologies," *Microelectronics Reliability*, vol. 51, no. 9-11, pp. 1508-1514, Sept.-Nov. 2011.
8. B. H. Calhoun, Y. Cao, X. Li, K. Mai, L. T. Pileggi, R. A. Rutenbar, and K. L. Shepard, "Digital circuit design challenges and opportunities in the era of nanoscale CMOS," *Proceedings of the IEEE*, vol. 96, no. 2, pp. 343-365, Feb. 2008.
9. A. Ramadan, "Compact model council's standard circuit simulator interface for reliability modeling," *International Reliability Physics Symposium*, pp. 2A.5.1-2A.5.6, 2013.
10. J. Hicks, D. Bergstrom, M. Hattendorf, J. Jopling, J. Maiz, S. Pae, et al., "45 nm transistor reliability," *Intel Technology Journal*, vol. 12, no. 02, pp. 131-144, Jun. 2008.
11. S. V. Kumar, C. H. Kim and S. S. Sapatnekar, "Adaptive techniques for overcoming performance degradation due to aging in digital circuits," *Asia and South Pacific Design Automation Conference*, pp. 284-289, 2009.
12. V. Reddy, "Impact of negative bias temperature instability in digital circuit reliability," *International Reliability and Physics Symposium*, pp. 248-253, 2002.
13. R. Zheng, et al., "Circuit aging prediction for low-power operation," *Customs Integrated Circuits Conference*, pp. 427-430, 2009
14. G. Chen, K. Y. Chuah, M. F. Li, Daniel SH Chan, C. H. Ang, J. Z. Zheng, Y. Jin, and D. L. Kwong, "Dynamic NBTI of PMOS transistors and its impact on device lifetime," *International Reliability Physics Symposium*, pp. 196-202, 2003.
15. B. C. Paul, K. Kang, H. Kufluoglu, M. A. Alam, and K. Roy, "Impact of NBTI on the temporal performance degradation of digital circuits," *IEEE Electronic Device Letters*, vol. 26, no. 8, pp. 560-562, Aug. 2005.
16. R. Vattikonda, W. Wang, and Y. Cao, "Modeling and minimization of pmos nbti effect for robust nanometer design," *IEEE/ACM Design Automation Conference*, pp. 1047-1052, Jul. 2006.
17. M. Agarwal, B. C. Paul, Ming Zhang, and S. Mitra, "Circuit failure prediction and its application to transistor aging," *VLSI Test Symposium*, pp. 277-286, 2007.
18. V. Reddy, A. T. Krishnan, A. Marshall, J. Rodriguez, S. Natarajan, T. Rost, and S. Krishnan, "Impact of negative bias temperature instability on digital circuit reliability," *International Reliability Physics Symposium*, pp. 248-254, 2002.
19. W. Wang, Z. Wei, S. Yang and Y. Cao, "An efficient method to identify critical gates under circuit aging," *Int. Conference on Computer Aided Design*, pp. 735-740, 2007.
20. A. T. Krishnan, F. Cano, C. Chancellor, V. Reddy, Q. Zhangfen, P. Jain, J. Carulli, J. Masin, S. Zuhoski, S. Krishnan, and J. Ondrusek, "Product drift from NBTI: Guardbanding, circuit and statistical effects," *Int. Electron Devices Meeting*, pp. 4.3.1-4.3.4, 2010.

21. W. Wang, S. Yang, S. Bhardwaj, R. Vattikonda, S. Vrudhula, F. Liu, Y. Cao, "The impact of NBTI effect on combinational circuit: Modeling, simulation, and analysis," *IEEE Transactions on VLSI Systems*, vol. 18, no. 2, pp. 173-183, 2010.
22. H. Sangwoo, K. Juho, "NBTI-aware statistical timing analysis framework," *IEEE International SOC Conference*, pp.158-163, 2010.
23. E. Maricau and G. Gielen, "Computer-aided analog circuit design for reliability in nanometer CMOS," *IEEE Transactions on Emerging and Selected Topics in Circuits and Systems*, vol. 1, no. 1, pp. 50-58, Mar. 2011.
24. A. R. Brown, V. Huard and A. Asenov, "Statistical simulation of progressive NBTI degradation in a 45-nm technology pMOSFET," *IEEE Transactions on Electron Devices*, vol. 57, no. 9, pp. 2320-2323, Sept. 2010.
25. S. Ogawa and N. Shiono, "Generalized diffusion-reaction model for the low-field charge-buildup instability at the Si-SiO2 interface," *Physical Review B*, vol. 51, no. 7, pp. 4218-4230, Feb. 1995.
26. G. Chen, K. Y. Chuah, M. F. Li, D. S. H. Chan, C. H. Ang, J. Z. Zheng, Y. Jin, et al., "Dynamic NBTI of PMOS transistors and its impact on device lifetime," *International Reliability Physics Symposium*, pp. 196-202, 2003.
27. S. Chakravarthi, A. T. Krishnan, V. Reddy, C. F. Machala and S. Krishnan, "A comprehensive framework for predictive modeling of negative bias temperature instability," *International Reliability Physics Symposium*, pp. 273-282, 2004.
28. M. A. Alam, S. Mahapatra, "A comprehensive model of PMOS NBTI degradation," *Microelectronics Reliability*, vol. 45, pp. 71-81, 2005.
29. V. Huard, C. R. Parthasarathy, C. Guerin, M. Denais, "Physical modeling of negative bias temperature instabilities for predictive extrapolation," *International Reliability Physics Symposium*, pp. 733-734, 2006.
30. S. V. Kumar, C. H. Kim, and S. S. Sapatnekar, "An analytical model for Negative Bias Temperature Instability (NBTI)," *International Conference for Computer Aided Design*, pp. 493-496, 2006.
31. W. Wang, V. Reddy, A. Krishnan, R. Vattikonda, S. Krishnan and Y. Cao, "Compact modeling and simulation of circuit reliability for 65 nm CMOS technology," *IEEE Transactions on Device and Materials Reliability*, vol. 7, no. 4, pp. 509-517, Dec. 2007.
32. R. Tu, E. Rosenbaum, W. Chan, C. Li, E. Minami, K. Quader, et al., "Berkeley reliability tools – BERT," *IEEE Transactions on Computer-Aided Design of Integrate Circuits and Systems*, vol. 12, no. 10, pp. 1524-1534, Oct. 1993.
33. *Reliability Simulation in Integrated Circuit Design*, Cadence, 2003.
34. *MOS Device Aging Analysis with HSPICE and CustomSim*, Synopsys, 2011.
35. *ELDO User's Manual*, Mentor Graphics, 2005.
36. V. Huard, M. Denais, "Hole trapping effect on methodology for DC and AC negative bias temperature instability measurements in pMOS transistors," *International Reliability Physics Symposium*, pp. 40-45, 2004.
37. T. Grasser, B. Kaczer, W. Goes, H. Reisinger, T. Aichinger, P. Hehenberger, et al., "The paradigm shift in understanding the bias temperature instability: from reaction-diffusion to switching oxide traps," *IEEE Transactions on Electron Devices*, vol. 58, no. 11, pp. 3652-3666, Nov. 2011.
38. G. I. Wirth, R. da Silva and B. Kaczer, "Statistical model for MOSFET bias temperature instability component due to charge trapping," *IEEE Transactions on Electron Devices*, vol. 58, no. 8, pp. 2743-2751, Aug. 2011.
39. J. B. Velamala, K. B. Sutaria, T. Sato, Y. Cao, "Aging statistics based on trapping/detrapping: silicon evidence, modeling and long-term prediction," *International Reliability Physics Symposium*, pp. 2 F.2.1-2 F.2.5, 2012.
40. T. Grasser, H. Reisinger, W. Goes, T. Aichinger, P. Hehenberger, P.-J. Wagner, M. Nelhiebel, J. Franco, B. Kaczer, "Switching oxide traps as the missing link between negative bias temperature instability and random telegraph noise," *International Electron Devices Meeting*, pp.1-4, 2009.

41. G. I. Wirth, J. Koh, R. da Silva, R. Thewes, and Ralf Brederlow, "Modeling of statistical low-frequency noise of deep-submicron MOSFETs," *Trans. on Electron Dev.*, vol. 52, pp. 1576-1588, 2005.
42. A.P. van der Wel, E.A.M. Klumperink, J.S. Kolhatkar, E. Hoekstra, M.S. Snoeij, C. Salm, H. Wallinga, B. Nauta, "Low-Frequency Noise Phenomena in Switched MOSFETs," *IEEE Journal of Solid-State Circuits,* vol. 42, no. 3, pp.540-550, March 2007.
43. M. J. Kirton and M. J. Uren, "Noise in solid-state microstructures: A new perspective on individual defects, interface states and sow-frequency (1/f) noise," *Advances in Physics*, vol. 38, p. 367-468, 1989.
44. G. Wirth, R. da Silva and R. Brederlow, "Statistical model for the circuit bandwidth dependence of low-frequency noise in deep-submicrometer MOSFETs," *Trans. on Electron Devices*, vol. 54, pp.340-345, Feb. 2007.
45. G. Wirth, R. da Silva, P. Srinivasan, J. Krick and R. Brederlow. "Statistical model for MOSFET low-frequency noise under cyclo-stationary conditions," *International Electron Devices Meeting*, p.30.5.1-4, 2009.
46. B. Kaczer, T. Grasser, Ph. J. Rousse, J. Martin-Martinez, R. O'Connor, B. J. O'Sullivan, and G. Groeseneken, "Ubiquitous relaxation in BTI stressing—new evaluation and insights," *International Reliability Physics Symposium*, pp. 20-27, 2008.
47. T. Grasser, H. Reisinger, P.-J. Wagner, F. Schanovsky, W. Goes and B. Kaczer, "The time dependent defect spectroscopy (TDDS) for the characterization of the bias temperature instability," *International Reliability Physics Symposium*, pp. 16-25, 2010.
48. J. B. Velamala, K. B. Sutaria, H. Shimizu, H. Awano, T. Sato, G. Wirth, Yu Cao, "Compact modeling of statistical BTI under trapping/detrapping," *IEEE Transactions on Electron Devices*, vol.60, no.11, pp.3645-3654, Nov. 2013
49. J. B. Velamala, K. Sutaria, H. Shimizu, H. Awano, T. Sato, Y. Cao, "Statistical aging under dynamic voltage scaling: A logarithmic model approach," *Custom Integrated Circuits Conference*, pp. 1-4, 2012.
50. K. Sutaria, J. Velamala, V. Ravi, G. Wirth, T. Sato, Y. Cao, "Multilevel reliability simulation for IC design," pp. 719-749, in *Bias Temperature Instability for Devices and Circuits*, edited by T. Grasser, Springer, 2014.
51. K. B. Sutaria, J. B. Velamala, C. Kim, T. Sato, Y. Cao, "Aging statistics based on trapping/detrapping: Compact modeling and silicon validation," *IEEE Transactions on Device and Materials Reliability*, vol. 14, no. 2, pp. 607-615, June 2014.
52. M. Denais, C. R. Parthasarathy, G. Ribes, Y. Rey-Tauriac, N. Revil, A. Bravaix, V. Huard, F. Perrier, "On-the-fly characterization of NBTI in ultra-thin gate oxide PMOSFET's," *International Electron Devices Meeting*, pp. 109-112, 2004.
53. H. Reisinger, O. Blank, W. Heinrigs, A. Muhlhoff, W. Gustin, C. Schlunder, "Analysis of NBTI degradation- and recovery-behavior based on ultra-fast VT-measurements," *International Reliability Physics Symposium*, 2006.
54. K. B. Sutaria, A. Ramakumar, R. Zhu, R. Rajeev, Y. Ma, Y. Cao, "BTI-induced aging under random stress waveforms: Modeling, simulation and silicon Validation," *Design Automation Conference*, pp. 1-6, 2014.

Chapter 7
Circuit Resilience Roadmap

Veit B. Kleeberger, Christian Weis, Ulf Schlichtmann, and Norbert Wehn

Abstract Technology scaling has an increasing impact on the resilience of integrated circuits. This growth is the result of (a) increasing sensitivity to various noise sources, and (b) an increase in parametric variability. This chapter examines the issue of circuit resilience by studying ongoing trends in technology scaling. Additional experiments with basic circuit blocks, such as memory or logic cells, reveal insights into their behavior for future technology generations and major threats for circuit resilience.

1 Introduction

Faults in integrated circuits can broadly result from four aspects: (a) *design bugs* which originate from mistakes in the system specification or implementation, (b) *manufacturing problems* which can range from variations of design parameters to complete changes in topology, such as resistive shorts or opens, (c) *aging problems* which arise over lifetime and also can range from a degradation of design parameters to changes in the circuit topology, and (d) *signal corruption* which results from intrinsic or extrinsic temporary disturbances.

All these faults may lead to errors and failures in the whole system. These we can again categorize broadly: (a) *permanent errors* which relate to changes in the topology, (b) *soft errors* which result from signal corruptions, such as a particle

V.B. Kleeberger (✉)
Infineon Technologies AG, Neubiberg, Germany
e-mail: veit.kleeberger@infineon.com; kleeberger@tum.de

C. Weis • N. Wehn
Microelectronic Systems Design Research Group, Technische Universität Kaiserslautern, Kaiserslautern, Germany
e-mail: weis@eit.uni-kl.de; wehn@eit.uni-kl.de

U. Schlichtmann
Institute for Electronic Design Automation, Technische Universität München, Munich, Germany
e-mail: ulf.schlichtmann@tum.de

R. Reis et al. (eds.), *Circuit Design for Reliability*,
DOI 10.1007/978-1-4614-4078-9_7

strike changing the contents of a storage element, and (c) *parametric errors*, where the circuit fails to meet its specification in terms of frequency of operation, power, or similar metrics.

In this context, a circuit is more resilient if it is able to tolerate increased occurrence of soft and parametric errors. For digital synchronous circuits, we can easily evaluate their resilience, using standardized measures and metrics for their performance and correctness. Consider for example a metric such as logic functionality: we can easily state that the steady-state output voltage of an inverter has to be the logic inverse of its input voltage. Similarly, for example a latch has to keep its stored value until the next latching clock pulse.

Traditionally, *hard* and *soft* errors were distinguished in the following way. Hard errors are associated with an incorrect behavior of the circuit under all conditions, which results from a topological change of the circuit. In contrast, soft errors are based on changes of electrical parameters of the circuit, which causes a faulty behavior only under specific conditions. For example, a short or open inside a CMOS inverter may cause its output to be *stuck at* logic one or zero, while a shift in gate oxide thickness behaves more subtle, as it changes the time the inverter takes to change its state given a switching input signal. This in turn, leads only to an error if this change causes a path to violate one of its timing constraints.

However, excessive parametric variability can cause the circuit to behave in a way which we would traditionally associate with a hard error. A prominent example for this in current technologies can be found in static random access memory (SRAM), where we use the smallest possible device due to high density requirements. These small devices are extremely susceptible to various sources of variability, such as *Random Dopant Fluctuations* (RDF) or *Line Edge Roughness* (LER) [1]. Such excessive variability can lead to scenarios, where a particular SRAM cell cannot be read or written under all practical operating conditions, although the underlying topology might still be correctly manufactured [13].

In addition to parametric variability, which we discussed above, circuits—as they get smaller—become more sensitive to various forms of noise. As the amount of charge being held in a storage element or transported by a switching signal decreases, the potential for a noise source, such as a striking particle, to disrupt this charge increases.

These problems already exist in memories for several years, which has lead to a huge variety of solutions and techniques to deal with them. Over the years, various kinds of redundancy, parity checking, and error correction, were proposed and are nowadays an essential part of any practical memory design. As we start to face similar problems now with the remaining logic parts of the design, also here several techniques have been proposed over the last years, such as RAZOR [2] or BISER [22] to name just two prominent examples.

Our goal in this chapter is to discuss and analyze the ongoing trend of continued technology scaling and its implications for circuit resilience. This continues earlier work on predicting circuit resilience conducted in [10] and [14]. To demonstrate this behavior on specific examples, we will focus on two basic circuit blocks. On the one hand we analyze the behavior and resilience of different SRAM architectures.

As already mentioned above, SRAM cells represent the most susceptible parts of a chip due to their minimal sizing. On the other hand we discuss trends of technology scaling in logic standard cells. Standard cells in logic designs, such as inverters, represent (compared to SRAM cells) the other extreme and are typically considered to be the most resilient parts of the design.

In the following, Sect. 2 discusses general trends in technology, focusing on manufacturing variability, aging degradation, and particle strikes. Section 3 discusses the behavior of SRAM performances for different architectures and technologies. Section 4 does the same for different logic standard cells. Section 5 concludes the chapter.

2 Trends in Technology

There exist several technology related effects, which influence the final resilience of the circuit and its individual components. For example, manufacturing variations, which may be in their nature either random, such as process variations, or systematic, such as chemical-mechanical polishing, cause a static deviation, which remains constant over the entire life-time of the chip. Aging effects cause an additional degradation of these characteristics which increases over life-time. Finally, effects such as particle strikes, disturb chip operation for small amounts of time, but this possibly quite frequently.

2.1 *Manufacturing Variations*

Due to imperfections of the semiconductor manufacturing process multiple parameters deviate from their designed nominal values. These variations can be either random or systematic and manifest in fluctuations of transistor parameters, such as channel length, oxide thickness or number of doping atoms. Additionally these variations may be spatially distributed, which leads to correlations of parameter variations of two transistors, depending on their proximity.

Channel length of transistors may deviate globally (i.e., all transistors on one die are effected in the same way) due to mask offset variations. Typically, this effect is assumed to be Gaussian distributed with a 3σ variation being equal to 10 % deviation from its nominal value [7]. Channel width variation typically exhibits a similar behavior. Gate oxide thickness also follows a Gaussian distribution, but this time due to variations in film thickness. A common assumption here is 3σ being equal to 5–10 % deviation of the nominal value.

To pattern transistors subwavelength lithography is used. As the utilized wavelength nowadays remains constant at 193 nm, there exists a lithographic gap between wavelength and minimum feature size of the transistors, which will increase with ongoing technology scaling until ultraviolet lithography becomes available.

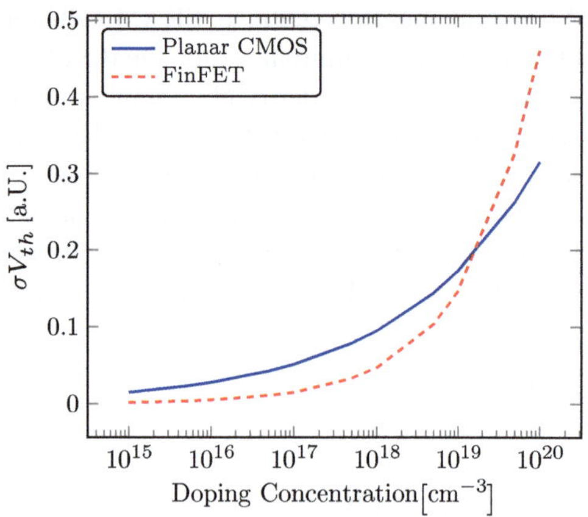

Fig. 7.1 Threshold voltage variability due to RDF dependent on doping concentration for a 20 nm planar CMOS and FinFET technology [11, 17]

This leads to line edge roughness (LER) which impacts for example the effective gate length of all transistors. LER does not scale with device scaling and is almost independent of the lithographic process.

Metal gates, which were introduced to overcome limitations of using polysilicon gate electrodes, are also subject to another variability source called metal gate granularity (MGG). Metal gate granularity refers to the random orientation of the different metal grains and leads to variations in the gate work function. The magnitude of this effect is primarily dependent on the number of metal grains inside the gate. As the size of metal grains is a material-specific property [3], the number of metal grains decreases with ongoing scaling, which increases the effect of MGG in future technologies.

The number and position of dopant atoms in the channel is another common source of process variability called *Random Dopant Fluctuations* (RDF). Due to decreasing numbers of dopant atoms in the channel, this effects tends to have more influence in CMOS technologies, as one dopant more or less means a large difference in resulting device properties.

With ongoing technology scaling transistors are not anymore produced in their common planar style, but more advanced techniques, such as FinFET evolved, especially due to problems resulting from RDF. Figure 7.1 shows the variation in threshold voltage dependent on doping concentration for a planar and a FinFET transistor produced in a similar technology.

As FinFET transistors are able to tolerate low channel doping, they can be produced with almost negligible amount of threshold voltage variation due to RDF. Traditional planar CMOS technologies require higher doping concentrations and, hence, impose a higher threshold variability due to RDF.

In FinFET the transistor channel is produced in a 3D fashion and the gate is folded around the channel, which enables a better control of the channel as well as low channel doping, but also introduces additional variation sources. Fin height

Table 7.1 Process parameter variations for different technology effects

Effect	Transistor parameter	Correlations	Amount
Mask offsets, etc.	Channel length	Global	$3\sigma/\mu = 10\,\%$ [7]
	Fin thickness	Global	$3\sigma/\mu = 10\,\%$ [7, 17]
	Fin height	Global	$3\sigma/\mu = 10\,\%$ [7, 17]
Film thickness variation	Oxide thickness	Global	$3\sigma/\mu = 5\,\%$ [7]
Line edge roughness	Channel length	Local	According to [7]
	Fin thickness	Local	According to [7]
Random dopant fluctuations	Threshold voltage	Local	According to [11, 17]
Metal gate granularity	Gate work function	Local	According to [3]

and fin thickness are additionally subject to global variations due to mask offsets. Additionally fin thickness also suffers from line edge roughness. In contrast, fin height is only dependent on film thickness and not on lithography and, hence, does not vary due to LER.

Table 7.1 summarizes the main process parameter variations which are modeled in this work. Global variations affect each transistor on the die in the same way, hence, each transistor pair is fully correlated in this regard. Local variations affect each transistor on the die individually, hence, each transistor pair is fully uncorrelated.

2.2 *Aging*

Aging changes transistor parameters after the manufacturing process. These changes are the consequence of a certain stress condition, which lead to changes in the physical structures of the transistor. For example, *Time Dependent Dielectric Breakdown* (TDDB) leads to additional conducting paths in the gate oxide, which increase the gate leakage current. TDDB can be divided into a soft breakdown phase, where this additional current leads to a degradation of the physical properties of the transistor. The soft breakdown is typically followed by a hard breakdown, which leads to a complete loss of oxide dielectric properties of the transistor. Another effect is hot carrier (HC) degradation, which is caused by a large electric field near the drain of the transistor channel. HC leads to a degradation of threshold voltage and carrier mobility of the transistor. Bias Temperature instability (BTI) has attracted much attention in the last years. As gate oxides become thinner than 4 nm, BTI-induced shift in V_{th} has become the dominant factor limiting device lifetime and circuit reliability [19]. BTI can be distinguished into *Negative Bias Temperature Instability* (NBTI), which occurs in p-channel transistors and *Positive Bias Temperature Instability* (PBTI), which occurs in high-k dielectric n-channel transistors. BTI requires that the transistor is operated in inversion, which means for a PMOS transistor that its gate is negatively biased with respect to its drain.

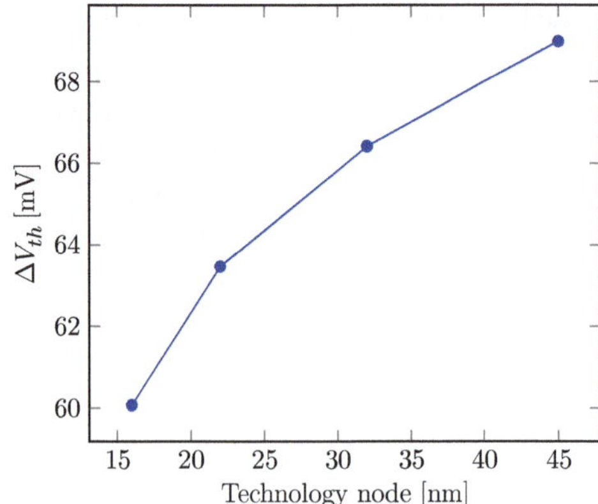

Fig. 7.2 Prediction of ΔV_{th} degradation in future technologies (T = 125 °C, t_{stress} = 9 years)

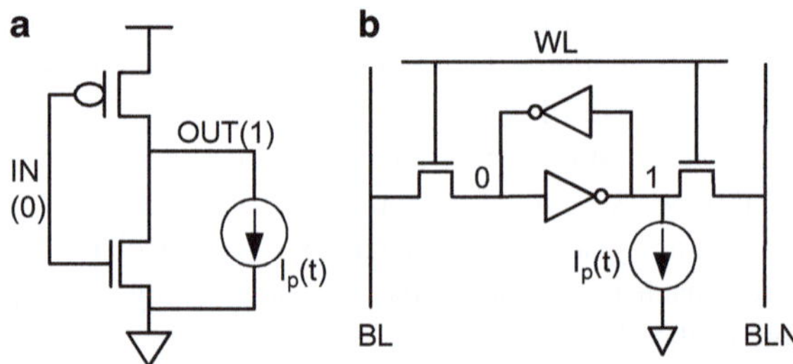

Fig. 7.3 Temporary current disturbance $I_p(t)$ by an SET in (**a**) an inverter and (**b**) a 6-T SRAM cell

The exact amount of V_{th} shift is a strong function of device parameters and circuit operation conditions such as the switching activity. Figure 7.2 presents a prediction of NBTI induced V_{th} degradation for future technologies based on a model from [9].

2.3 Particle Strikes

Logic upsets and value changes in high density memories can be induced by charge collection from ion tracks [21]. Soft error transients (SETs) are momentary current or voltage disturbances in a circuit caused by energetic particle strikes (see Fig. 7.3).

Such an SET may propagate through subsequent circuitry, eventually reaching a memory element or latch and causing a logic failure. The sensitivity of digital ICs to SETs is rapidly increasing with aggressive technology scaling. Due to decreasing parasitic capacitances and operating voltage, the critical charge Q_c, which is required to upset a storage cell also decreases. Figure 7.4 shows the dependence of Q_c on supply voltage for different technology nodes in a 6T SRAM cell, which can be found using a binary search and the simulation setup from Fig. 7.3. When entering the fC region for the critical charge, as in current logic and SRAM devices, lighter particles such as alpha and proton particles become

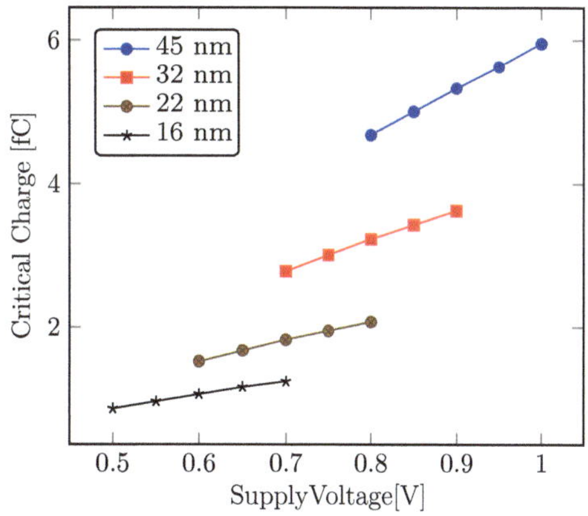

Fig. 7.4 Critical charge dependence on technology node and supply voltage for a 6T SRAM cell

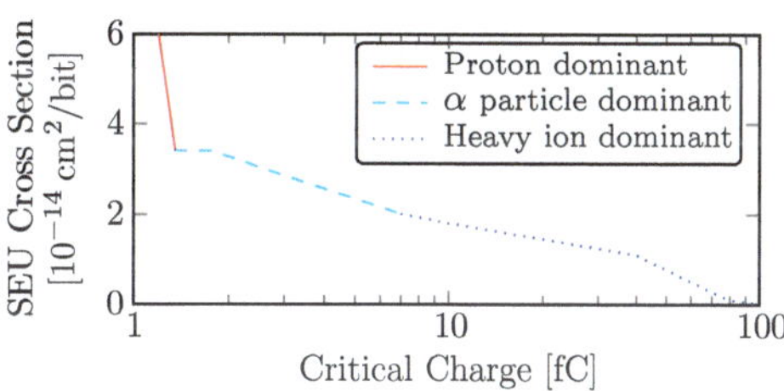

Fig. 7.5 Particle dominance dependent on critical charge (adapted from [6])

dominant (see Fig. 7.5). This increases not only error rates, but also their spread, as the range of lighter particles is much longer compared to residual nucleus [6].

For a scaled transistor, the surroundings of the drain junction are the most sensitive sites to SET, e.g., the drain junction of the inverter and internal storage nodes in a SRAM cell. For alpha particles, the charge deposition is modeled by a double-exponential current pulse (i.e., funneling current) [12]:

$$I_p(t) = I_0(e^{-t/\tau_\alpha} - e^{-t/\tau_\beta}) \tag{7.1}$$

where I_0 is the maximum charge collection current; τ_α is the charge collection time constant and τ_β is the time constant for the rising edge of the current pulse, which is mainly determined by the energy of the particle.

Assuming the device is a bulk CMOS transistor, the dependence of I_0 and τ_α on device parameters is expressed as:

$$I_0 = q\mu N E_0 \tag{7.2}$$

and

$$\tau_\alpha = k\varepsilon_0/q\mu N \tag{7.3}$$

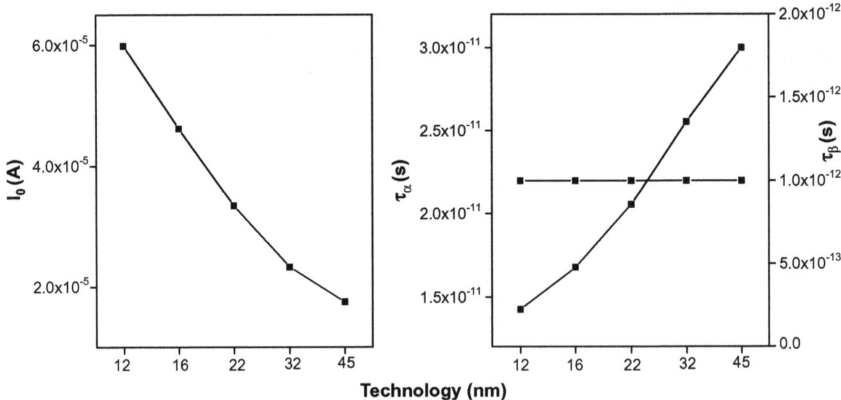

Fig. 7.6 Maximum charge collection current I_0, and collection time constants τ_α and τ_β as a function of CMOS technology

where N is the doping concentration and μ is the effective mobility. With the scaling of CMOS technology, the product of $N\mu$ increases, even though the effective mobility decreases due to higher channel doping concentration. Hence, I_0 increases and τ_α decreases when the device moves to a smaller technology node, as shown in Fig. 7.6. However, τ_β is relatively independent on device parameters (Fig. 7.6). It is also much smaller compared to τ_α [12].

3 Trends in Memory

In this section, we first analyze different design options for SRAM cells based on a 14 nm FinFET technology extracted from TCAD simulation. Second, we demonstrate resilience trends for FinFET-based SRAM architectures down to 7 nm with the help of Predictive Technology Models (PTM) [16, 18]. For this purpose we have chosen two common SRAM architectures, namely the 6-transistor (6T) and 8-transistor (8T) bit cell shown in Fig. 7.7.

3.1 CMOS SRAM

The SRAM is well known to already have high failure rates in current technologies. Figure 7.7 shows both architectures we analyzed. For the 6T architecture we have as design choices the number of fins for the pull-up transistors (PU), the number of fins for the pull-down transistors (PD) and the number of fins for the access transistors (PG). The resulting architecture choice is then depicted by *6T_(PU:PG:PD)*. For

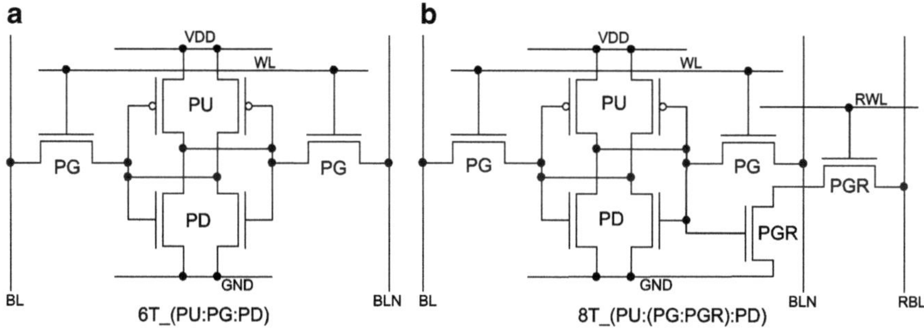

Fig. 7.7 Circuit schematics for standard 6T (**a**) and 8T (**b**) SRAM bit cells

the 8T architecture we have additionally two transistors for the read access (PGR). Hence, the corresponding architecture choice is named *8T_(PU:(PG:PGR):PD).*

As any other circuit an SRAM cell can fail in many different ways, for example:

- **Read delay failure**: An SRAM cell cannot be read within a specified time.
- **Write Trip Voltage (WTV) failure**: The voltage swing during a write is not high enough at the SRAM cell.
- **Static Voltage Noise Margin (SVNM) failure**: An SRAM cell can be flipped unintentionally, when the voltage noise margin is too low (stability).
- **Soft Error/Single Event Upset (SEU) failure**: If the critical charge Q_{crit} is low, the susceptibility to a bit flip caused by radiation is higher.

For the design parameters we have extended this list by:

- **Write delay**: Write a new content into the cell within a specified time.
- **Dynamic power (SW)**: The power measured during a new content write (switching).
- **Leakage power**: Standby leakage with zero and one content averaged (word line set to zero).

We measure the write delay in our experiments, by first initializing the SRAM content with a one, setting the bit lines such that they will attempt to write a zero into the cell, pulse the word line, and measure the delay until the internal cell node takes on the new value.

For the complete SRAM array architecture we assumed 128 cells connected to one column (nrows = 128).

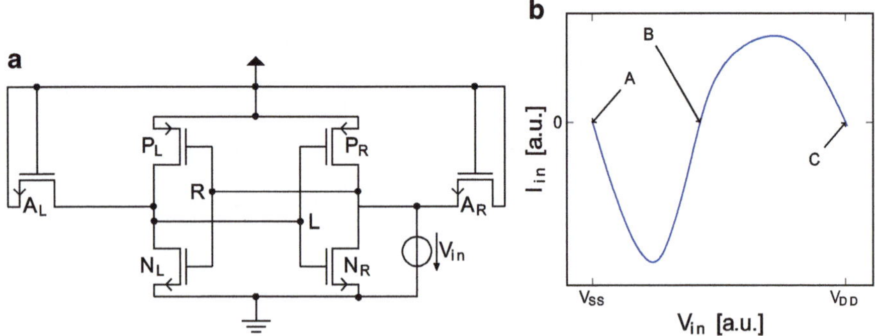

Fig. 7.8 N-Curve extraction circuit (**a**) and resulting N-Curve (**b**)

3.2 Resilience Key Parameters

We identified four parameters, namely Read delay, WTV, SVNM and Q_{crit} as resilience key parameters.

To measure the SVNM and WTV we used the *N-Curve* metric [20]. This metric finds the minimal amount of voltage that is required to change the cell's content. The extraction procedure for the N-curve is as follows (see also Fig. 7.8):

1. The bitlines BL and BR and the wordline WL are clamped to V_{DD}
2. A DC voltage source V_{in} is attached to the node R
3. The voltage of V_{in} is swept from V_{SS} to V_{DD} and the current I_{in} through the voltage source is measured.

This then results on the curve shown in Fig. 7.8b. The voltage difference between the points A and B in the curve is the maximum tolerable DC voltage noise or the static voltage noise margin (SVNM) [5]. Similarly, the voltage difference between the points C and B is the voltage drop required to flip the internal node of the cell, also called the write-trip voltage (WTV).

For measuring the read delay, we initialize the SRAM bit with a one, initialize both bit lines to the supply voltage, leave them floating (high-Z), and pulse the wordline. Then we measure the delay as the time the bit line node requires to drop to 2/3 of the supply voltage.

Figure 7.9a shows the impact of the number of rows used in one column on the read delay for various 6T and 8T architectures. The 6T and 8T one fin for all transistors designs are very similar (in fact their read delays are equal) and show the largest impact, while the double sized 8T cell is almost as good as the high-performance 6T from Intel (6T_(1:2:3)) [8]. Figure 7.9b shows the impact of V_{DD} on the SVNM. We see that 8T cells show for both models PTM 20 nm and TCAD 14 nm a very similar trend as they have nearly identical P/N ratios. The SVNM of the 8T cell relies only on the intrinsic feedback of the storage inverter. The 6T cell

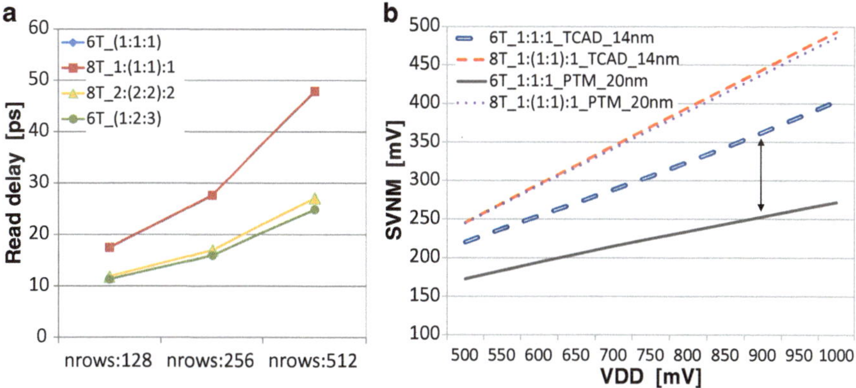

Fig. 7.9 Impact of number of rows on read delay for a 14 nm TCAD model (**a**) and SVNM comparison between 6T and 8T cells using different models (**b**)

behavior differs between TCAD based model and PTM. This is due to the 100 mV lower threshold voltage (V_{th}) of the PTM models, which reduces the SVNM.

For Q_{crit} measurements we use the setup shown in Fig. 7.3b. Basically, we measured the amount of charge needed to *flip* the cell, i.e., to corrupt the data.

3.3 Design Trends and Considerations

We observed that there is no single cell, which provides the best solution under any requirement specification. Therefore, also Intel presented three different sizings for the 6T cell [8]. There is a need for at least a high-density cell and a high-performance cell. So we included in our study the three flavors of Intel's 6T cell and several 8T design alternatives to show what might be achievable.

First, we compared various 6T and 8T architectures by an exhaustive analysis to quantify the design influence.

In Fig. 7.10 different SRAM design options are evaluated for the resilience key parameters normalized to the respective values of a 6T_(1:1:1) minimal sized cell. In general, higher values indicate a performance improvement (e.g. +10 %). Obviously, it is not good idea to use the design options 6T_(1:2:1) and 6T_(2:2:1), as they decrease the SVNM by 30 %. For 8 fin designs the 8T_(1:(1:1):1) cell might be a much better solution, which offers 22 % SVNM improvement. But also the 6T_(1:1:2) used by Intel as high density low voltage design has an increased SVNM (10 %). When it comes to high performance cell options, the Intel variant 6T_(1:2:3) shows the fastest read delay, followed by many 8T designs and some 6T cells. However, if we want to lower the soft error rate, despite of increasing the voltage supply, the best SRAM design option is the 8T_(2:(2:2):2) cell, which unfortunately occupies a very large area.

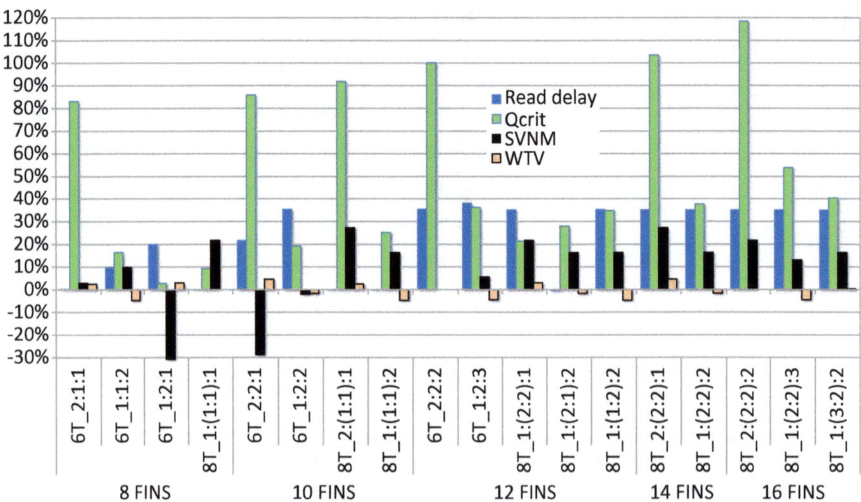

Fig. 7.10 Comparison of read delay, SVNM, WTV and Q_{crit} for different design variants normalized to the results of the 6T_(1:1:1) bit cell

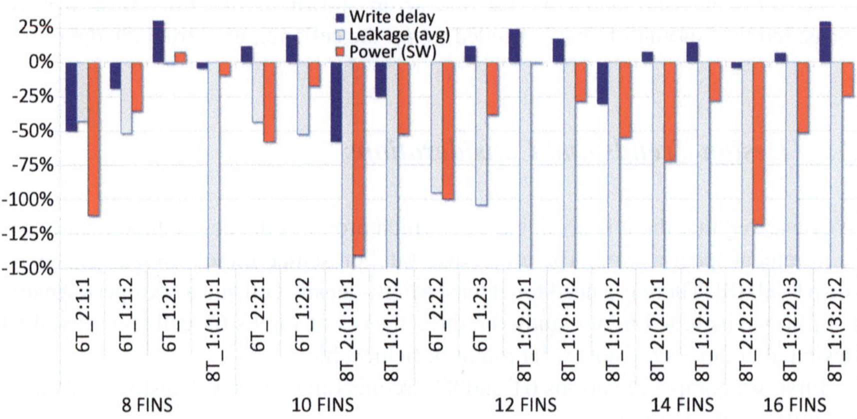

Fig. 7.11 Comparison of write delay, leakage, and switching power (SW) for different design variants normalized to the results of the 6T_(1:1:1) bit cell

In Fig. 7.11 we plotted the additional design parameters. We see that all 8T cells have much increased leakage power, while Intel's 6T_(1:2:3) offers a 10 % decreased write delay with an moderate increase of dynamic power consumption. However, the leakage power increase is for this option also above 100 % of a 6T_(1:1:1) cell. An interesting alternative design option with 12 fins could be the 8T_(1:(2:2):1) cell, as it shows 25 % decreased write delay with no impact on dynamic power. The leakage power increase of this variant could be controlled by lowering V_{DD} in standby or by active backward body biasing.

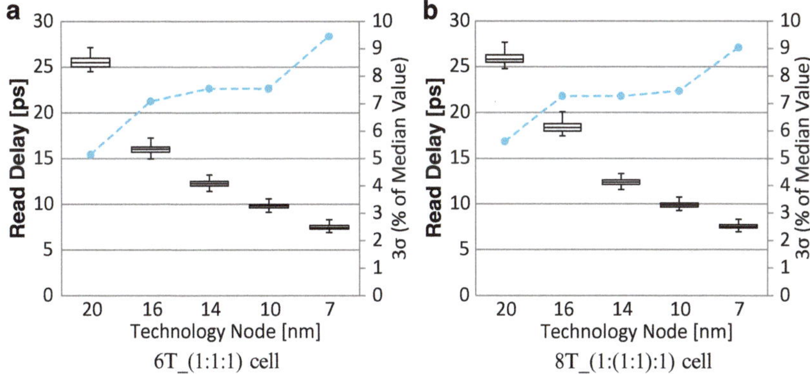

Fig. 7.12 Impact of V_{DD} variation ($\pm 10\,\%$) in absolute values (*box plots*) and relative increase compared to the median (*dashed line*) for a 6T_(1:1:1) high density (**a**) and a standard 8T_(1:(1:1):1) cell (**b**)

From a design perspective there is no clear winner for all evaluated design variants. The best choice depends on the individual requirements and priorities, which can be either high density, high robustness or high performance.

3.4 SRAM Scaling Trends

In this section we want to analyze different scaling related trends for SRAM cells using PTM SPICE models from [18]. First of all we want to analyze the impact of V_{DD} variations. Afterwards, we quantify trends considering transistor technology variations as described in Sect. 2.1.

To evaluate the different technology nodes we run Monte-Carlo (MC) simulations with 1,000 samples to get an impression of the general trend.

We quantified the impact of supply voltage noise by assuming a maximal V_{DD} variation of $\pm 10\,\%$ for a 3σ yield. Figure 7.12 shows the corresponding results of the respective Monte Carlo simulations. We recognize, as expected, an increasing influence of V_{DD} on the read delay for the smaller nodes (up to 9.4 % for the 7 nm node). Additionally, we observe a very similar trend for the 6T and 8T architectures.

To evaluate the scaling trends of the chosen resilience key parameters when scaling down to the 7 nm node, we run MC simulations considering process variations. Figures 7.13 and 7.14 show box plots and the percentage deviation for a 3σ yield with respect to median values for a 6T_(1:1:1) high density and an 8T_(1:(1:1):1) SRAM cell, respectively.

Although the read delay decreases with technology scaling (see Figs. 7.13a and 7.14a), which theoretically enables a higher working frequency, its relative 3σ variation becomes as high as 50 % at the 7 nm node. This compromises its robustness and cripples the possible increase in frequency.

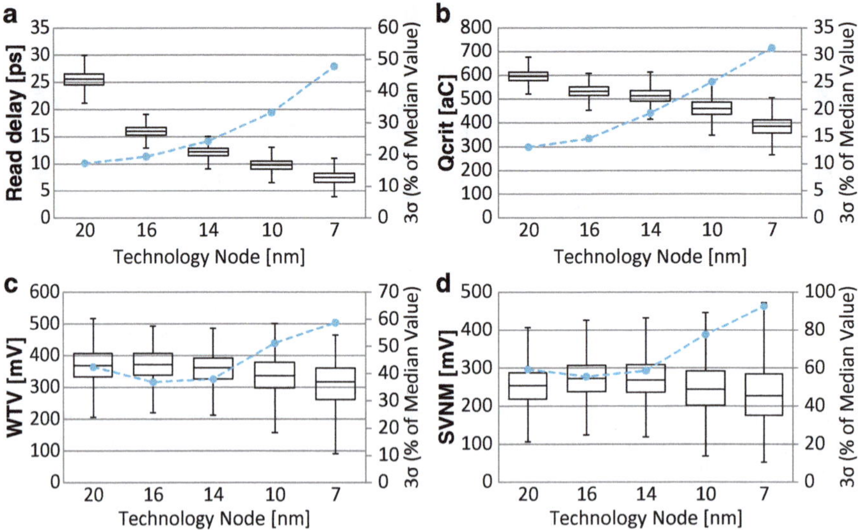

Fig. 7.13 Scaling trends for a 6T_(1:1:1) high density SRAM cell: (**a**) read delay, (**b**) Q_{crit}, (**c**) WTV and (**d**) SVNM

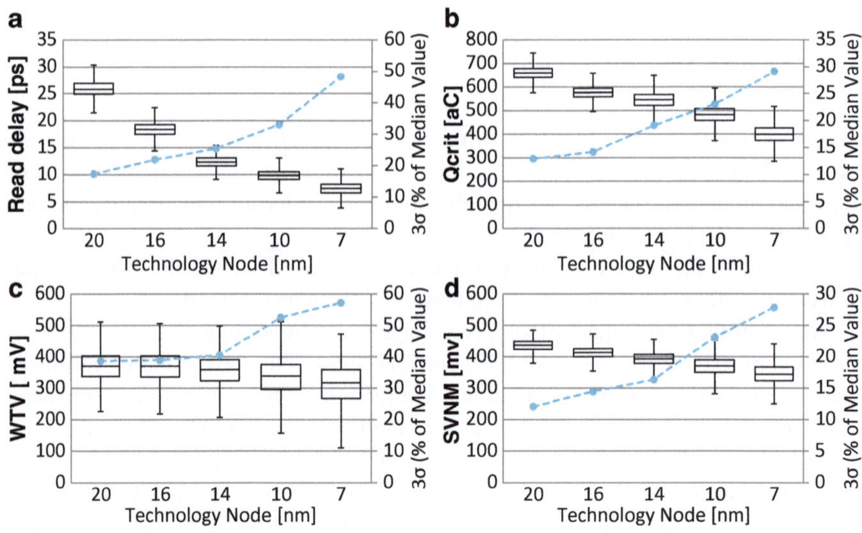

Fig. 7.14 Scaling trends for a 8T_(1:(1:1):1) SRAM cell: (**a**) read delay, (**b**) Q_{crit}, (**c**) WTV and (**d**) SVNM

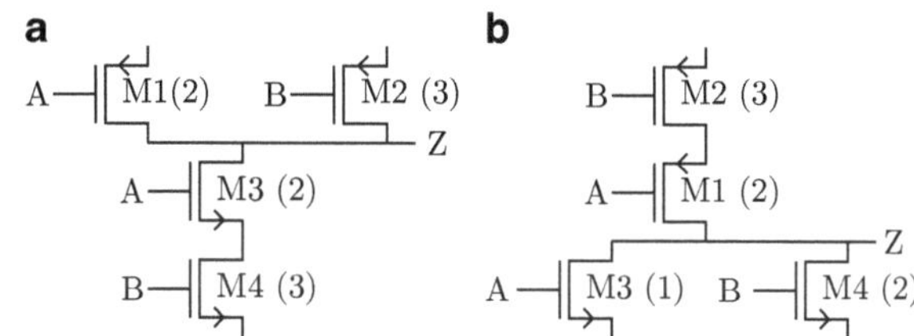

Fig. 7.15 (**a**) NAND and (**b**) NOR cells with obtained sizings (=fin numbers) in *parentheses*

The critical charges Q_{crit} shows a similar trend, where its decrease with technology scaling is accompanied by a noticeable increase in 3σ variability (around 30 % at 7 nm). Nevertheless, the 8T_(1:(1:1):1) design provides an approximately 10 % higher Q_{crit} than the 6T_(1:1:1), and a slightly smaller 3σ variability. Regarding WTV, a quick examination of Fig. 7.7 shows that both designs are identical in terms of writing procedure, which is exemplified in Figs. 7.13c and 7.14c. In both cases their 3σ variability increases from 40 % at 20 nm to nearly 58 % at 7 nm. Such high variations have to be taken into account to guarantee enough voltage swing during a write procedure to flip the cell's content as expected. Perhaps the biggest advantage of the 8T_(1:(1:1):1) design over the 6T_(1:1:1) version with respect to the resilience key parameters presented here, is its much improved SVNM. Not only reaches the 8T design an approximately 22 % higher SVNM than the 6T design, but it is also much more robust in terms of 3σ variability (28 % for 8T 7 nm compared to 90 % for 6T 7 nm).

This overall analysis of the resilience key parameters presented here (read delay, Q_{crit}, WTV, SVNM) shows clearly that the variability increases rapidly as technology is scaled down, which has to be taken into account when designing robust SRAM cells.

4 Trends in Logic

Before making predictions regarding the scaling trend of FinFET standard cells, we first want to evaluate in detail how process variations influence the behavior of FinFET-based logic.

4.1 *Influence of Process Variations*

To evaluate the influence of process variations we utilize the a 14 nm PTM model, which has a nominal channel length of 18 nm [18]. As characteristic representatives we choose a NAND and a NOR cell, which are shown in Fig. 7.15 together with their obtained sizings using a discrete sizing algorithm from [15].

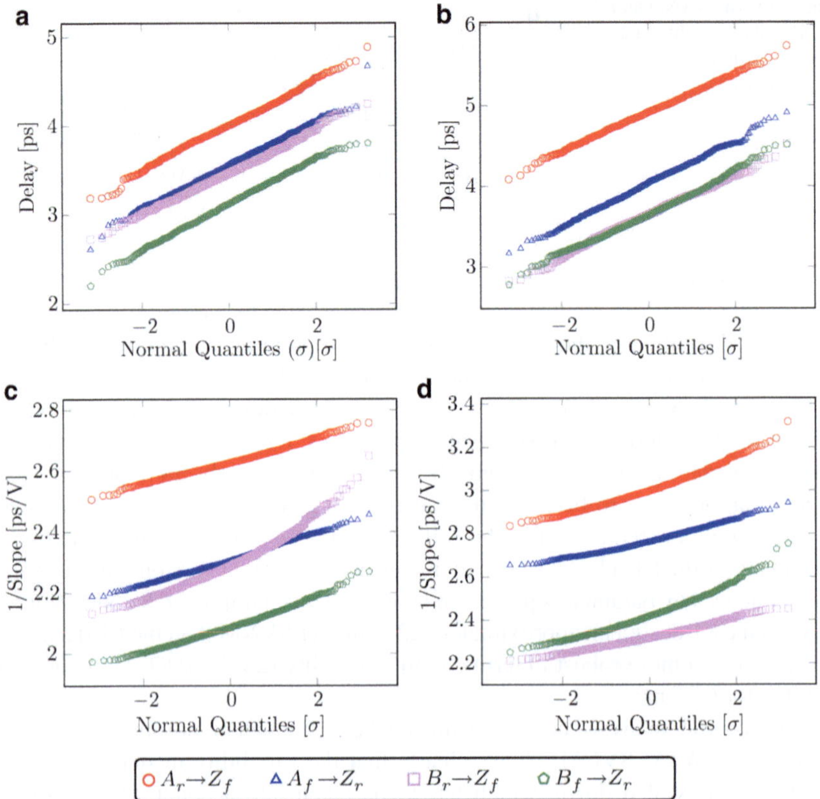

Fig. 7.16 Q-Q plot—delay and slope. (**a**) NAND—delay. (**b**) NOR—delay. (**c**) NAND—slope. (**d**) NOR—slope

To evaluate these cells we run a Monte-Carlo simulation with 1,000 samples and measure delay, slope, and leakage power for every input timing arc. Figures 7.16 and 7.17 show the corresponding Q-Q plots.

Obviously, delay is very well standard normal distributed as we already expect it to be from bulk technologies. For the slope inverse (i.e., switching time) this holds also for most timing arcs. Only for the timing arcs where the entire transistor stack is involved (e.g., $B_r \rightarrow Z_f$ for the NAND) this property does not hold. Figure 7.17 shows the lognormal Q-Q plot for leakage power. Leakage variability fits very well a lognormal distribution for most samples. However, the tail seems to not fulfill this assumption, hence, for accurate yield predictions more advanced distribution types will probably be required here.

Additionally we conduct a worst case analysis as described by [4]. Based on the gradient pointing to the worst-case point in the process parameter space, we

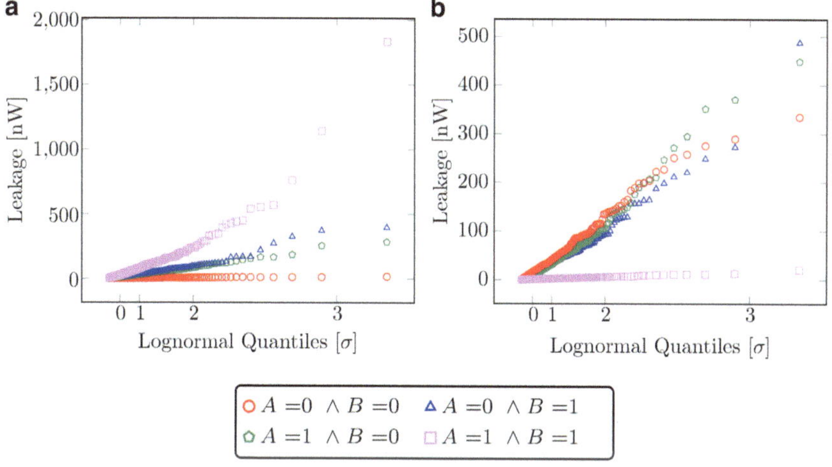

Fig. 7.17 Q-Q plot—leakage. (**a**) NAND. (**b**) NOR

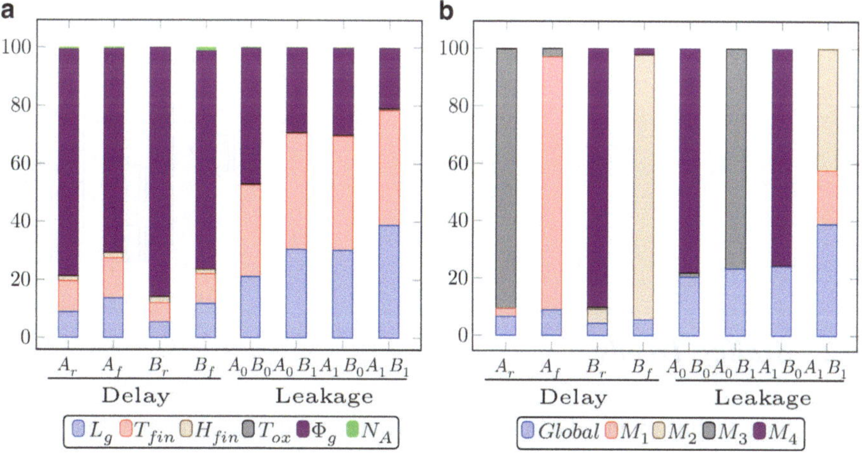

Fig. 7.18 Delay and leakage variability—NAND. (**a**) Variability sources. (**b**) Global vs. local

compute the contribution of the different variability sources as well as the single transistors compared to global variability (Figs. 7.18 and 7.19).

In both cell types, metal gate granularity is the major variability source for delay and leakage. Besides, fin thickness also contributes considerably, especially for leakage. The contribution of random dopant fluctuations is well suppressed due to the low channel doping. Most variability is caused by local variations, which can again be attributed to the dominance of metal gate granularity, followed by line edge roughness effects.

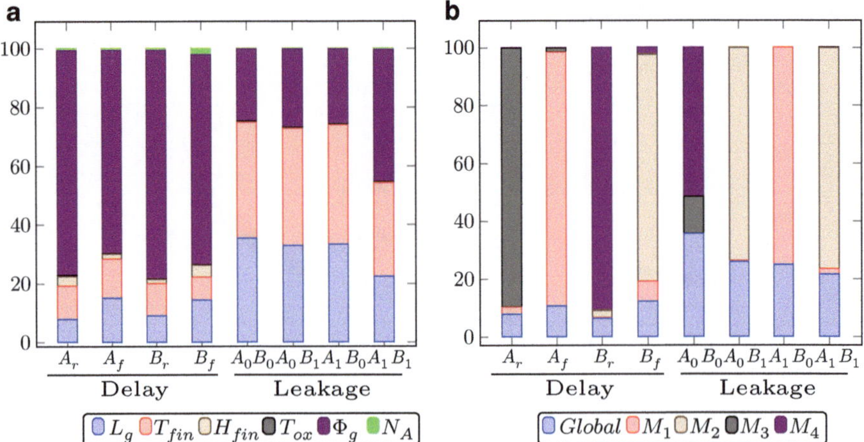

Fig. 7.19 Delay and leakage variability—NOR. (**a**) Variability sources. (**b**) Global vs. local

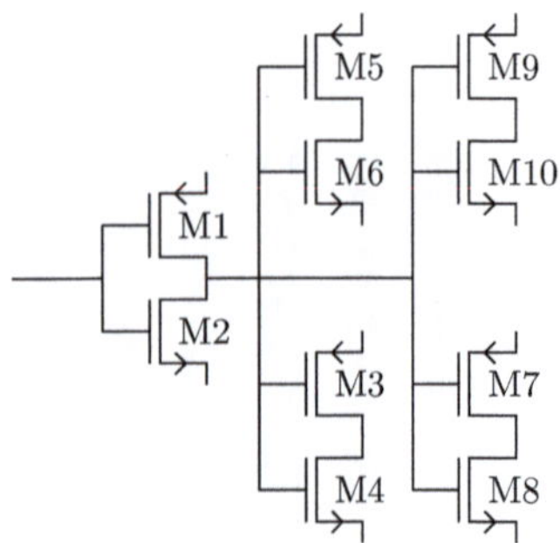

Fig. 7.20 FO4 circuit

4.2 Scaling Trends

To evaluate scaling trends of nominal performances we utilize a fanout of four (FO4) inverter circuit as depicted by Fig. 7.20. The driving inverter is represented by the transistors M1 and M2, while the remaining transistors M3 to M10 represent its load. We assume that all p-channel transistors (i.e., M1, M3, ...) are equally sized and that all n-channel transistors (i.e., M2, M4, ...) are equally sized. The transistors were sized so that the rise and fall delay as well as the rise and fall output slope of the driving inverter (composed of M1 and M2) are approximately equal. Additionally an approximately equal input and output slope was targeted.

We sized this circuit for different technology nodes using models from [18] and [23] and extracted characteristic performances such as P/N sizing ratio, output slope, delay and equivalent input capacitance of the driving inverter (Fig. 7.21).

While in planar CMOS it is required to size P- and N-transistors differently to obtain similar behavior (i.e., equal rise and fall delay), in FinFET P- and N-transistors behave more equally. This leads to an equal number of fins for both

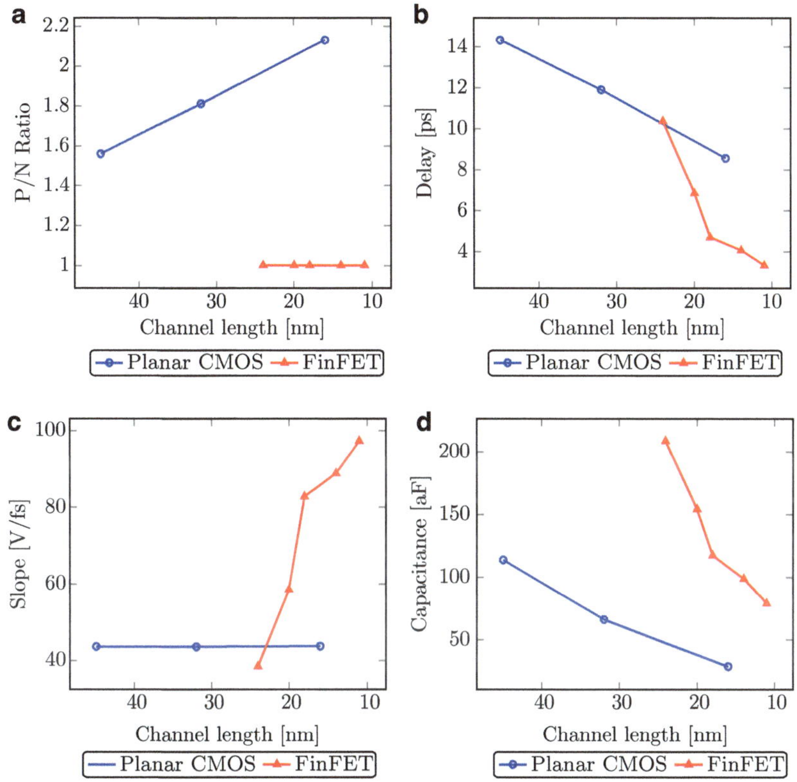

Fig. 7.21 Scaling of minimal-sized inverter performances. (**a**) P/N ratio. (**b**) Delay. (**c**) Slope. (**d**) Input capacitance

transistor types. The delay of the FO4 inverter initially scales with a very steep slope, but after some generations this trend quickly reverts to an less steep scaling as we know from planar technologies. Interestingly, slope scales very steeply in FinFET technology. Input capacitance is in general much higher for FinFET compared to planar technologies in the same technology node. Both, the scaling of slope and the value of the capacitance, can be very well explained by the three-dimensional gate controlling the channel.

To evaluate the scaling trend in FinFET technologies in the presence of variability, we size a mid-performance inverter and evaluate its behavior under different specifications, which are listed in Table 7.2. Specification 0 represents nominal conditions without any variability. In each subsequent specification one additional variation source is added: Process variations in spec 1, device temperature variation in spec 2, and supply voltage variation in spec 3.

Table 7.2 Specifications for performance evaluation

Spec-Nr.	Yield	V_{DD}	T (°C)
0	Nominal	Nominal	27
1 (P)	3σ	Nominal	27
2 (PV)	3σ	$\pm 10\,\%$	27
3 (PVT)	3σ	$\pm 10\,\%$	0–60

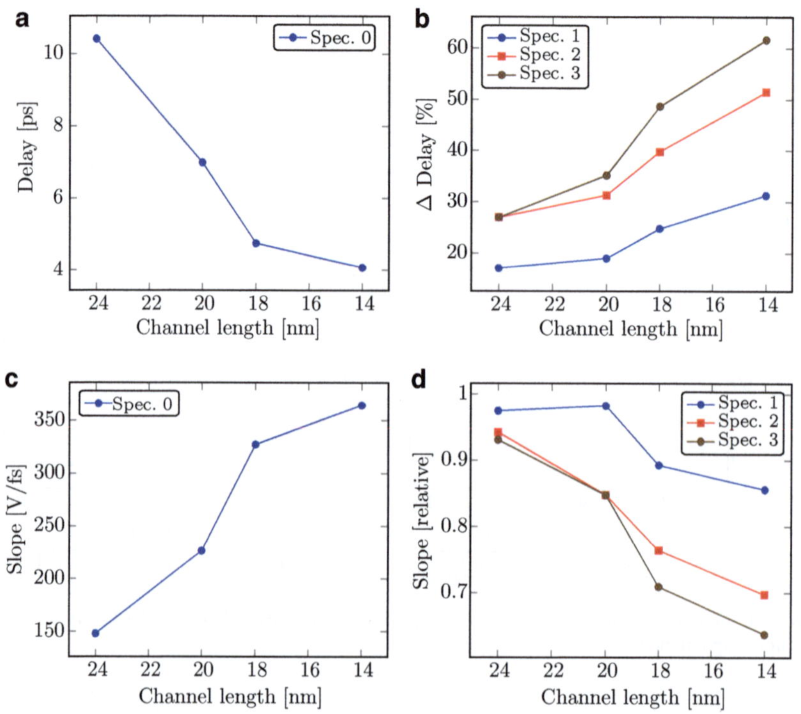

Fig. 7.22 Scaling trends for timing performances of a medium-sized inverter. (**a**) Nominal delay. (**b**) Delay degradation. (**c**) Nominal slope. (**d**) Slope degradation

Figure 7.22 shows the change of timing performances (delay and slope) over the respective technology nodes, while Fig. 7.23 shows the change in power performances (leakage power and switching energy).

Delay degradation due to process variations increases from 17 % up to 30 % for a 3σ-design. A detailed analysis reveals that from this increase about 5 % can be accounted to an increase of sensitivity to process variation due to technology scaling. The remaining larger increase results mostly from the increase of gate work function variability due to metal gate granularity. This corroborates that metal gate granularity is the major source of process variability in FinFET as metal grain size does not scale. Another interesting insight is that the influence of temperature increases over the studied technologies. While we observed in this study almost no

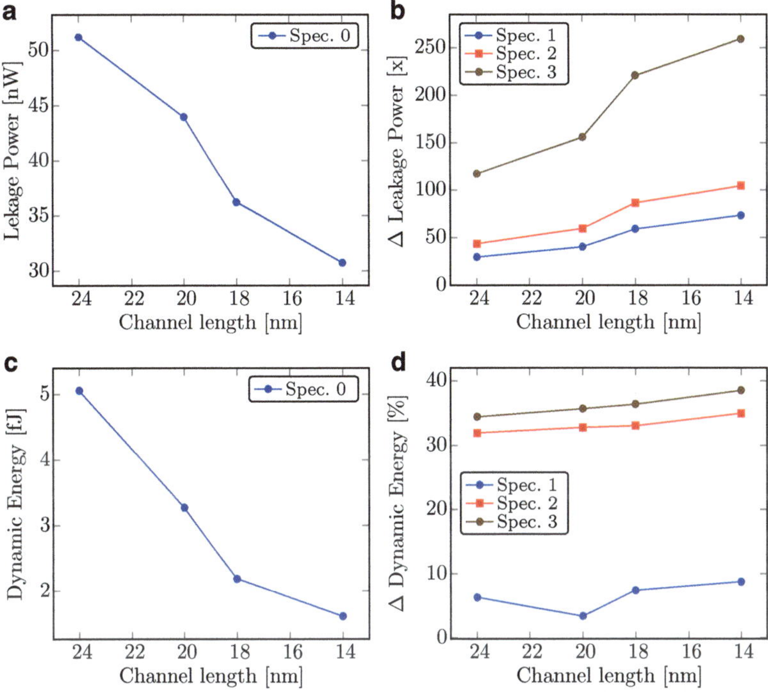

Fig. 7.23 Scaling trends for power performances of a medium-sized inverter. (**a**) Nominal leakage. (**b**) Leakage degradation. (**c**) Nominal switching energy. (**d**) Switching energy degradation

influence of temperature at 24 nm channel length, the additional increase in delay at 14 nm channel length is about 10 percentage points.

Leakage power shows temperature and process variation as its major variability contributors, while supply voltage mainly determines the variation of switching energy.

5 Conclusion

We presented in this chapter an analysis of ongoing scaling related trends of integrated circuits. For this purpose, we discussed major trends observed in process technology, as well as in memory and logic designs. While newer technology styles, such as FinFETs provide necessary advances to enable further technology scaling, we can observe that there exists also a trend for reduced resilience with ongoing scaling. Our hope is that with this roadmap, which concentrates on FinFET-based designs, we are able to sensitize the research community to present and near future problems associated with this technology. Additionally, we analyzed different architectures for SRAM memories to demonstrate the different design choices available and tradeoffs that have to be made.

Acknowledgements The authors want to gratefully acknowledge the contributions of Sani Nassif, Nikhil Mehta, Helmut Graeb, and Javier Varela to this work. Veit Kleeberger was employed at the Institute for Electronic Design Automation (TUM) during this work. This work was supported in part by the German Research Foundation (DFG) as part of the priority program "Dependable Embedded Systems" (SPP 1500—http://spp1500.itec.kit.edu).

References

1. Asenov, A.: Random dopant induced threshold voltage lowering and fluctuations in sub-0.1 m mosfet's: A 3-d atomistic simulation study. IEEE Trans. Electron Devices **45**(12), 2505–2513 (1998)
2. Austin, T., Blaauw, D., Mudge, T., Flautner, K.: Making typical silicon matter with Razor. IEEE Computer **37**(3), 57–65 (2004)
3. Dadgour, H., Endo, K., De, V., Banerjee, K.: Modeling and analysis of grain-orientation effects in emerging metal-gate devices and implications for SRAM reliability. In: IEEE International Electron Devices Meeting (IEDM), pp. 1–4 (2008)
4. Graeb, H.: Analog Design Centering and Sizing. Springer Verlag (2007)
5. Grossar, E., Stucchi, M., Maex, K., Dehaene, W.: Read stability and write-ability analysis of SRAM cells for nanometer technologies. IEEE Journal of Solid-State Circuits **41**(11), 2577–2588 (2006)
6. Ibe, E., Chung, S.S., Wen, S., Yamaguchi, H., Yahagi, Y., Kameyama, H., Yamamoto, S., Akioka, T.: Spreading diversity in multi-cell neutron-induced upsets with device scaling. In: IEEE Custom Integrated Circuits Conference (CICC), pp. 437–444 (2006)
7. International technology roadmap for semiconductors (2013)
8. Jan, C.-H. et al.: A 22nm SoC Platform Technology Featuring 3-D Tri-Gate and High-k/Metal Gate, Optimized for Ultra Low Power, High Performance and High Density SoC Applications. In: IEEE International Electron Devices Meeting (IEDM), pp. 44–47 (2012)
9. Kleeberger, V.B., Barke, M., Werner, C., Schmitt-Landsiedel, D., Schlichtmann, U.: Compact Model for NBTI Degradation and Recovery under Use-Profile Variations and ist Application to Aging Analysis of Digital Integrated Circuits. Microelectronics Reliability 54(6–7), pp. 1083–1089 (2014)
10. Kleeberger, V.B., Graeb, H., Schlichtmann, U.: Predicting Future Product Performance: Modeling and Evaluation of Standard Cells in FinFET Technologies. In: ACM/IEEE Design Automation Conference (DAC) (2013)
11. Lu, D.D., Lin, C.H., Niknejad, A.M., Hu, C.: Compact Modeling of Variation in FinFET SRAM Cells. IEEE Design & Test of Computers **27**(2), 44–50 (2010)
12. Messenger, G.C.: Collection of charge on junction nodes from ion tracks. IEEE Trans. Nuclear Science, vol. 29, issue 6, pp. 2024–2031 (1982)
13. Nassif, S.R., Kleeberger, V.B., Schlichtmann, U.: Goldilocks failures: Not too soft, not too hard. In: IEEE International Reliability Physics Symposium (IRPS) (2012)
14. Nassif, S.R., Mehta, N., Cao, Y.: A resilience roadmap. In: IEEE Conference on Design, Automation & Test in Europe (DATE) (2010)
15. Pehl, M.: Discrete Sizing of Analog Integrated Circuits. Ph.D. thesis, Technische Universität München (2012)
16. Predictive technology model (ptm). available at http://www.eas.asu.edu/~ptm

17. Saha, S.K.: Modeling process variability in scaled CMOS technology. IEEE Design & Test of Computers **27**(2), 8–16 (2010)
18. Sinha, S., Yeric, G., Chandra, V., Cline, B., Cao, Y.: Exploring sub-20nm FinFET design with predictive technology models. In: ACM/IEEE Design Automation Conference (DAC) (2012)
19. Wang, W., Reddy, V., Krishnan, A.T., Vattikonda, R., Krishnan, S., Cao, Y.: Compact modeling and simulation of circuit reliability for 65nm cmos technology. IEEE Trans. on Device and Materials Reliability **7**(4), 509–517 (2007)
20. Wann, C., Wong, R., Frank, D., Mann, R., Ko, S.B., Croce, P., Lea, D., Hoyniak, D., Lee, Y.M., Toomey, J., et al.: SRAM cell design for stability methodology. In: International Symposium on VLSI Technology (VLSI-TSA-Tech), pp. 21–22 (2005)
21. Wirth, G., Vieira, M., Neto, E., Kastensmidt, F.: Generation and propagation of single event transients in cmos circuits. In: Design and Diagnostics of Electronic Circuits and systems, 2006 IEEE, pp. 196–201 (2006)
22. Zhang, M., Mitra, S., Mak, T., Seifert, N., Wang, N.J., Shi, Q., Kim, K.S., Shanbhag, N.R., Patel, S.J.: Sequential element design with built-in soft error resilience. IEEE Transactions on Very Large Scale Integration (VLSI) Systems **14**(12), 1368–1378 (2006)
23. Zhao, W., Cao, Y.: New generation of predictive technology modeling for sub-45nm early design exploration. IEEE Trans. Electron Devices **53**(11), 2816–2823 (2006)

Chapter 8
Layout Aware Electromigration Analysis of Power/Ground Networks

Di-an Li, Malgorzata Marek-Sadowska, and Sani R. Nassif

Abstract In this chapter, we briefly introduce physical foundations of electromigration (EM) and present two classical EM-related theories. We discuss physical parameters affecting EM wire lifetime and we introduce some background related to the existing EM physical simulators. In our work, for EM physical simulation we adopt the atomic concentration balance-based model. We discuss the simulation setup and results. We present VEMA—a variation-aware electromigration (EM) analysis tool for power grid wires. The tool considers process variations caused by the chemical–mechanical polishing (CMP) and edge placement error (EPE). It uses a compact model that features critical region extraction and variation coefficient calculation. VEMA is a full-chip EM analysis tool; it extracts the effective jL product values and performs a via-centric EM lifetime calculation on ideally manufactured EM-mortal wires. It analyzes process variation effects on EM reliability and reports variation tolerances of EM-sensitive power grid wires.

1 Introduction to Electromigration

1.1 Basic Concepts

Electromigration (EM) is the transport of material caused by the gradual movement of ions in a conductor due to the momentum transfer between conducting electrons and diffusing metal ions that make up the lattice of the interconnect material [1, 2]. The tensions caused by accumulating ions will gradually nucleate a void or an

D. Li
Qualcomm Inc, 5775 Morehouse Dr, San Diego, CA 92121, USA
e-mail: dianl@qti.qualcomm.com

M. Marek-Sadowska (✉)
University of California, Santa Barbara, CA 93106, USA
e-mail: mms@ece.ucsb.edu

S.R. Nassif
Radyalis, 7000 North Mopac Expressway, Suite 200, Austin, TX 78731, USA
e-mail: srn@radyalis.com

R. Reis et al. (eds.), *Circuit Design for Reliability*,
DOI 10.1007/978-1-4614-4078-9_8

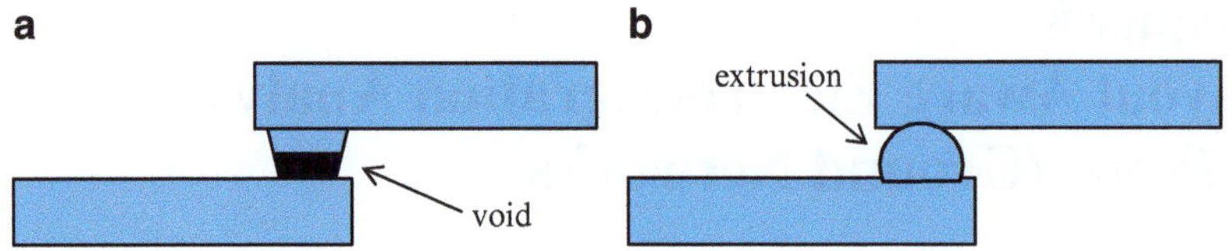

Fig. 8.1 EM failures. (**a**) Void (**b**) extrusion

extrusion which will grow in size over time and eventually form an open or short circuit as shown in Fig. 8.1.

In reality, void formation [3] occurs much faster than extrusion growth, which is very unlikely to occur [4]. Therefore voids are considered the major EM-caused failures. In this chapter, we only focus on the EM-induced void damage of interconnects.

The ion movement is caused by several forces, the two main ones are: (1) The direct electrostatic force $\mathbf{F_e}$ caused by the electric field, and (2) The force from the momentum transfer with moving electrons $\mathbf{F_p}$, which is in the opposite direction to the electric field [5]. The resulting force $\mathbf{F_{res}}$ acting on an ion is given by (8.1).

$$F_{res} = F_e - F_p = q\left(Z_e + Z_p\right)E = qZ^* j\rho \tag{8.1}$$

In (8.1), q is the electron charge, Z_e is the nominal valence of the metal, Z_p accounts for the magnitude and direction of the momentum exchange between the conducting electrons and the metal ions, Z^* is the effective valence, E is the electrical field, j is current density and ρ is the metal resistivity.

1.2 *Classical Theories*

1.2.1 Black's Equation

Black's Equation is a mathematical model for the mean time to failure (MTTF) of a wire due to electromigration [6]. The equation is given by (8.2).

$$MTTF = Aj^{-n}e^{\left(\frac{E_a}{kT}\right)} \tag{8.2}$$

In (8.2), A is an experimental constant, j is the current density, n is a model parameter, E_a is the activation energy, k is Boltzmann constant, T is the absolute temperature in K.

The model is abstract, not based on a specific physical model, but it flexibly describes the failure rate dependence on the temperature, the electrical stress, and the specific technology and materials. The value of Black's equation is that it maps

Fig. 8.2 Back stress buildup

experimental data taken at elevated temperature and stress levels in short periods of time to the expected component failure rates under actual operating conditions.

1.2.2 Blech Length Effect

For electromigration to occur, there is also a lower limit on the interconnect length [2]. It is known as “Blech length” [7], and any wire that has a length below this limit (typically on the order of 10–100 μm) will not fail by electromigration. Here, the mechanical stress buildup causes a reversed migration process which reduces or even compensates the effective material flow towards the anode (Fig. 8.2).

Although known as “Blech length” effect, the actual criterion to determine if a wire is EM-immortal or not, is the product of its current density j and length L. Specifically, a conductor line is not susceptible to electromigration if the product of the wire’s current density j and length L is smaller than a process technology-dependent threshold value $jL_{threshold}$. Below, we briefly explain the phenomena behind the Blech length effect. The electromigration induced drift velocity is determined by (8.3).

$$v_d = \frac{D_a \left|Z^*\right| e\rho j}{kT} \tag{8.3}$$

I.A. Blech observed that the metal ions stopped moving, when the wire length was reduced to a certain length. Also, he observed that no ion drift could be detected below a threshold current density.

These observations can be explained by considering the flux due to electromigration and the gradient of the chemical potential via a gradient of mechanical stress [7–10] according to (8.4)

$$J_v = \frac{D_v C_v}{kT}\left(\left|Z^*\right| e\rho j - \Omega\frac{\partial\sigma}{\partial x}\right) \tag{8.4}$$

where Ω is the atomic volume, and σ is the hydrostatic stress. This equation shows that a gradient of mechanical stress acts as driving force against electromigration. Thus, electromigration stops, when the opposing stress gradient, commonly referred to as “back stress” [11], equals the electromigration driving force, so that $J_v = 0$. This steady-state condition is the so-called “Blech Condition”, given by (8.5)

$$\frac{\partial\sigma}{\partial x} = \frac{\left|Z^*\right| e\rho j}{\Omega} \tag{8.5}$$

Integrating (8.5) over the interconnect line length yields (8.6)

$$\sigma(x) = \sigma_0 + \frac{|Z^*| e\rho j}{\Omega} x \tag{8.6}$$

where σ_0 is the stress at $x = 0$. This equation shows that the stress varies linearly along the line, when the backflow flux equals the electromigration flux. Given that the maximum stress a conductor line can withstand is σ_{th}, the critical product for electromigration failure can be stated as (8.7)

$$(jL)_c = \frac{\Omega\left(\sigma_{th} - \sigma_0\right)}{|Z^*| e\rho} \tag{8.7}$$

This is the so-called "Blech Product". The critical product provides a measure of the interconnect resistance against electromigration failure and several experimental works have reported that the critical product for modern copper interconnects is in the range from 2,000 to 10,000 A/cm [12–15].

1.3 Affecting Factors

1.3.1 Current Density

Current density of a wire is a ratio of the total current flowing through that wire, I, and its cross-section area, A (8.8).

$$j = \frac{I}{A}. \tag{8.8}$$

According to Black's equation, MTTF of a wire is inversely proportional to its current density. And their correlation depends on the current exponent n. There is some controversy across EM literature regarding the exact value of n. A common agreement is that the typical value of n is between 1 and 2. Interestingly, the most important factor that determines the value of n is the current density itself. The value of n is close to 1 when the current density is small, and it gradually increases to two as the current density increases. For example, in [16], the authors report $n = 1$ for $j \leq 0.1$ MA/cm^2, $n = 1.5$ for 0.1 MA/cm$^2 < j < 1$ MA/cm^2 and $n = 2$ for $j \geq 1$ MA/cm^2; in [17], the authors report $n = 1.1 \pm 0.2$ for $j \leq 2.5$ MA/cm^2 and $n = 1.8$ for $j > 2.5$ MA/cm^2. For $n = 1$, the time-average current density can be used in (8.2) and the current waveform shape does not affect EM lifetime; else if $n > 1$, transient current density should be considered and the shape of a current waveform matters.

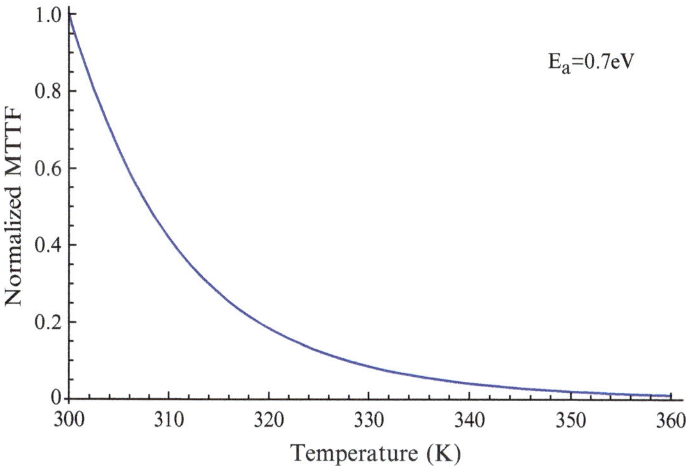

Fig. 8.3 MTTF–temperature dependency

1.3.2 Temperature

According to Black's equation, MTTF of a wire has an exponential dependency on wire temperature. Figure 8.3 depicts the MTTF trend when temperature changes. In Fig. 8.3, MTTFs are normalized to MTTF at 300 K. A sharp MTTF decrease (0.5X) can be observed when temperature rises from 300 to 310 K. Some papers report huge temperature rise (hundreds of K) on wires [18]. We believe that these results are exaggerated due to unrealistic assumption of extremely high current wire densities. However, in practice, 10 K wire temperature rise above the substrate temperature is common, which can still cause a non-negligible effect on wire EM MTTF.

1.3.3 Wire Length

Wire length affects two parameters: j_{limit} and j_{effect}. According to Blech length effect, there is a wire length limit that any wire shorter than that limit is considered EM-immortal. From a different perspective, it can be also interpreted as given a wire length, there is a current density limit that prevents the EM on the wire, as stated in (8.9).

$$j_{\mathrm{limit}} = \frac{\Omega\,(\sigma_{th} - \sigma_0)}{|Z^*|\,e\rho L} \tag{8.9}$$

Since the Blech length effect is caused by the mechanical stress buildup with a reverse migration process, even if the current density on a wire is greater than j_{limit}, the reverse stress still exists. Therefore, the effective current density j_{effect} that is

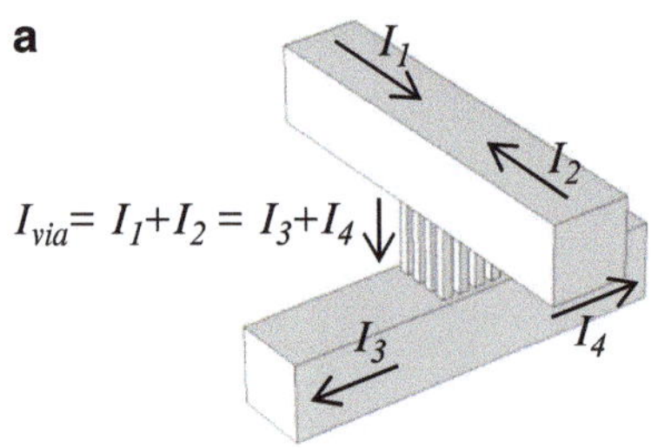

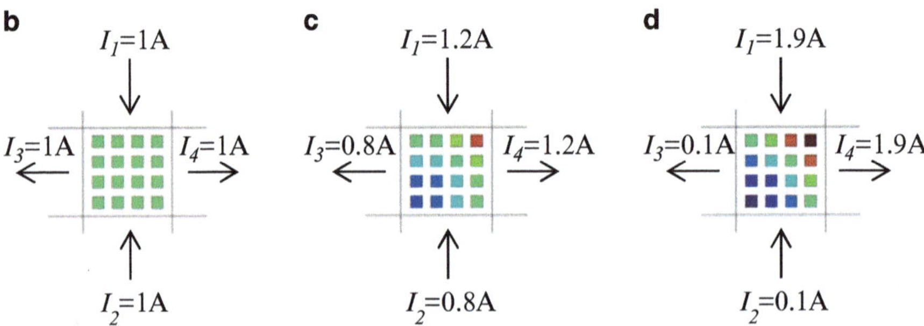

Fig. 8.4 Via-array current distribution. (**a**) 3D orthogonal wire structure with 4-by-4 via-array. (**b**) Even distribution (**c**) uneven distribution (**d**) extreme uneven distribution

used in Black's equation is modified by $j - j_{limit}$ as in (8.10) [19].

$$MTTF = A(j - j_{\text{limit}})^{-n} e^{\left(\frac{E_a}{kT}\right)} \tag{8.10}$$

1.3.4 Interconnect Structure

The interconnect structure that affects EM mainly refers to the via-array design. For a single via, all current is flowing through the via and its current density can be easily obtained. However in a via-array, the current is not evenly distributed into each via. References [20, 21] discuss this effect.

In Li and Guan [22], a 3D orthogonal wire structure with a 4-by-4 via-array is built as shown in Fig. 8.4a. Several current configurations are applied to the structure, and current densities through the horizontal cross-sectional plane of multiple vias are recorded as shown in Fig. 8.4b–d. It can be observed that when all wire currents are identical as in Fig. 8.4b, current distribution into the vias is also uniform. Greater differences among the wire segment currents cause the current split into the vias to be more non-uniform.

This effect can be explained by the differences among path resistances through different vias. A SPICE simulation on a pure resistive mesh validates this observa-

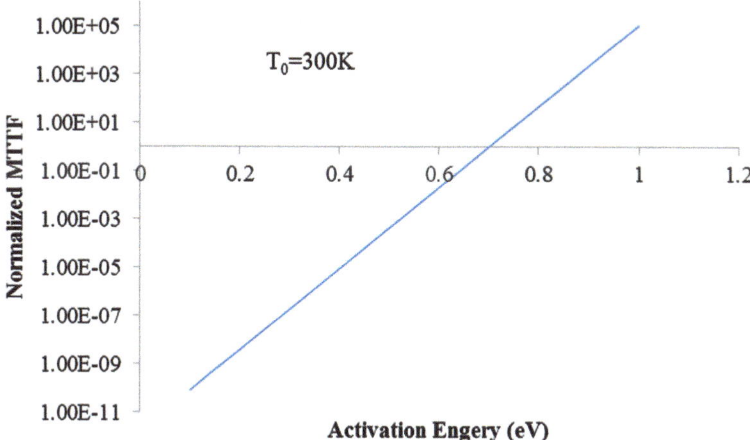

Fig. 8.5 MTTF-activation energy dependency

tion. Based on the resistive mesh model, the current distribution into each via can be determined and the EM lifetime of the via-array can be estimated. A detailed discussion can be found in [22].

1.3.5 Activation Energy

Activation energy is the minimum energy required to cause thermal diffusion of metal ions. It plays an important role due to its exponential effect on MTTF [23]. Figure 8.5 shows the MTTF trend when activation energy changes, while temperature is fixed at 300 K. In Fig. 8.5, MTTF is normalized to its value at $E_a = 0.7$ eV. The exponential dependency can be observed.

Interconnect material is the dominant factor that determines the value of E_a. Aluminum was the typical material used for interconnects before copper was first introduced in 1997 [24]. Due to its generally higher activation energy, copper has a significantly better electromigration resistance than aluminum does. In this chapter, we focus our discussion on copper interconnect EM reliability.

1.3.6 Diffusion Paths

In a homogeneous crystalline structure, because of the uniform lattice structure of the metal ions, there is hardly any momentum transfer between the conducting electrons and the metal ions. However, this symmetry does not exist at the grain boundaries and material interfaces, so here the momentum is transferred much more vigorously [25]. Since in these regions, the metal ions are bonded more weakly than in a regular crystal lattice, once the electron wind has reached certain strength,

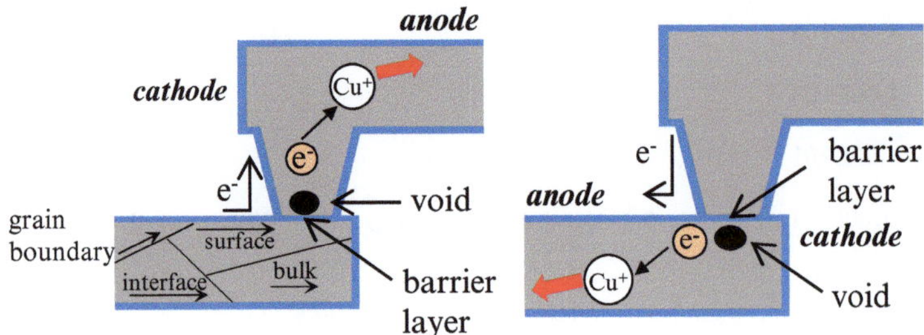

Fig. 8.6 Dual-damascene copper EM diffusion paths

Table 8.1 Copper diffusion paths and the activation energy

Diffusion path	E_a (eV)
Surface	0.5–0.7
Interface	0.8–1.25
Grain boundary	1.2–1.25
Bulk	2.1

atoms become separated from the grain boundaries/interfaces and are transported in the direction of the current. Figure 8.6 shows the EM diffusion paths in a dual-damascene copper interconnect structure, which is the main metallization process in modern semiconductor fabrication [26].

In a dual-damascene structure, only a single metal deposition step is used to simultaneously form the main metal line and the via beneath it. After the via and trench recesses are etched, the via is filled in the same metal-deposition step that fills the trench. Therefore, vias are always in the same copper (Cu) body as the upper level metal, but are separated from the lower level metal by a barrier layer which prevents copper diffusion into the dielectric as shown in Fig. 8.6. Typical barrier layer materials are Ta, TaN, TiN and TiW. The diffusion processes caused by electromigration can be divided into grain boundary diffusion, bulk diffusion and surface diffusion. In general, surface diffusion is dominant in copper interconnects. The diffusion path differences can be explained by their different activation energies listed in Table 8.1.

2 Physical Simulation of EM

2.1 Background

One branch of the EM-related research is focused on its physical simulation. Classical theories such as Black's equation and Blech length effect provide macro-level analysis tools for EM but are limited in explaining micro-level EM behavior. The purpose of physical simulation is to better understand the physics behind EM phenomena and more precisely predict interconnect EM behavior at microstructural level. With such a complex system, typically a finite element method (FEM) tool is used for computation and simulation. In [27], the author develops an EM simulation method based on flux divergence caused by driving forces including electromigration, thermo-migration, stress-migration and atomic concentration. In [28], an alternative concept of EM modeling is proposed based on the system energy reduction as the simulation criterion. In this section, we mainly apply the atomic concentration model from [27] to simulate interconnect structures. As EM theory becomes more complete, we expect the physical simulation to be closer to the real situation.

2.2 *Balance of Atom Concentration*

The governing equation which describes the atom concentration evolution throughout an interconnect segment, is the conventional mass balance (continuity) equation in (8.11) [1].

$$\frac{\partial N\left(x,t\right)}{\partial t} + \nabla \cdot J = 0 \tag{8.11}$$

In (8.11), $N(x, t)$ is the atom concentration at the point with coordinates $x = (x, y, z)$ at the moment of time t, and J is the total atomic flux at this location. The total atomic flux J is a combination of the fluxes caused by the different atom migration forces. The major forces are induced by the electric current, and by the gradients of temperature, mechanical stress and concentration: $J = J_E + J_{th} + J_S + J_C$.

To define the fluxes mentioned above, we have (8.12).

$$\begin{cases} \vec{J}_E = \frac{N}{kT} e Z^* j \rho D_0 e^{-\frac{E_a}{kT}} \\ \vec{J}_{th} = -\frac{N Q^* D_0}{kT^2} e^{-\frac{E_a}{kT}} gradT \\ \vec{J}_s = -\frac{N \Omega D_0}{kT} e^{-\frac{E_a}{kT}} grad\sigma_H \\ \vec{J}_c = -D_0 e^{-\frac{E_a}{kT}} gradN \\ \frac{dN}{dt} + div\left(\vec{J}_E + \vec{J}_{th} + \vec{J}_s + \vec{J}_c\right) = 0 \end{cases} \tag{8.12}$$

In (8.12), $\boldsymbol{J_E}$, $\boldsymbol{J_{th}}$, $\boldsymbol{J_S}$ and $\boldsymbol{J_C}$ are atomic fluxes induced by electrical, thermal, stress and atomic concentration forces, respectively. N is the atomic concentration, Q^* is the specific heat of transport of metal, and σ_H is the local hydrostatic stress. EM lifetime is considered to be proportional to atomic flux divergence ($AFD = div\boldsymbol{J}$), so we have here MTTF $\propto AFD$.

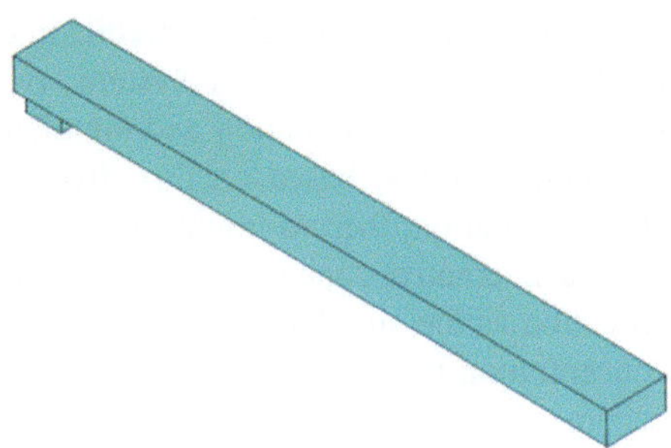

Fig. 8.7 3D view of a simulated wire structure

2.3 *Simulation Setup and Result*

The 3D interconnect structure used for simulation is a copper wire segment with a via below the wire. The wire geometry is specified by its 5 μm width, 100 μm length and 1.5 μm height; the via is of 2.5 μm size and is 1.5 μm off the wire edge; the initial temperature is set to 300 K; the activation energy is set to 0.7 eV for the copper body and is set to 1.25 eV along the surface to mimic the behavior of the interface between the copper body and the cap layer. The current is flowing from the via bottom up to the wire, then to the other end of the wire. A 3D view is shown in Fig. 8.7.

We build an EM physical simulator using ANSYS [29] and a C++ program. ANSYS is a multi-physics FEM-based tool, which provides AFD distribution for specified physical parameters including temperature, current density and stress. The whole simulation time is divided into many small time steps Δt. During each time step, ANSYS simulates the wire structure and C++ program calculates the AFD using (8.12), then the AFD is fed back into ANSYS for simulation of the next time step. This ANSYS/C++ loop continues until an EM failure state is reached (e.g. 10 % resistance increase).

Figure 8.8 shows the AFD contour map of the interconnect structure. The wire end that connects to the via below has large AFD, while small AFD is found at the other end. The result matches well with the theoretical analysis and experimental observations from real EM tests in literature [30, 31].

3 Variation-Aware Compact MTTF Model

3.1 *Variation Sources*

3.1.1 CMP Dishing

Chemical Mechanical Polishing/Planarization (CMP) is a process of smoothing surfaces with the combination of chemical and mechanical forces [32]. Copper and

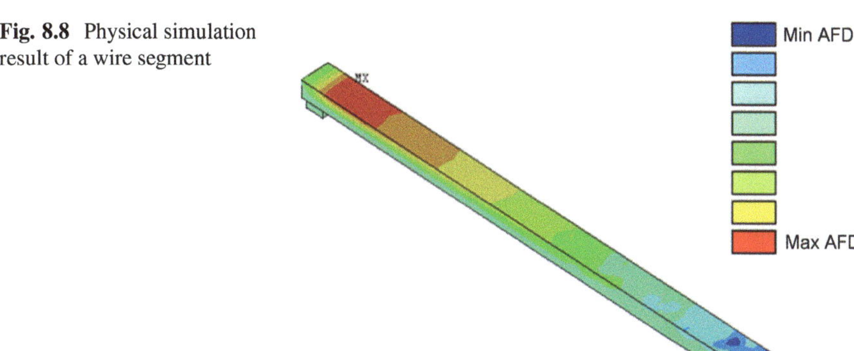

Fig. 8.8 Physical simulation result of a wire segment

the adjacent dielectric are removed from the wafer at different rates during CMP, creating surface anomalies and a varying topography [33]. Many factors, including pattern geometry (e.g., line density), affect the material removal rates. The CMP process strives to achieve flat topography to improve yield. Dishing is a copper surface anomaly caused by CMP. Dishing occurs when copper recedes below the level of the adjacent dielectric. Although dummy metal fills are added in order to achieve a flat topography, dishing still occurs (Fig. 8.9).

3.1.2 EPE

As technology continues to scale down, lithography no longer produces ideal geometric shapes. The most typical error is the Edge Placement Error (EPE) [34]. EPEs outside the feature are considered positive errors (bumping) and EPEs inside the feature are negative errors (necking) as shown in Fig. 8.10. Although techniques such optical proximity correction (OPC) are applied to reduce EPE, these variations still exist.

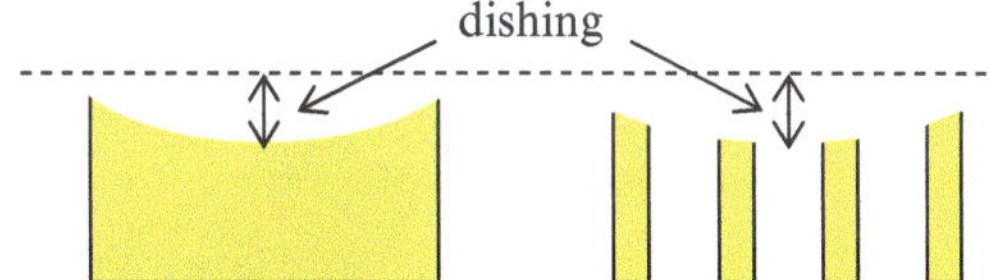

Fig. 8.9 CMP dished copper wires

3.2 Compact Model

3.2.1 Observations

Figure 8.11 shows an example of AFD simulation result by ANSYS of a CMP dished wire. A maximum AFD occurs at the cathode and a minimum AFD at the anode end of the wire.

We make several observations based on the simulations with CMP/EPE variations:

(1) A bumping on a wire does not relax the overall EM stress, whereas necking reduces the EM lifetime;
(2) On a wire, there exists a critical region where maximum AFD occurs. Necking in this region affects EM significantly; necking elsewhere affects EM much less. The position of a critical region depends on the wire length, current density, and current direction;
(3) When CMP dishing affects the critical region, the resulting EM lifetime is proportional to the square of the wire height;
(4) When necking affects the critical region, the resulting EM lifetime is a function of the necking size and location.

These observations inspire us to develop a compact model that captures dependencies between wire geometry imperfections and EM lifetime. We run simulations with various configurations, and develop a reasonably accurate model within a practical range of geometric variations.

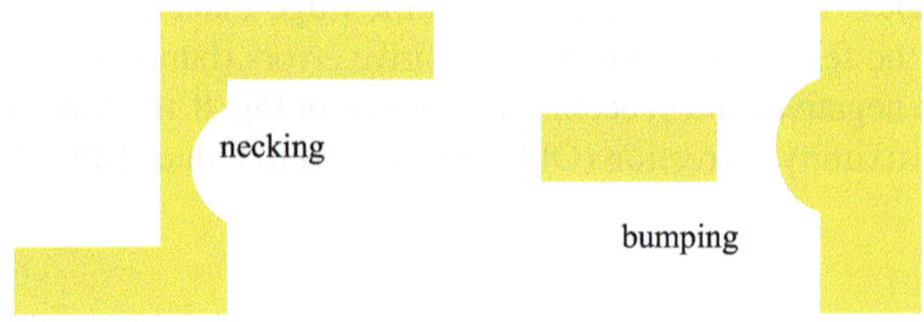

Fig. 8.10 EPE affected wires

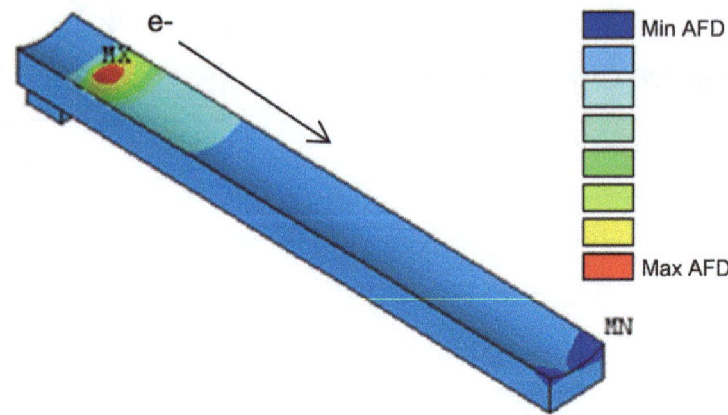

Fig. 8.11 EM physical simulation of a CMP dished wire

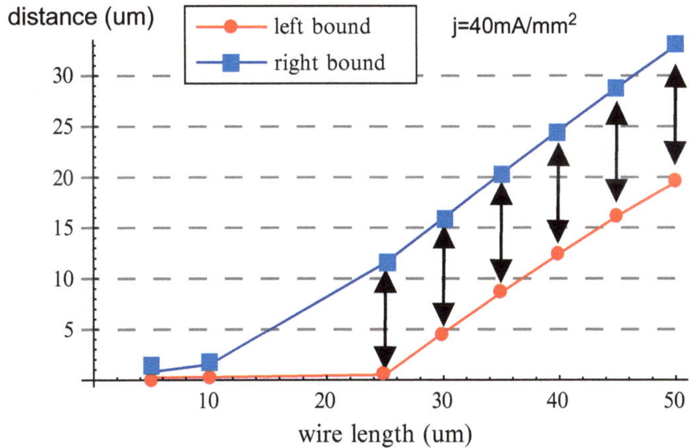

Fig. 8.12 Critical region boundaries

3.2.2 Critical Region

We define the critical region of a wire as a region where the AFD is greater than 10 % of AFD_{max} (the maximum value of AFD along the entire wire). The critical region boundary position is found to be a piecewise function of wire length and current density as shown in (8.13), where $k_1 = 0.85$, $k_2 = 1.65\text{e-}12$, $C_R = 0.85\text{e}6$, $C_L = 1.02\text{e}16$ and $C_C = 0.0915$ are constants, l is the wire length and j is the current density.

$$B_L = \begin{cases} 0 & 0 \le l < \frac{C_L}{k_1 j} \\ k_1 l - \frac{C_L}{j} & l \ge \frac{C_L}{k_1 j} \end{cases}$$
$$B_R = \begin{cases} k_2 (j + C_C) \cdot l & 0 \le l < \frac{C_R}{k_1 j^2} \\ k_1 l - \frac{C_R}{j^2} & l \ge \frac{C_R}{k_1 j^2} \end{cases} \tag{8.13}$$

A sample curve for current density 40 mA/mm^2 is shown in Fig. 8.12. In this figure, the square- and circle-marked lines represent distances from the cathode via to the right and left boundaries of the critical region. The double arrowed lines mark the lengths of critical regions. It can be observed that for a fixed current density, the length of critical region does not change.

To quantify the effects of process variation we introduce the variation coefficient.

Definition 1

Variation coefficient C_{var} is the ratio of an ideal wire EM lifetime t_{life0} and EM lifetime t_{life} of the corresponding wire with geometric variations.

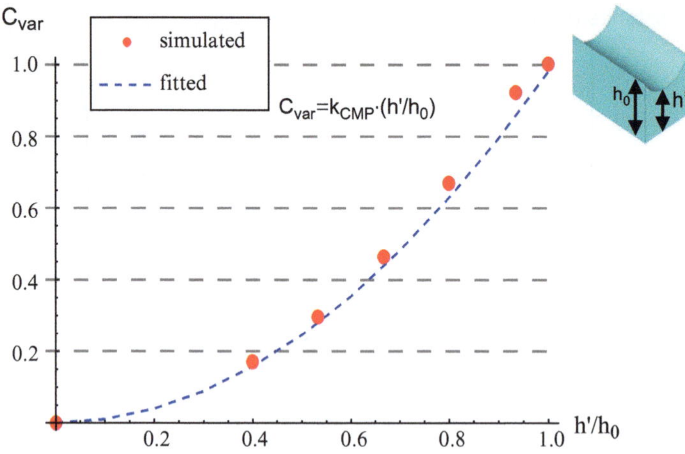

Fig. 8.13 Compact model curve for C_{var} due to CMP

$$C_{\text{var}} = \frac{t_{life_0}}{t_{life}} \tag{8.14}$$

Unlike for process variation induced global effects on EM lifetime, when considering local effects, we assume that the amount of charge passing through a wire does not change.

3.2.3 Compact Model

There are many factors that affect the EM lifetime of a wire with CMP/EPE variations. However, instead of investigating those detailed physical effects, we try to extract the major components that are most important to EM lifetime with variations and develop a simplified model that can be easily applied to full-chip interconnect EM analysis.

The ratio between an imperfect and ideal wire height (h'/h_0) is the dominant factor for C_{var} of CMP dishing. This dependency is given by (8.15), and shown in Fig. 8.13, for $k_{CMP} = 1$ and $m = 2$.

$$C_{\text{var}} = k_{CMP}\left(\frac{h'}{h_0}\right)^m \tag{8.15}$$

To determine C_{var} of EPE necking, we first consider the case where a single circular dent is present on a wire. The dent can be completely described by its depth d, length l and location x, and C_{var} is a function of these three parameters captured by (8.16), where b and w are the length of the critical region and the width of a wire;

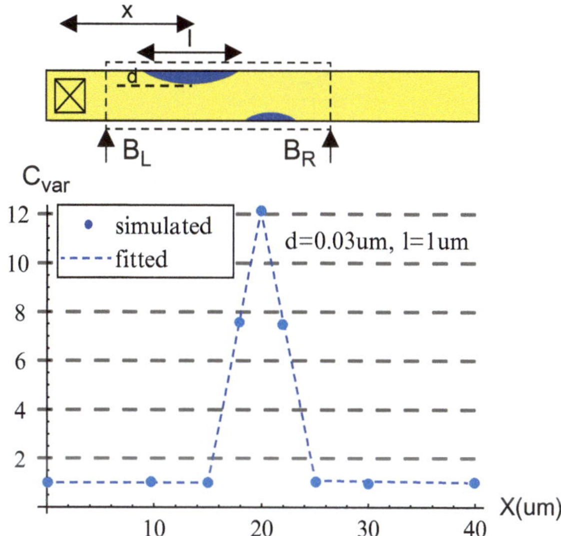

Fig. 8.14 Compact model curve for C_{var} due EPE

k_{EPE} is a fitted parameter equal to 7.2e18. Figure 8.14 shows a sample plot for C_{var} of EPE necking with $d = 0.03$ μm and $l = 1$ μm.

$$C_{\mathrm{var}} = f\left(\frac{d}{w}, \frac{l}{b}, x\right) = \begin{cases} 1 x \leq B_L \ or \ x \geq B_R \\ 1 + k_{EPE}\frac{l}{b}\frac{d}{w}(x - B_L)\, B_L < x \leq (B_L + B_R)/2 \\ 1 - k_{EPE}\frac{l}{b}\frac{d}{w}(x - B_R)\,(B_L + B_R)/2 < x < B_R \end{cases} \tag{8.16}$$

When multiple dents are present on a wire, their cumulative effect is found to be the product of C_{var} caused by each dent, given by (8.17).

$$C_{\mathrm{var}} = \prod_i C_{\mathrm{var},i} \tag{8.17}$$

This variation-aware EM lifetime model provides a mapping methodology from an assumed variation to the resulting EM lifetime. A more useful way to apply this model is to estimate the variation tolerance of EM-sensitive wires. For example, we may want to know how much CMP dishing or how much EPE is allowed for a given wire. These are later referred to as variation criticalities, and will be given to designers.

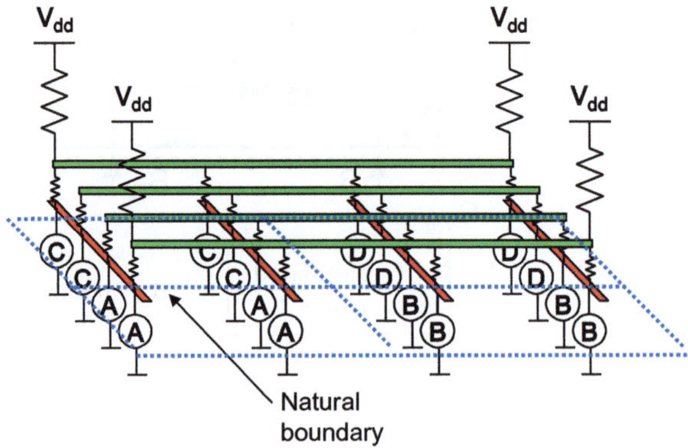

Fig. 8.15 Circuit-level power grid model

4 Full-Chip EM Analysis

4.1 Power Grid Model

Power grid of a VLSI chip carries unidirectional currents of large magnitudes and it is the most EM-vulnerable part of the on-chip interconnect network. In this chapter, we focus on power grid wires only.

In Fig. 8.15, we show the circuit-level model of power grid used in our work. We assume that power is delivered through a controlled collapse chip connection (C4) package. Wires and vias are modeled as resistors, external voltage supplies are modeled as ideal voltage sources, and switching transistors are modeled as current sources.

There are many different power grid EM analysis flows depending on data format types used to describe the power grid. In this chapter, we use the SPICE format. The link between the SPICE netlist naming and numbering scheme for the circuit and the original geometry of the power grid is described below.

- Node name:

  n<net-index>_<x-location>_<y-location>

- Data associated with each layer starts from:

  * layer: <name>,<net>_net: <net-index>
  Each layer/net combination is associated with a unique net-index.

- Vias starts from:

  * vias from: <net-index> to <net-index>
  Vias are implemented as resistors or as zero voltage sources.

- Current source:

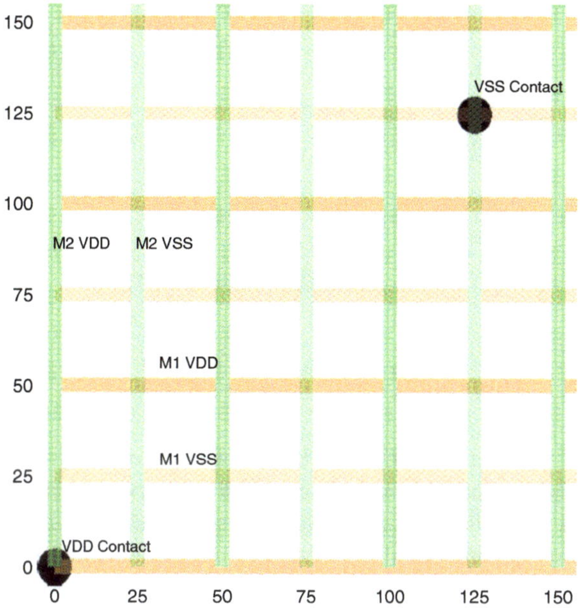

Fig. 8.16 Small power grid sample

iB<block-number> <node> 0 <value> and iB<block-number> 0 <node> <value>iB

Each current source is split into two components: from VDD to ideal ground and from ideal ground to VSS.

Figure 8.16 shows a sample small power grid. Reference [35] provides more details of this model.

4.2 *Effective jL Product Extraction*

Blech length effect can be used to quickly filter out a great number of EM-immortal wires which do not require further consideration. The remaining EM-mortal wires will be processed by a more time-consuming EM lifetime calculation procedures. The studies of Blech length effect usually consider simple straight point-to-point wires, for which the values of j and L can be easily determined. For complex interconnect topologies the accumulated jL product along the longest possible path is usually taken. However, in [36], the authors show that it is non-trivial to extract jL product in a complex interconnect topology, and they develop an effective jL extraction strategy based on EM physics, as stated in (8.18). In (8.18), j_k and L_k are the current density and length of a wire segment. The sum is taken over all possible paths in the network, with $(jL)_{eff}$ being the maximum of these sums. The effective

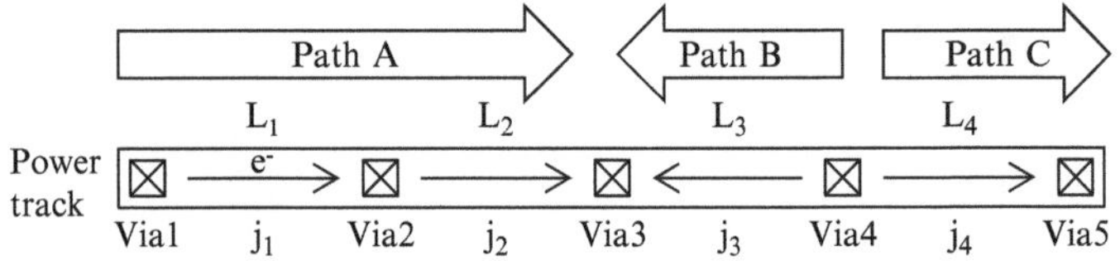

Fig. 8.17 Effective *jL* extraction

jL product matches well with experimental results in [37].

$$(jL)_{eff} = \max\left(\sum_k j_k L_k\right) \tag{8.18}$$

The effective *jL* product extraction (8.18) can be applied to any complex interconnect structure. But instead of computing all paths in a power grid, the effective *jL* can be obtained as follows. For any power wire segment, find the longest consistent current path that contains it and compute the sum of individual *jL* products of all wire segments along the path. For example, in Fig. 8.17, paths A, B and C are longest consistent current paths in the given power track, so we have (8.19).

$$\begin{aligned} jL_{1,eff} &= jL_{2,eff} = jL_{3,eff} = jL_{pathA} = j_1L_1 + j_2L_2 + j_3L_3 \\ jL_{4,eff} &= jL_{5,eff} = jL_{pathB} = j_4L_4 + j_5L_5 \\ jL_{6,eff} &= jL_{7,eff} = jL_{pathC} = j_6L_6 + j_7L_7 \end{aligned} \tag{8.19}$$

4.3 *Analytical Lifetime Calculation*

The Black's equation given by (8.2) has its limitations, since it ignores some physical factors that can substantially affect EM lifetime such as critical stress for void nucleation, material effective modulus and diffusivity. Many physics-based EM lifetime calculations are developed in literature. We adopt the method presented in [36], which obtains EM lifetime based on the void nucleation and growth model. It also provides EM lifetime calculation for complex interconnect topologies, which can be applied to a power grid.

Here, we briefly explain how to calculate EM lifetime of a power grid wire segment. Detailed discussion can be found in [36]. In Fig. 8.17, arrows denote electron flow in a wire segment. First, each wire segment EM failure is associated with its cathode via. So we have (8.20).

$$\begin{aligned} t_{L1} &= t_{via1} \\ t_{L2} &= t_{via2} \\ t_{L3} &= t_{L4} = t_{via4} \end{aligned} \tag{8.20}$$

Table 8.2 Physical parameters

Parameters	Values
σ_{void}	40 MPa
Z^*	1
B	28 GPa
Ω	$1.18 \times 10^{-29}\ \text{m}^{-3}$
ρ	2.1μΩ cm (metal 5, 6), 3.9μΩ cm (metal 2, 3, 4), 4.8μΩ cm (metal 1)

Next, EM lifetime of a via depends on the metal atomic flux divergences contributed by all the connected wire segments. Due to the flux divergence, a void is nucleated when threshold stress is reached. The time for a void nucleation comprises the first part of (8.21). After a void is nucleated, it keeps growing until it reaches a critical length such that EM failure occurs. The time for void growth comprises the second part of (8.21).

$$t_{life} = \left(\frac{\sigma_{void}\Omega}{\rho e Z^*} \sqrt{\frac{\pi}{4}} \sqrt{\frac{kT}{B\Omega}} \frac{\sum_i \sqrt{D_i}}{\sum_i D_i j_i} \right)^2 + \frac{L_c kT}{\rho e Z^*} \frac{1}{\sum_i D_i j_i} \tag{8.21}$$

In (8.21), σ_{void} is the critical stress required to nucleate a void, j_i and D_i are the current density and the effective diffusivity of each wire segment i connected to a via, B is the effective modulus of the materials surrounding the metal, L_c is the critical void length, and T is the temperature. In reality, if the actual chip temperature profile is available, it should be used for EM lifetime calculation. However, here we assume temperature is uniformly fixed to 300 K. The effective diffusivity D is computed using equation (8.22):

$$D = D_0 \cdot e^{-\frac{E_a}{kT}} \cdot \frac{w}{h} \tag{8.22}$$

In (8.22), w is the wire width and h is its height. When all D_i values are the same, which is typical in a power grid, the lifetime of a wire segment and thus the lifetime of its cathode via is inversely proportional to the current density through the via. The parameter values we used are listed in Table 8.2.

Wires are classified into three subgroups based on their analytical lifetime: EM-safe, EM-sensitive and EM-weak. The classification criteria are listed below.

EM-safe: $t_{life} > (1+\beta)\, t_{req}$;
EM-weak: $t_{life} < t_{req}$;
EM-sensitive: $t_{req} \leq t_{life} \leq (1+\beta) t_{req}$,

where t_{req} is the required lifetime, assumed here to be 10 years. β is the lifetime slack threshold; its value is selected according to the maximum lifetime reduction found by the variation-aware analysis. The wires with lifetime $t_{life} > (1 + \beta)\ t_{req}$ are EM-safe and should not fail even if they experience the worst process variation effects. Wires with $t_{req} \le t_{life} \le (1 + \beta)t_{req}$ are considered EM-sensitive and their status is determined by the variation-aware analysis using the compact model. A user can either use a pre-estimated upper bound of β, or run several iterations of our flow to tune the value of β. For example, if initially the value of β is set to 10 %, and later variation-aware analysis reports a 12 % maximum lifetime reduction, then the user would adjust β value to a number greater than 12 %. In practice, β is within a range of 10–15 %, thus the convergence of tuning β is not considered a problem in our work. All EM-sensitive wires will then be processed by the variation-aware EM analysis.

4.4 Global Current Redistribution

CMP/EPE variations not only locally affect EM lifetime, but also cause global power grid current redistribution from full-chip analysis point of view. However, we need to know the exact CMP/EPE variations on the power grid to study those redistribution effects.

For experimental purpose, to create the environment for studying global current distribution changes due to resistance change, we apply Monte-Carlo simulation. Monte-Carlo simulation samples are generated by first randomly throwing some darts on the P/G network. Each such selected resistor, referred to as *c-resistor*, is a variation center. Its own resistance and those of a few wires around it are changed. Each *c-resistor* randomly picks a value according to a Gaussian distribution $N(\mu,\sigma^2)$, where μ is mean and σ^2 is variance. This variation then propagates to its neighbors with decreasing magnitude as the distance to the *c-resistor* increases. The assumption is that wires in close layout proximity have similar imperfections due to similar process conditions. Figure 8.18 gives an example.

In Fig. 8.18, the nominal value of each horizontal resistor is 5Ω and of vertical resistor is 4Ω. *A* is an EM-sensitive wire. *B* and *C* are *c-resistors* in the perturbed network. The perturbed resistance values are shown in Fig. 8.18b. An EM-sensitive wire segment may also become a *c-resistor*.

Currents of the EM-sensitive wires have to be computed in the perturbed grid. Solving Kirchhoff's Current Law (KCL) or Kirchhoff's Voltage Law (KVL) equations for the whole system each time a perturbed network is generated is infeasible due to a huge computational cost. Here, knowing the node voltages in the nominal network, we apply random walk method [38] to compute current flows in its perturbed variant. Random walk is a technique that allows localized computation; its principle is briefly described below using Fig. 8.19.

A random walk starts from any node *x*, and then travels to some adjacent node according to a certain probability as in (8.23) which depends on electrical conductance between the two nodes. In (8.23), V_x is the voltage of the node *x*, degree(*x*) is the number of nodes adjacent to *x*, g_i is the electrical conductance between node i and x, I_x is the current load connected to x ($I_x = 0$ if no load is connected to x). The travelling continues until a boundary node (a ground node or an ideal voltage source node) is reached. Multiple travels are executed from a given node, and then the average value is taken as its node voltage. So the voltages of those nodes of interest (nodes of EM-sensitive wires) in a perturbed network can be obtained without solving the whole system.

$$V_x = \sum_{i=1}^{\text{degree}(x)} \frac{g_i}{\sum_{j=1}^{\text{degree}(x)} g_j} V_i - \frac{I_x}{\sum_{j=1}^{\text{degree}(x)} g_j} \tag{8.23}$$

Lifetimes of all EM-sensitive wires are then recalculated for each perturbed network. Two distributions are obtained: lifetime distribution of each EM-sensitive wire and cumulative distribution function (CDF) of the full-chip EM reliability.

4.5 VEMA Flow

The variation-aware EM analysis tool VEMA includes four major steps: fast jL filter, analytical EM lifetime calculation, global current redistribution and local variation effect analysis. Figure 8.20 shows a flowchart for our variation-aware EM analysis tool VEMA.

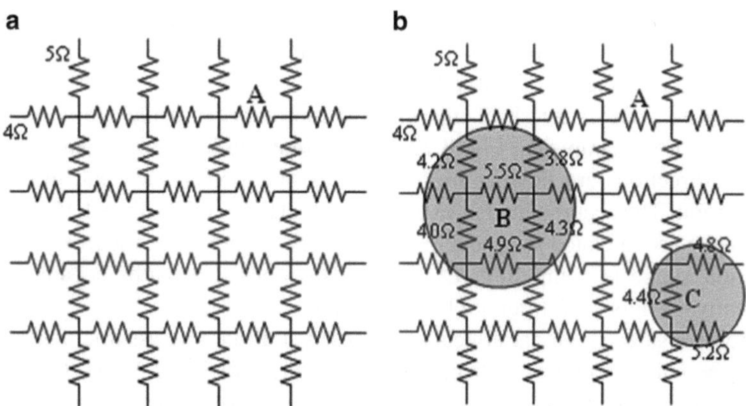

Fig. 8.18 P/G network. (**a**) Original (**b**) perturbed

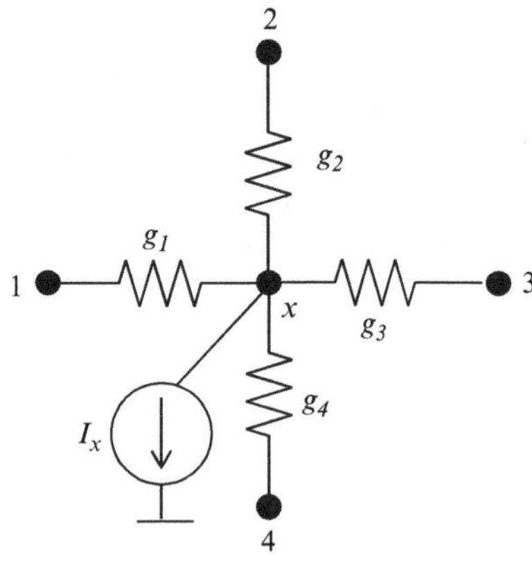

Fig. 8.19 Random walk from node x

5 Experimental Results

We implemented our tool in C++ and tested it with several publicly released IBM P/G benchmarks [35]. Experiments were carried on GUN/LINUX server with Intel Xeon E5440 2.83 GHz CPU and 16 GB memory. The specification required EM lifetime (referred to as t_{req}) was set to 10 years and the temperature was assumed to be 300 K for all experiments.

In Fig. 8.21, we compare *jL* filter results from the accumulated and effective *jL* product extraction for metal three layer in IBM PG1 benchmark. An obvious reduction in EM-mortal wires can be observed in effective *jL* product distribution. This is because the accumulated *jL* extraction uses the longest possible path for an entire power track, thus a large number of wires are unnecessarily classified as EM-mortal. This imposes an onerous burden on further EM analysis and significantly diminishes the efficiency of fast *jL* filter.

Moreover, there is a risk of missing EM-mortal wires using the accumulated *jL* product. For example, in Fig. 8.22, two currents with the same magnitude but opposite directions are flowing from two ends of a power track and meet at the middle point. The accumulated *jL* extraction will classify this wire as EM-immortal, because the accumulated *jL* product would be zero, whereas the effective *jL* product extraction suggests computing *jL*s for the two consistent paths separately. The quantitative results of *jL* filter are listed in Table 8.3.

When considering global effects of process variations, we assume that 10 % of wires experience variations, with each *c-resistor* R_i following $N(R_i, 0.001R_i^2)$ distribution. We simulated the global effects of process variation on a randomly selected EM-sensitive wire, named W_2, taken from IBM PG2 benchmark. Figure 8.23 shows the lifetime histogram with a trend line for W_2. The solid vertical line denotes t_{sp} and the dotted line marks MTTF for W_2. It can be observed that with probability 0.84, W_2 is safe from EM failure taking into account global effects of process variations; $P_{safe,w2} = 0.84$.

Table 8.3 Experimental results

Benchmarks	Total #wires	Mortal			Worst MTTF (year)	#EM-sensitive	Avg. prob. no failure (%)	Avg. CMP tolerance (%)	Avg. EPE tolerance (%)
		A	E	Miss					
IBM PG1	30027	25196	10032	623	4.07	46	32.7	15.74	8.50
IBM PG2	208325	111245	23124	77	4.94	63	89.7	16.24	15.41
IBM PG3	1401572	33872	33309	546	4.90	917	82.0	14.37	12.04
IBM PG4	1560645	198103	66226	0	5.53	554	74.6	12.53	11.06

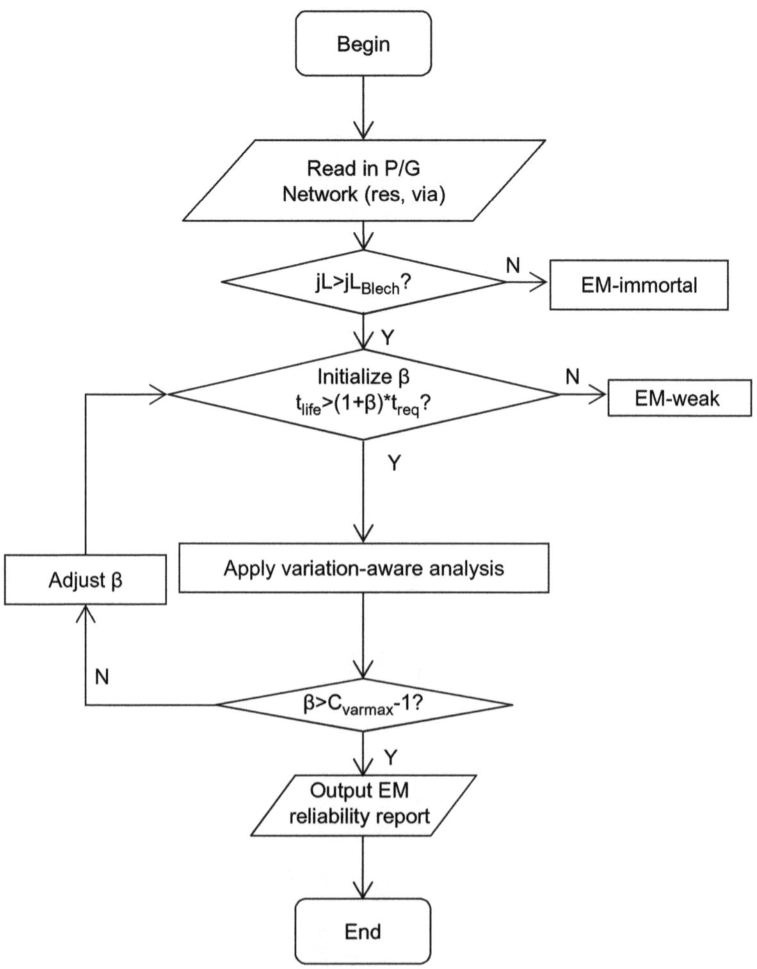

Fig. 8.20 VEMA flowchart

We determine an overall EM failure cumulative distribution function (CDF) using (8.24) considering global effects from process variation. In (8.24) $P_{CDF}(q)$ is the probability of no more than q EM-sensitive wire failures, p is the total number of EM-sensitive wires, $P_{safe,i}$ is the EM-safe probability of i_{th} wire. EM-sensitive wires are sorted according to the order of their increasing EM-safe probability.

$$P_{CDF}(q) = \prod_{i=0}^{p-q} P_{safe,i} \tag{8.24}$$

Figure 8.24 shows the CDF curve for IBM PG2 benchmark. Point *A* on the CDF curve corresponds to probability 0.95 of no more than eight EM-sensitive wire failures, and the gray-shaded region beneath point *A* is highly sloped, elsewhere it is gently sloped. This can be interpreted that strengthening the worst eight EM-sensitive wires can boost EM-safe probability to 0.95; but beyond that due to gently sloped region, the benefit of strengthening more EM-sensitive wires will be very small.

Using the variation-aware compact EM lifetime model, we perform EM analysis considering CMP/EPE variations. In Fig. 8.25, we show the EM lifetime distribution for a randomly selected EM-sensitive wire W_4 from IBM PG4 benchmark. For W_4, we assume 10 % EPE-caused area loss within the critical region. Local effects caused by CMP and EPE are similar and manifest themselves as a shift of EM lifetime distribution which reduces the ideal EM lifetime $MTTF_0$ to $MTTF'$ and causes a higher EM failure probability. The maximum achievable distance between $MTTF_0$ and t_{req} indicates how critical EPE effect on MTTF is.

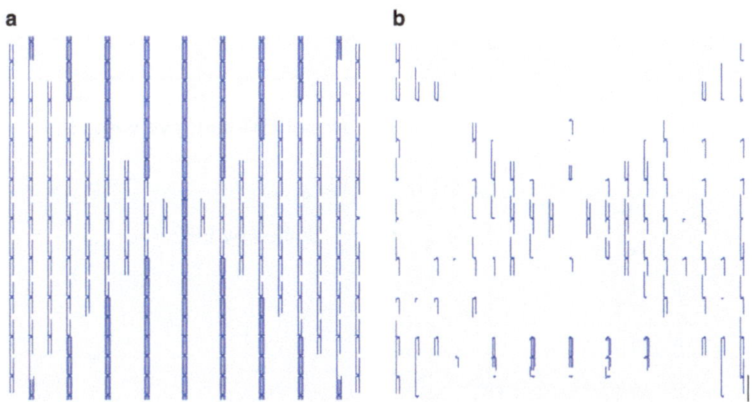

Fig. 8.21 EM-mortal map for IBM PG1. (**a**) Accumulated mortal map (**b**) effective mortal map

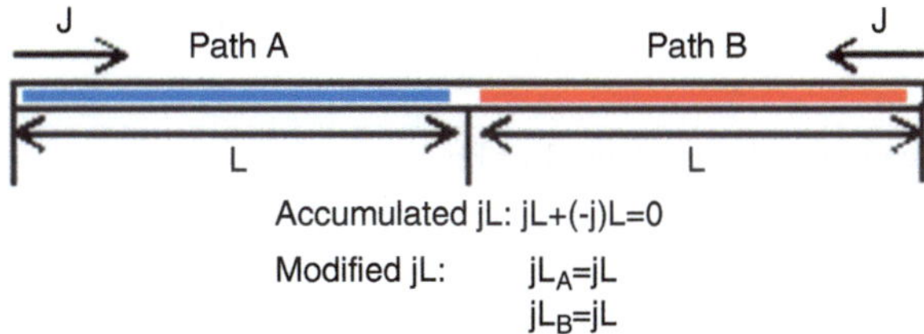

Fig. 8.22 *jL* extraction comparison

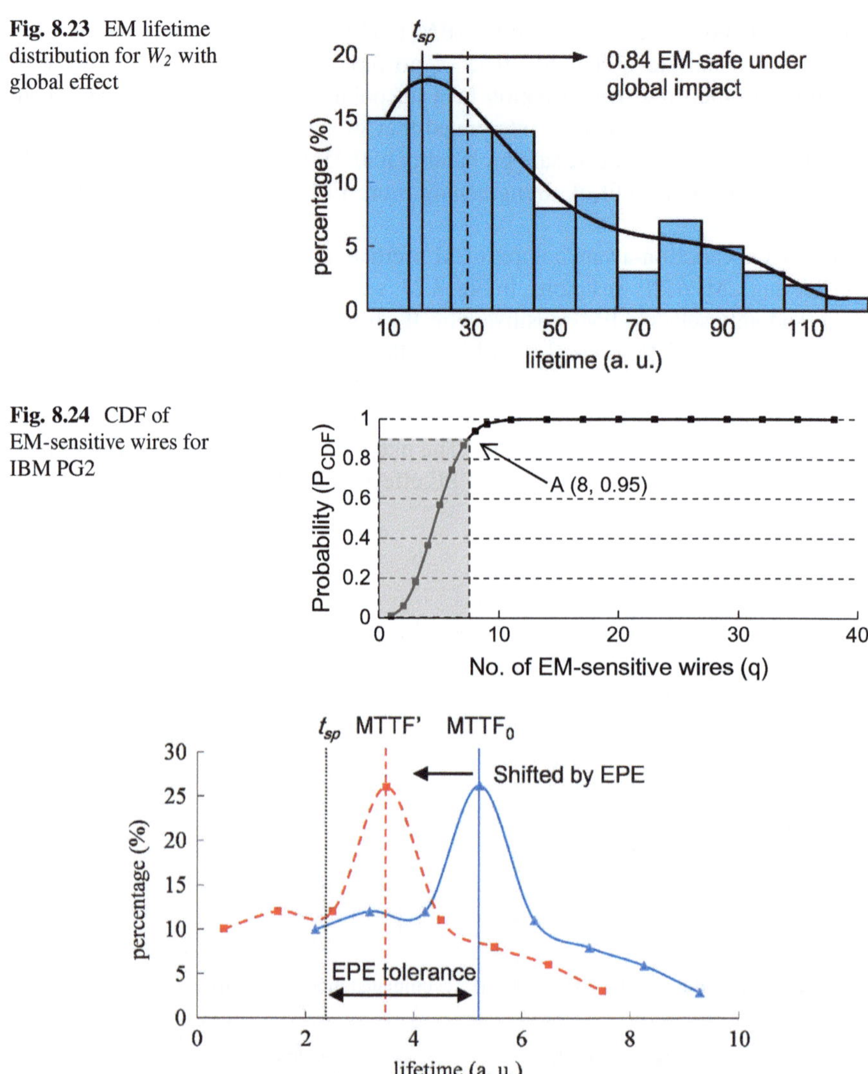

Fig. 8.23 EM lifetime distribution for W_2 with global effect

Fig. 8.24 CDF of EM-sensitive wires for IBM PG2

Fig. 8.25 EM lifetime distribution shifted by EPE

Definition 2

CMP criticality is defined as a ratio of the original wire height over the maximum wire height loss that a wire can tolerate while keeping $t_{life} > t_{req}$. EPE criticality is defined as the ratio of original total critical region area over the maximum area loss that a wire can tolerate while maintaining its lifetime greater than the required limit, $t_{life} > t_{req}$.

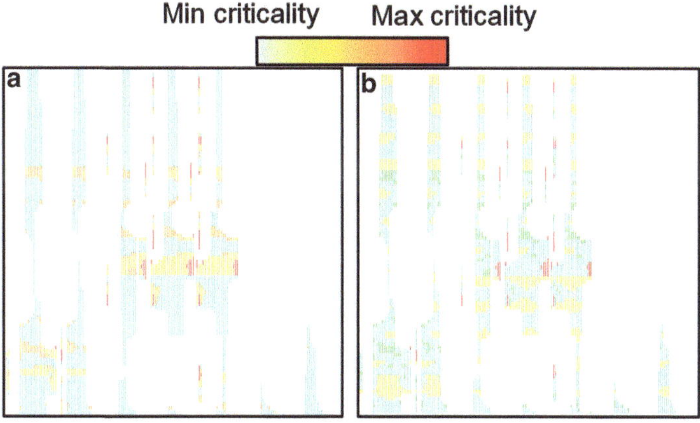

Fig. 8.26 Variation criticality map for IBM PG4. (**a**) CMP criticality (**b**) EPE criticality

$$CMP\_criticality = \frac{h_0}{\max(h_0 - h')}$$
$$EPE\_criticality = \frac{A_c}{\max(A_{loss})} \tag{8.25}$$

In Fig. 8.26, we report CMP/EPE criticality map for metal five layer in IBM PG4 benchmark. For high criticality EM-sensitive wires process variation induced printing imperfections should be minimized by techniques such as better OPC or more careful insertion of dummy metal fill.

All these results are quantitatively summarized in Table 8.3. In Table 8.3, the sub column labeled *A* under column *Mortal* provides the accumulated *jL* extraction values, the sub column labeled *E* refers to effective *jL* extraction and the sub column labeled *Miss* gives the number of EM-mortal wires missed by the accumulated *jL* extraction. The column labeled *Avg. prob. no fail* is the average probability that EM-sensitive wires will not fail.

Conclusions

As technology scales down, EM is becoming progressively a more serious problem. It posts huge challenge on analysis tools and limits the application of conventional EM theories such as Black's equation and Blech length effect. Moreover, the process variation brings in more uncertainty to EM analysis.

In this chapter, we introduced the basic background knowledge of EM, and described the tool VEMA, which is a post-design EM analysis and reliability evaluator. Fast *jL* filter and ideal EM lifetime calculation are performed to quickly filter out EM-immortal wires and classify all EM-mortal wires into three subgroups: EM-weak, EM-safe and EM-sensitive. Variation-aware EM analysis is applied on EM-sensitive wires. The variation effects mainly refer

(continued)

to the changes of wire geometry caused by CMP and EPE that affect the EM lifetime of individual wires. A compact variation-aware EM lifetime model is developed. Using this model, variation tolerance of each EM-sensitive wire can be fed back to designers. As technology is scaling down, process variation on wires becomes increasingly more significant. Our work shows that its effect on EM reliability is non-negligible, and a variation-aware EM analysis such as VEMA may provide a more realistic assessment of the EM reliability of power grids.

Acknowledgments Research support was provided in part by SRC through IBM Corporation and in part by NSF grant CCF-1115663.

Many people have contributed at various stages to this book chapter. In particular we would like to thank our collaborator from IBM, Dr. Baozhen Li for motivating this work and for support. We also thank Xiang Qiu, Zhong Guan and Ali Abbasinasab from the University of California, Santa Barbara for their helpful discussions.

There also have been many discussions with professors at the University of California, Santa Barbara. In particular, we would like to thank Professor Tim Cheng and Professor Li-C Wang.

References

1. http://en.wikipedia.org/wiki/Electromigration.
2. J. Lienig, "Invited Talk: Introduction to Electromigration-Aware Physical Design," *International Symposium on Physical Design*, 2006, pp. 39-46.
3. F. Wei, C. L. Gan, T. L. Tan, et.al, "Electromigration-induced extrusion failures in Cu/low-k interconnects," *Journal of Applied Physics*, 104, 023529 (2008).
4. R. S. Sorbello, "Theory of the Direct Force in Electromigration," *Phys. Rev. B*, vol. 31, no. 2, pp. 798-804, 1985.
5. J. R. Black, "Electromigration - A Brief Survey and Some Recent Results," *IEEE Trans. on Electron Devices*, Vol. ED-16 (No. 4), pp. 338-347, April 1969.
6. I. A. Blech, "Electromigration in thin aluminum films on titanium nitride," *J. Appl. Phys.*, vol. 47 (1976), pp. 1203–1208.
7. I. A. Blech and C. Herring, "Stress Generation by Electromigration," *Appl. Phys. Lett.*, vol. 29, no. 3, pp. 131-133, 1976.
8. I. A. Blech and K. L. Tai, "Measurement of Stress Gradients Generated by Electromigration," *Appl. Phys. Lett*, vol. 30, no. 8, pp. 387-389, 1977.
9. C. Herring, "Diffusional Viscosity of a Polycrystalline Solid," *J. Appl. Phys.*, vol. 21, pp. 437-445, 1950.
10. http://www.iue.tuwien.ac.at/phd/orio/node26.html.
11. E. T. Ogawa, A. J. Bierwag, K.-D. Lee, H. Matsuhashi, P. R. Justinson, and et al., "Direct Observation of a Critical Length Effect in Dual-Damascene Cu/Oxide Interconnects," *Appl. Phys. Lett.*, vol. 78, no. 18, pp. 2652-2645, 2001.
12. D. Ney, X. Federspiel, V. Girault, O. Thomas, and P. Gergaud, "Stress-Induced Electromigration Backflow Effect in Copper Interconnects," *Trans. Dev. Mater. Reliab.*, vol. 6, no. 2, pp. 175-180, 2006.
13. L. Doyen, E. Petitprez, P. Waltz, X. Federspiel, L. Arnaud, and Y. Wouters, "Extensive Analysis of Resistance Evolution due to Electromigration Induced Degradation," *J. Appl. Phys.*, vol. 104, p. 123521, 2008.

14. A. S. Oates and M. H. Lin, "Void Nucleation and Growth Contributions to the Critical Current Density for Failure in Cu Vias," *Proc. Intl. Reliability Physics Symp.*, pp. 452-456, 2009.
15. J. W. McPherson and P. B. Ghate, "A methodology for the calculation of continuous dc electromigration equivalents from transient current waveforms," in *Proc. Symp. on Electromigration of Metals*, New Orleans, LA, pp. 64-74, Oct. 7-12, 1984.
16. C. K. Hu, R. Rosenberg, H. S. Rathore, et.al, "Scaling Effect on Electromigration in On-Chip Cu Wiring," *International Conference on Interconnect Technology*, 1999, pp. 267-269.
17. N. Srivastava, K. Banerjee, K. E. Goodson, "Scaling Analysis of Multilevel Interconnect Temperatures for High-Performance ICs," *IEEE Trans. on Electron Devices*, volume 52, issue 12, 2005, pp. 2710-2719.
18. B. Li, C. Christiansen, C. Burke and et al., "Short Line Electromigration Characteristics and their Applications for Circuit Design," *International Reliability Physics Symposium*, 2013, pp 3.F.2.1-3.F.2.5.
19. M. Lin, N. Jou, W. Liang and K. C. Su, "Effect of Multiple Via Layout on Electromigration Performance and Current Density Distribution in Copper Interconnect," *International Reliability Physics Symposium*, pp. 844-847, 2009.
20. N. Raghavan and C. M. Tan, "Statistical Modeling of Via Redundancy Effects on Interconnect Reliability," *International Symposium on the Physical and Failure Analysis of Integrated Circuits*, pp. 1-5, 2008.
21. D. Li, Z. Guan, M. Marek-Sadowska and S. R. Nassif, "Multi-Via Electromigration Lifetime Model," International Conference on Simulation of Semiconductor Process and Devices, pp. 308-311, 2012.
22. http://en.wikipedia.org/wiki/Activation_energy.
23. http://en.wikipedia.org/wiki/Copper_interconnect.
24. R. L. D. Orio, "Electromigration Modeling and Simulation," doctoral dissertation, Institute for Microelectronics, TU Wien, 2010.
25. D. Li, M. Marek-Sadowska and S. R. Nassif, "A Method for Improving Power Grid Resilience to Electromigration-Caused Via Failures," to appear in *IEEE Trans. on VLSI Systems*.
26. D. Dalleau, "3-D Time-depending Simulation of Void Formation in Metallization Structures", Doctoral thesis, University of Hannover, 2003.
27. W. Li, C. M. Tan and N. Raghavan, "Dynamic simulation of void nucleation during electromigration in narrow integrated circuit interconnects," *Journal of Applied Physics* 105, 014305 (2009).
28. http://www.ansys.com.
29. M. Lin, N. Jou, W. Liang and K. C. Su. Effect of Multiple Via Layout on Electromigration Performance and Current Density Distribution in Copper Interconnect. *IEEE 47th Annual International Reliability Physics Symposium*, Montreal, 2009, pp. 844-847.
30. B. Li, J. Gill, C. J. Christiansen, et. al. Impact of Via-Line Contact on Cu Interconnect Electromigration Performance. *IEEE 43rd Annual International Reliability Physics Symposium*, San Jose, 2005, pp. 24-30.
31. http://en.wikipedia.org/wiki/Chemical-mechanical_planarization.
32. F. M. Serry, D. Dawson, "Minimizing Dishing and Erosion in Copper CMP," http://www.veeco.com/pdfs/database_pdfs/minimizing_de_in_copper_cmp_45.pdf.
33. https://www.si2.org/openeda.si2.org/dfmcdictionary/index.php/Edge_Placement_Error.
34. S. R. Nassif, "Power grid analysis benchmarks," *Asia and South Pacific Design Automation Conference*, 2008, pp. 376-381.
35. S. P. Hau-Riege, "New Methodologies for Interconnect Reliability Assessments of Integrated Circuits," Doctoral thesis, Massachusetts Institute of Technology, 2000.
36. C. W. Chang, Z. –S. Choi, C. V. Thompson, et.al, "Electromiration resistance in a short three-contact interconnect tree," *Journal of Applied Physics* 99, 094505 (2006).
37. H. Qian, S. R. Nassif, and S. S. Sapatnekar, "Power Grid Analysis Using Random Walks," *IEEE Trans. on CAD*, vol. 24, no. 8, August 2005, pp. 1204-1224.
38. S. M. Alam, C. L. Gan, C. V. Thompson, et al, "Reliability computer-aided design tool for full-chip electromigration analysis and comparison with different interconnect metallizations," *Microelectronics Journal* 38 (2007) 463-473.

Chapter 9
Power-Gating for Leakage Control and Beyond

Andrea Calimera, Alberto Macii, Enrico Macii, and Massimo Poncino

Abstract The need of reliable nanometric integrated circuits is driving the EDA community to develop new automated design techniques in which *power consumption* and *variability* are central objectives of the optimization flow.

Although several *Design-for-Low-Power* and *Design-for-Variability* options are already available in modern EDA suites, the contrasting nature of the two metrics makes their integration extremely challenging. Most of the approaches used to compensate and/or mitigate circuit variability (e.g., Dynamic Voltage Scaling and Adaptive Body Biasing) are, in fact, intrinsically power inefficient, as they exploit the concept of redundancy, which is known to originate power overhead.

In this work, we introduce possible solutions for concurrent leakage minimization and variability compensation. More specifically, we propose *Power-Gating* as a mean for simultaneously controlling static power consumption and mitigating the effects induced by two of the most insidious sources of variability, namely, *Process Variations* (PV) due to uncertainties in the manufacturing and *Transistor Aging* due to Negative Bias Temperature Instability (NBTI).

We show that power-gating, when implemented through the insertion of dedicated switches (called *sleep transistors*), has a double effect: On one hand, when sleep transistors are enhanced with tunable features, it acts as a natural supply-voltage regulator, which implements a control knob for PV compensation; on the other hand, during the idle periods, it makes the circuits immune to NBTI-induced aging.

We describe optimization techniques for the integration of a new concept of power-gating into modern sub-45 nm design flows, that is, *Variation-Aware Power-Gating*. The experimental results we have obtained are extremely promising, since they show 100 % timing yield under the presence of PV and circuit lifetime extension of more than 5× in the presence of NBTI.

A. Calimera • A. Macii • E. Macii (✉) • M. Poncino
Dip. di Automatica e Informatica, Politecnico di Torino, 10129 Torino, Italy
e-mail: enrico.macii@polito.it

R. Reis et al. (eds.), *Circuit Design for Reliability*,
DOI 10.1007/978-1-4614-4078-9_9

1 Introduction

MOS devices are approaching the size of atoms, which is a fundamental barrier for the scaling process of the bulk technology. The research community is thus actively pursuing alternative materials, fabrication processes, devices and architectures to be adopted in mainstream circuit and system manufacturing. While several new solutions have shown good potential (e.g., carbon nanotubes and graphene, memristors, spin-based devices, ferromagnetic logic, atomic switches, NEMS), the debate of which of them will prevail is still open; therefore, nanometer CMOS will remain the dominant technology on the electronics market for a few years.

The 2009 ITRS Roadmap [17] reported that *static power consumption* and *variability* are the most serious concerns for the design of nanometer ICs below 45 nm. Static power due to internal leakage mechanisms represents the main source of power consumption in modern CMOS circuits [28], which have shown to be power-hungry even when not switching. Variability, instead, refers to the marked tendency of a manufactured circuit to show a deviation from its nominal behavior. Main sources of variability include random and systematic process variations [4], environmental condition variations [33] (e.g., temperature and V_{dd} fluctuations) and aging effects (e.g., Negative Bias Temperature Instability and Hot Carrier Injection) [1]. While static power translates to low energy efficiency, variability originates lower reliability and lower fabrication yield; both factors make electronic circuits less reliable.

Static power and variability are not new in the EDA field, and various options for their management are already available. However, while in the past considering the two as independent variables in the design space was accurate enough to obtain reasonable results, advanced technology nodes are showing the need of holistic design strategies that could provide *concurrent* static power optimization and variability compensation. Unfortunately, this challenge is complicated by the fact that most of the design solutions for compensating variability are intrinsically power inefficient: Fault-tolerance approaches, such as "fail and correct", or adaptive strategies, such as "dynamic voltage regulation" or "forward body biasing", are based on the concept of redundancy, which is in contrast with low-power requirements.

In this work, we address this critical issue and we propose the use of power-gating [2] as the enabling technology for achieving simultaneous leakage optimization and variability compensation.

Power-gating (PG) is based on the insertion of a dedicated MOS switch, called *sleep transistor*, between the gated block and the actual ground rail. This provides the circuit with two power modes: A low-power mode, during which the sleep transistor is turned-off and the leakage power is reduced by more than one order of magnitude, and an active mode, during which the sleep transistor, which is turned-on, guarantees a normal connection to the global power-rails. What is interesting to note is that, in terms of variability, the sleep transistor shows important and unique properties, that is: During the active periods, when enhanced with tunable-width features, it can act as a natural supply-voltage regulator usable to implement adaptive control schemes for process-variation mitigation [14]; during the low-power mode, its effect is to make the gated circuit immune by the aging mechanisms (NBTI and HCI) [9].

Based on these observations, we claim that PG can represent a viable solution for the implementation of low-power, aging-free reliable ICs. Obviously, the beneficial effects that PG can provide must be weighted against the amount of overhead it introduces. A direct Sleep Transistor Insertion (STI) methodology, in fact, may originate excessive timing, area and power overhead, which can off-set the beneficial effects that PG offers in terms of aging and variability compensation.

To overcome this drawback, we propose new optimization techniques based on the concept of *Variation-Aware Clustered PG*, which consists of a methodology for clustering and power-gating critical cells only, that is, to apply variation-aware PG only to cells whose process variation-induced and/or aging-induced variations have a direct impact on the overall performance of the circuit. To enable such a strategy, we opted for an STI flow in which sleep transistors are inserted in layouts with row granularity. This allows a finer control of variability/aging compensation, while reducing the design overhead. Experimental results performed on a set of benchmark circuits and mapped to an industrial 45 nm CMOS library prove the effectiveness of the proposed solutions, as well as their integrability into industrial design flows.

The remainder of the chapter is organized as follows. In Sect. 2, we discuss the main challenges to be faced during nanometric IC design; we introduce models for sub-threshold leakage power consumption, also describing the sources of variability in scaled nanometric technologies. Section 3 addresses the key design issues related to power-gating, with particular emphasis on automated solutions for physical sleep-transistor insertion. In the last two sections, we provide detailed background and models for process variation and NBTI effects and we present automated power-gating strategies for process variation compensation (Sect. 4) and NBTI mitigation (Sect. 5). Finally, section "Conclusions" gives some concluding remarks.

2 Design Issues for Nanometric CMOS Circuits

2.1 *Sub-threshold Leakage Power Consumption*

Among all the leakage current mechanisms induced by Short Channel Effects (SCEs), the *sub-threshold current* I_{sub-th} has proven to be the major contributor to the total static power consumption [28].

I_{sub-th} is defined as the drain-to-source current which flows when the transistor operates in the weak inversion region, i.e., when the gate voltage V_g is below the threshold voltage V_{th}. Under this condition, the channel shows a small, but non-zero concentration of minority carriers that are diffused from the drain to the source terminal whenever a potential greater than 0 is applied between drain and source, i.e., $V_{ds} > 0$.

A well-known model for the I_{sub-th} of a single nMOS transistor is given by the following equation [28]:

$$I_{sub-th} = \mu C_{ox} \frac{W}{L} (m-1) v_T^2 \cdot e^{\frac{V_g - V_{th}}{m v_T}} \cdot (1 - e^{\frac{-V_{ds}}{v_T}}) \tag{9.1}$$

with

$$m = 1 + \frac{C_{dm}}{C_{OX}} \tag{9.2}$$

where $v_T = KT/q$ is the thermal voltage, C_{OX} is the gate oxide capacitance; μ is the carrier mobility; m is the sub-threshold swing coefficient, with C_{dm} representing the capacitance of the depletion layer.

The magnitude of the sub-threshold current is a function of several parameters, such as the operating temperature (I_{sub-th} increases as T increases), the supply voltage (I_{sub-th} increases for larger V_g), and the device size (I_{sub-th} increases as the transistor gets larger and shorter). However, the parameter that affects most (i.e., exponentially) I_{sub-th} is the threshold voltage V_{th}: Decreasing the V_{th} by 100 mV increases the leakage current by a factor of 10. That is why scaled CMOS technologies (characterized by ever smaller V_{th}) suffer from sensible sub-threshold leakage.

2.2 Sources of Variability

With variability we commonly refer to the marked tendency of a manufactured CMOS circuit to show a deviation from its nominal behavior. The sources of variation can be broadly classified according to their nature (statistical vs. deterministic), their spatial reach (local vs. global), and their temporal rate of change (static vs. dynamic). Figure 9.1 summarizes the typical variations arising in nano-scale CMOS circuits and systems.

Under the label *statistical* it is possible to include all those variations that are induced by stochastic events; they differ from *deterministic* variations, which can be somehow predicted at design time. *Global* variations affect all the transistors on the die, while *local* variations are limited to a few transistors in the immediate vicinity of each other. Finally, the classification between *static* and *dynamic* depends on the actual rate of change with time. Static variations, e.g., process variations, remain effectively invariant over the entire lifetime of the manufactured chips, while dynamic variations change over the lifetime of the chips. The changes can manifest on a large time-sale (that is the case of slow-variations like aging effects: NBTI, HCI and TDDB) or in a short time-scale (fast-variations like IR-drop, clock jitter, coupling noise, temperature and V_{dd} variations). Although all these sources of variability have deleterious effects on the reliability of CMOS digital circuits, two of them have been recognized as particularly critical, thus worth specific consideration: Process variations and aging.

		Local	Global
Statistical	Static	**Intra-die (WID) random process variations**	**Inter-die (D2D) random process variations**
	Dynamic	-	-
Deterministic	Static	**Systematic process variations**	**Lifetime degradation (NBTI, HCI, TDDB), Systematic process variations**
	Dynamic	**IR drop, clock jitter, coupling noise (capacitive and inductive)**	**Temperature and Vdd fluctuations**

Fig. 9.1 Types of variability in CMOS-based circuits and systems

Process variations (PV) [4, 5] are mostly due to random fluctuations of dopant atoms, which result in the mismatching of the electrical characteristics of transistors in the same die, and to systematic or non-systematic impreciseness of the manufacturing process (like lithography, etching, and chemical-mechanical polishing). Process variations have a significant impact both on the power dissipation and performance of a design: 20× variation in leakage power for a 1.5× variation in delay between fast and slow dies has been reported in the literature. Given the power/performance tradeoff, those figures translate into an increase of dies that must be discarded because either too slow or too power consuming. Therefore, as a relevant side effect, increased variability decreases yield, with important cost implications.

Aging, instead, includes all those wear-out mechanisms that induce *time-dependent* degradation of the operating characteristics of devices [1]. Two are the main sources of aging in active devices: Bias Temperature Instability (BTI), and Hot Carrier Interface (HCI) [26]. Both these physical/chemical effects result in the generation of interface traps at the silicon/oxide interface and cause a drift of the threshold voltage over time. These irreversible effects, and BTI in particular, have traditionally been regarded as "reliability issues", and have only recently received some consideration in the CAD community as a factor affecting performance of digital circuits. Traditional VLSI design, in fact, bypasses the analysis and optimization of such dynamic networks by approximating the problem to the optimization of uniform static networks with certain guard-band. This may induce unacceptable design overheads.

3 Power-Gating for Leakage Power Reduction

3.1 Power-Gating Basics

Power-gating has proven to be a very effective approach to reduce standby leakage, while keeping high speed in the active mode. It is based on the principle of adding devices, called *sleep transistors*, in series with the pull-up and/or the pull-down of logic gates, and turning them off when the circuit is idle, thereby decreasing the leakage power component due to I_{DS} sub-threshold currents. When a nMOS sleep transistor is used on the pull-down path, a SLEEP signal controls its active/standby mode (i.e., SLEEP = 1 during standby and SLEEP = 0 during active mode). In the standby mode, the sleep transistor is off, thus disconnecting its insertion point, called *virtual ground*, from the physical ground. In active mode the gated circuit operates normally, but it incurs a delay degradation due to the series resistance of the sleep transistor. Figure 9.2 shows a logic block with a nMOS sleep transistor connected.

The source terminals of the logic gates in the logic block are connected to the virtual ground which is, in turn, connected to the drain terminal of the sleep transistor.

Effective use of power-gating requires a proper sizing of the sleep transistor, since that affects the performance of all the gates connected to it. While a small transistor unacceptably slows down the circuit in active mode due to its high resistance, a large one implies a significant overhead in area and a non-negligible energy (i.e., power and delay) for ON↔OFF transitions. One additional difficulty is that sleep transistor sizing is determined by the *maximum* current injected by the circuit, which leads to maximum drop across the sleep transistor drain-source path and, as a consequence, causes worst-case delay degradation during active operation.

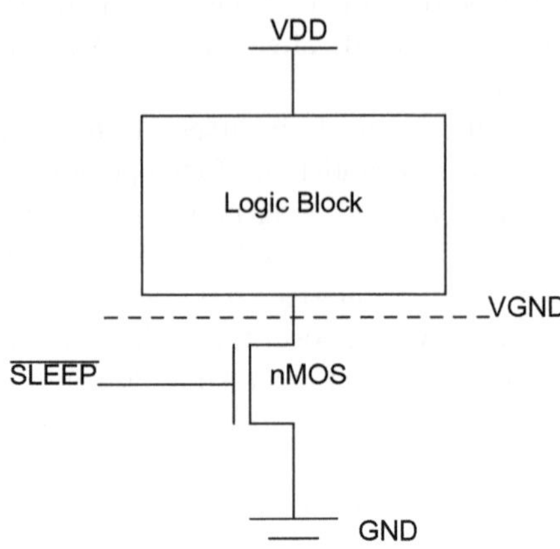

Fig. 9.2 A logic block with nMOS sleep transistor

Several power-gating styles have been proposed, differing in the *granularity* of the blocks to which sleep transistors are applied. Granularity may range from individual cells (the *fine-grained* sleep transistor insertion approach [21]) to large chip sub-units, (the *block-level power-gating* scheme), in which very large sleep transistors are placed on the root of the power distribution networks of large chip areas. While the fine-grained approach suffers from high area overhead and an excessive buffering of the sleep signal due to very high capacitance load to be driven by this signal, block-level power-gating has the disadvantage of having long transition delays between sleep and active state, caused by the large RC time constant of the sub-unit's power distribution network.

Moreover, and this is the most important aspect, its coarse granularity reduces the degrees of freedom available to the designer. If the decision of power-gating a block is taken, then all gates in the block are gated. This may not be desirable if, for instance, there is a critical subset of gates for which parametric variations (typically due to fabrication process or aging mechanisms) induce speed degradation of the whole circuit. In this case, a finer control of the sleep transistor insertion could guarantee a perfect match between leakage savings and design overhead, and, as we demonstrate afterwards, it also allows to exploit the beneficial effects of power-gating in terms of variability only where necessary, i.e., only for those subsets of gates that are more critical.

We refer to this finer strategy as the *clustered sleep transistor insertion*, a very effective solution where multiple subsets of cells (i.e., the clusters) are connected to dedicated sleep transistors distributed across the layout. It is worth mentioning that clustered power-gating is not new in the low-power design domain, and many clustering solutions have been proposed. In [20], the authors propose a solution in which gates having mutually exclusive current discharge patterns are grouped together; the resulting clusters allow optimal sleep transistor sizes. Instead, in [31], the authors show an effective timing-driven clustering strategy that is able to handle simultaneously timing and area constraints. Differently from previous works, in Sects. 4 and 5 we propose variation-aware clustering methods in which gates are grouped based on variation-induced timing criticality. Independently of the clustering algorithm, physical design details as well as constraints posed by the adoption of industrial EDA frameworks need to be considered while developing a strategy for sleep transistor insertion.

Using layout rows as atomic clustering objects greatly simplifies the physical-level management of virtual ground distribution, which is the major practical obstacle in clustering approaches that work on a cell-by-cell basis. In fact, having a mix of power gated and normal cells on the same row imposes drastic changes in power routing within a single row. The ensuing disruption of routing regularity makes it very difficult to control congestion and to ensure fast design convergence. Furthermore, since all the sleep transistors are placed in dedicated rows the sleep transistor placement is simplified and their overhead can be easily estimated and, consequently, traded for leakage reductions.

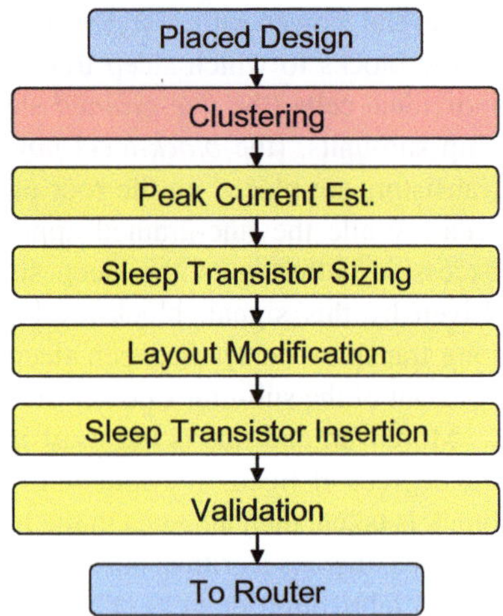

Fig. 9.3 Power-gating design flow

In order to fully integrate clustering and sleep transistor insertion with a state-of-the art physical design flow, placement and routing information have to be taken into account within the core clustering algorithm that selects gates to be power-gated, as well as in sleep transistor sizing and insertion.

Section 3.2 describes a common sleep transistor insertion flow suitable for the variation-aware clustered strategies proposed in this work.

3.2 Clustered Row-Based Sleep Transistor Insertion Methodology

State-of-the-art power gating methodologies follow a well known design flow, depicted in Fig. 9.3.

The entry point is a standard-cell placed design. The first step is *clustering*, in which cells are grouped together in order to be controlled by the same sleep transistor. Next, maximum current estimation is performed for each cluster. This information is essential to drive the selection of the appropriate sleep cells to be connected to the various clusters (*sleep transistor sizing*). The layout is then modified in order to accommodate the sleep transistor cells and the routing of the sleep signals. Finally, the modified layout is validated before it is fed to the routing tool.

This flow is fully compliant with industry standard back-end tools and it supports various power gating strategies (i.e., different sleep transistor insertion types), differing in the *granularity* of the blocks to which sleep transistors are applied, as described in Sect. 3.2.3.

3.2.1 Peak Current Estimation

This section describes a methodology for estimating the maximum current drawn by a cell cluster [30]. We define as Switching Window, T, of a gate under an input pattern the interval between the arrival time and the output transition of the gate. There is one switching window for each path through a particular gate. The width of the switching window of a gate is equivalent to the propagation delay of the gate for a rising or falling transition. For a given gate in the cluster, its Maximum Switching Window (MSW) is the time interval encompassing all its possible switching windows. If a gate has n switching windows, the set $T_1, \ldots, T_n$ is called the Full Switching Window (FSW) of the gate.

The peak current estimation algorithm consists of two steps. The first one computes the MSW for each gate in the cluster. The motivation behind extracting the MSW and not all the switching times of a gate is that, for a complex circuit and for a gate very deep in the logic, there is possibly a very large number of switching time intervals, each one corresponding to a path being activated through a gate. Moreover, it is very time consuming to extract all these windows for all the gates, which form a cluster. Conversely, it is very fast to extract the MSW for each gate. In fact, it only requires the calculation of the first (earliest) switching window T_1 and of the last (latest) one T_n; the MSW is simply obtained as $(T_1 \bigcup T_n)$. We thus need to extract only two switching time intervals for each gate in the cluster, which can be accomplished very fast.

The second step consists of the construction of a current plot over time which records the gates that switch at a particular time interval and, hence, the total discharge current in that time interval. This plot is the superposition over time of rectangles whose bases correspond to the MSWs of the individual gates of the cluster (which will, in general, partially overlap), and whose heights correspond to the currents drawn by the gates (e.g., derived by a technology library). From this current plot, the time interval during which the maximum current discharge occurs and the gates that contribute to this current discharge are obtained.

Figure 9.4 shows an example of current plot, in which five MSWs (from a to e) are shown. The interval in which the current drawn is maximum corresponds to the overlapping of the c, d, and e MSWs.

The current plot allows the identification of the time interval during which the maximum current discharge could occur. To tighten this upper bound, we need to extract the detailed FSW information only for those gates which contribute to this maximum value, which are normally a very small percentage of the gates in the cluster.

The algorithm enters then an iterative phase; it proceeds by finding, based on the current plot, the subset of gates that contribute to this maximum current value. Gate ordering on this subset of gates is executed first, so as to maximize the probability of non-overlapping switching of these gates. After ordering the gates, gate-by-gate FSW extraction is performed on the set of gates which contribute to the maximum

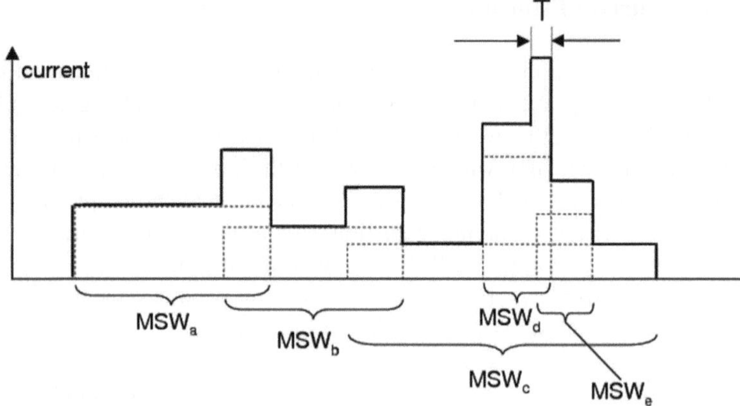

Fig. 9.4 Example of current plot

current. Once all the possible switching time intervals for this small subset of gates are extracted, the current plot is updated and the process is repeated until convergence is reached.

Convergence is guaranteed since, in the worst case, we have to extract the FSW for all the gates in the cluster and compute their maximum currents. When full FSW extraction is computationally too expensive, we may terminate the iteration early, for example by using a time bound. Since the maximum current estimate is monotonically non-increasing as the iteration proceeds (i.e., the upper bound is progressively tightened), the use of a time bound provides us with a fine-grained control of the accuracy vs. time trade-off.

3.2.2 Sleep Transistor Sizing

An effective use of power-gating requires a proper sizing of the sleep transistor. In fact, while a small sleep transistor may unacceptably slow down the circuit in the active mode due to its high resistance, a larger one implies a large area and a significant energy cost to drive it [15]. Designers usually define an IR-drop threshold (e.g., 10 % of V_{DD}) that is used as a constraint that must be met when sizing the sleep transistor resistance.

The maximum sleep transistor channel resistance can be computed using Eq. (9.3):

$$R_{st} = \frac{V_{DD} \cdot \alpha_{drop}}{I_{on}} \tag{9.3}$$

where $V_{DD} \cdot \alpha_{drop}$ is the allowed voltage drop across the sleep transistor, expressed as a percentage of the supply voltage, while I_{on} is the maximum discharge current that power-gated cells inject into the sleep transistor during the active mode. The estimation of the active current is not a trivial task. In fact, an erroneous estimation of I_{on} translates to a sub-optimal sleep transistor sizing, which may result in area and power increase or, even worse, in timing violations during the active mode [30]. Considering that the sleep transistor operates in the resistive region and knowing R_{st}, its size can be properly evaluated using Eq. (9.4):

$$\left(\frac{W}{L}\right)_{st} = \frac{1}{R_{st} \cdot \mu_{st} C_{OX}(V_{DD} - V_{th_{st}} - V_{DD} \cdot \alpha_{drop})} \tag{9.4}$$

where W/L is the ratio between width and length of the transistor channel, μ_{st} is the carrier mobility, C_{OX} is the oxide capacitance and $V_{th_{st}}$ is the threshold voltage.

The size of the sleep transistor is strongly influenced by its physical implementation, where carrier mobility, threshold voltage and gate length are key parameters. Hence, for a proper sleep transistor sizing, indicating the type of MOS device is mandatory: pMOS (i.e., *header*) or nMOS (i.e., *footer*). Although both devices can be used as power-switches without distinction (i.e., maintaining the same leakage reduction), nMOS transistors are usually more suitable. In fact, pMOS devices are less leaky than nMOS ones, but they show a lower carrier mobility, which limits their ON-current. As a result, in order to guarantee a certain current capability, pMOS sleep transistors require more silicon area compared to nMOS.

In a typical power-gating approach, switch devices are high-threshold-voltage transistors (i.e., HVT). Since a HVT transistor is less leaky, this choice helps in reducing the total static consumption when the circuit is in stand-by mode. Unfortunately, a larger threshold voltage reduces the current capability of the transistor and, as demonstrated by Eq. (9.4), more area is required to achieve the optimal resistance. On the contrary, with a low-threshold-voltage (i.e., LVT), the silicon area taken by the switch device can be sensibly reduced, but its leakage becomes significant.

Another design parameter that should be taken into account is the gate length. Usually, the sleep transistors are designed with minimum channel length, but for today's nanometric technologies, this implies high leakage currents due to Short Channel Effects (i.e., SCEs) and more power consumption. Increasing the gate length allows to reduce the power consumption [15], but the current sinking capability of the sleep transistor is drastically reduced and more area is required. Moreover, since the majority of the fab processes are optimized for a single channel length (typically, the minimum length), using devices with multiple lengths may drastically increase process variation and sensibly reduce fabrication yield.

As discussed above, the sleep transistor has to be sized in order to limit the IR-drop on the virtual rail when the maximum discharge current of the power-gated cells is injected into the virtual ground. Under this IR-drop constraint, designers can trade off channel length L and threshold voltage V_{th} (see Eq. (9.4)) to achieve area and power optimization. In principle, maximum area efficiency (i.e., minimum W) is obtained using transistors with the largest current density capability, namely, low-V_{th} transistors with minimum gate length. Obviously, this causes larger leakage consumption due to an increase of the sub-threshold current. On the other hand, in order to achieve minimum power overhead, high-V_{th} devices with larger gate length are the best choice. This helps in reducing leakage power consumption at the cost of a larger area. It is worth emphasizing that by increasing the equivalent sleep transistor area, also the load capacitance that the driving circuit has to charge and discharge increases, thus negatively impacting the dynamic power. Clearly, depending on the system constraints, designers can play with the values of W and V_{th} to achieve the required power/performance constraints. Most recent works propose optimum sleep transistor synthesis in which, for a given IR-drop and area constraints, the threshold voltage is selected to minimize the power [29]. To achieve the optimum V_{th}, both body-bias and multi-V_{th} transistors can be exploited.

3.2.3 Power-Gating Strategies

As anticipated in Sect. 3.1, when power gating is applied to a generic logic block, an important dimension of the problem concerns whether the gating is applied to: (1) The entire logic block (block-level or full power-gating); (2) a subset of the cells, typically, the non-timing critical ones (partial or clustered power-gating); (3) all the cells individually (cell-level power-gating). The block-level approach has the drawback of having an high reactivation period since it may have long transition delay between sleep and active state. The cell-level strategy is characterized by high area overhead and huge sleep signals buffering. The most promising power-gating style is then the clustered one, in particular the row-based style, which offers a reduced overhead with respect to the cell-level approach and limited active/sleep transition times with respect to the block level strategy.

3.2.4 Layout Modification

The modifications to the layout required to accommodate sleep transistor insertion depend on the chosen clustering granularity. Figure 9.5 shows an example of a modified layout in the case of row-based clustering; the sleep transistor is placed in a dedicated row indicated as ST row in the figure. Open channels indicate where the ground lines of adjacent rows are split to accommodate for clustering rows which share a common ground line.

In the case of block-level power-gating, the layout must be modified so that the rows are extended on both sides of the layout to make space for the sleep transistor cells. All side pins must be moved and an extra pin must be added to connect the SLEEP signal with the outside. For row-based power-gating, the clustering phase returns a set of rows to be disconnected from the ground and connected to Virtual Ground. If those rows share the ground line with any non-power-gated row they need to be spaced. If two rows sharing the ground line are both to be clustered, we can simply connect the ground line to the Virtual Ground. Besides opening channels,

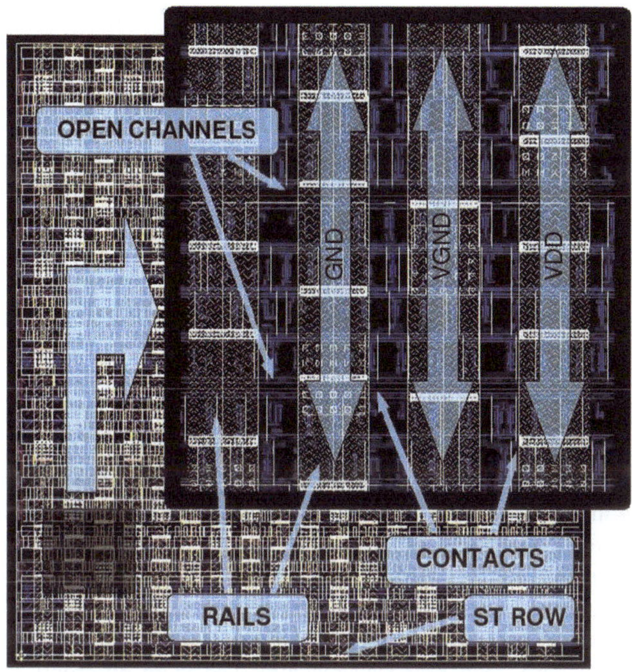

Fig. 9.5 Example of a complete layout after a row-based sleep transistor insertion

extra rows needed to host the sleep transistor cells are added. As in the block-level case the pins have to be moved and the SLEEP pin added.

For any clustering strategy, the power grid and its connection to the standard cells are created. The smallest region that can be power-gated is limited by the distance from the V_{gnd} power rail. The maximum distance from the V_{gnd} is equal to the distance between two power stripes of the same type divided by two. To avoid having regions of different sizes, the rows are always cut in the same points and those are determined by the position of the vertical stripes. Each cut, if needed, is done at the maximum distance from the V_{gnd} lines. To avoid wrong substrate polarizations the distance between two stripes of the same type has been kept equal to the one without power gating, while the width of the stripes has not been changed in order to avoid the IR drop increase.

In the remainder of this chapter, we will consider row-based sleep transistor insertion as the reference power-gating strategy. However, the solutions we present could be successfully adapted to other kinds of sleep transistor insertion approaches.

4 Clustered Tunable Power-Gating for PV Compensation

Parametric variations introduced by manufacturing represent the main cause of performance variability and yield degradation in modern VLSI circuits.

As discussed in Sect. 2, process variations may range from few percent to several orders of magnitude, following a deterministic scheme or a random distribution, also depending on the spatial-scale they reach: Wafer-to-wafer (W2W), die-to-die (D2D) and with-in-die (WID) variations. While the binning method has been successfully adopted for tackling W2W and D2D variations, considering WID variations (which are random, more localized and thus harder to manage) require dedicated counter-measures that must be taken since the early phases of the design flow.

Several design methodologies have been proposed in the last years, from those based on *Design for Manufacturability* approaches [13, 18, 25] (like litho-friendly and Restricted Design Rules layout design), where variability of a given design is either mitigated or amortized, to *Adaptive* strategies, which attempt to solve the variability issues by sensing and correcting the desired parameters using various knobs that affect them. These schemes are also called *Monitor & Control* (M&C) strategies, to emphasize their analogy with closed-loop control systems.

M&C approaches have proven to be extremely efficient and they are usually preferred to other strategies for their flexibility. Different embodiments of the M&C paradigm have been proposed recently. The main differentiating factor is represented by the strategy used to control the circuit. Most solutions use Dynamic Voltage Scaling (DVS) [35], or Adaptive Body Biasing (ABB) [34] as control knobs. Although both DVS and ABB are effective in adjusting circuit performance, as a side-effect they have a dramatic impact on the power consumption of the circuit; in fact, dynamic power is quadratically related to the supply voltage, while sub-threshold leakage current shows an exponential relationship to the body voltage. This makes their implementation energy inefficient, thus less effective for leakage-dominated CMOS technologies.

In this section, we show how power-gating may represent a viable solution to achieve concurrent power reduction and performance control for mitigating process variations (and random WID variation in particular) and increase the timing yield. We show that, when enhanced with tunable features, the sleep transistors can act as natural supply-voltage regulators for the power-gated circuit, thus providing a low-power, yet low-cost, control knob.

4.1 Modeling Process Variations and Timing Yield

WID variations are mostly due to systematic and random variations of several physical device parameters, such as the concentration of doping atoms in the substrate (N_i), the effective channel length (L_{eff}) and width (W_{eff}) and the oxide thickness (T_{ox}). The resulting effect is the shift of the electrical characteristics of

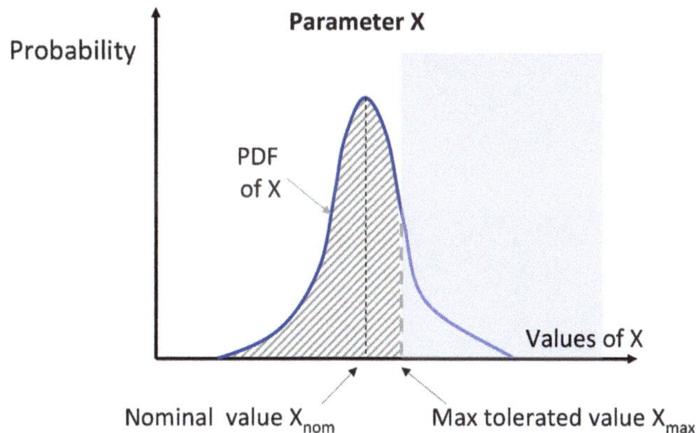

Fig. 9.6 Effects of parameter drifts due to process variations

the transistors, like the threshold voltage (V_{th}) and the maximum current density (I_{ds}/W), with a significant impact on the power dissipation and the performance of a manufactured circuit [6].

Under these conditions, device parameters (and design metrics) have to be treated as random variables, whose probability distribution function (PDF) depends on the actual fabrication process. As pictorially summarized in Fig. 9.6, deviations from the ideal case (i.e., where a given parameter X has a nominal, deterministic value) are represented by a PDF with a given shape with average value $avg(X)$ coincident with the nominal value X_{nom} of the parameter; variability is related to the variance of the PDF (e.g., the range of values between $\pm 3\sigma$). Under the typical assumption that values slightly exceeding the nominal value (X_{max} in the figure) can be considered as acceptable, we define the timing yield as the probability that $X < X_{max}$, which can be immediately obtained by the cumulative distribution function of X (area below the PDF and delimited by X_{max}; striped region in the figure).

4.2 Controlling Performance with Tunable Sleep-Transistors

The active current I_{on}, drained by a circuit from the power supply during normal operation, may represent a suitable metric for quantifying the performance degradation induced by process-variation. In fact, the imperfections of the fabrication process may affect several electrical parameters (such as channel dimensions and threshold voltage), which in turn alter the actual current capability of the active transistors, and thus, the intrinsic speed of the devices. Therefore, the larger the speed degradation due to process variations, the smaller the resulting I_{on}.

During the active periods, the sleep transistor, that is turned-on and that operates in the linear region (in Fig. 9.7, R_{on} represents the channel resistance in the linear region), behaves as a current-to-voltage transducer that transforms the flowing current I_{on} into a voltage drop (V_{drop} in Fig. 9.7). The virtual-ground rail is now

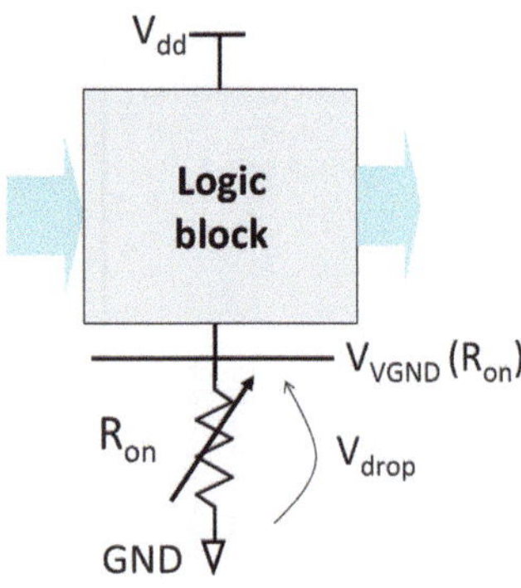

Fig. 9.7 Voltage drop across the sleep transistor during active periods

Table 9.1 Quantitative effect of sleep transistor sizing on delays

	d_{WC} (ps)				
d_{nom} (ps)	W	$2W$	$4W$	$8W$	$16W$
130.2	152.7	147.1	140.8	138.7	134.9

at a potential higher than ground, and the circuit operates at a scaled supply voltage ($V_{dd} - V_{drop}$). By modulating R_{on}, and thus V_{drop}, it is therefore possible to change the operating point of the entire circuit: A lower R_{on} implies a faster circuit; a larger R_{on}, on the contrary, will increase V_{drop} thus making the circuit slower.

In summary, circuits made slower by PV can recover their speed by means of R_{on} reduction, where R_{on} is proportional to the width of the sleep transistor.

The key aspect of this strategy is that the voltage regulation obtained through R_{on} appears only temporarily while the circuit is switching. Once all the internal switchings are completed, I_{on} falls to zero and so does V_{drop}. In this sense, the supply voltage V_{dd} is automatically adapted to the load, and there is no need of an external voltage regulator. A quantitative evaluation of the above dependency is given in Table 9.1, which reports delay values of a sample design (10 power-gated chains in parallel, each consisting of 10 inverters) as a function of the total sleep transistor width, in the presence of process variations.

Column d_{nom} reports the nominal delay when the circuit is power gated with a transistor of a given size W (in the example 20 "fingers" of 0.8 μm each for a total of 16 μm) and ignoring the presence of variations. Columns d_{WC} report worst case delay values for different values of W. By worst case, we mean the extreme value of the delay distribution obtained by randomly selecting 1,000 different instances of the circuit with Monte-Carlo. The column d_{WC}–W shows then the slowest circuit instance (152.7 ps) when a transistor of size W is used for gating, i.e., the same conditions for which d_{nom} is measured. This instance is approximately 17 % slower than the nominal one. If we then upsize the sleep transistor for such circuit instance, we see that, as expected, we can progressively speed up the instance, asymptotically reaching the nominal delay d_{nom}.

4.3 Design Issues and Architectures

4.3.1 Design Issues

As mentioned above, R_{on} modulation allows to compensate delay variations with a simple mechanism and with an arbitrarily fine granularity. However, such powerful control mechanism does not come for free, and designers must take into consideration the amount of overhead introduced by a tunable power-gating architecture.

Upsizing the sleep transistor may result in a significant area overhead, which causes extra static and dynamic power consumption. In fact, the sleep transistor itself leaks when turned off, and its leakage is proportional to its size. Moreover, driving a huge sleep transistor during power-mode transitions would imply additional load capacitance and larger logic effort. Such overhead may nullify the leakage power savings obtained by gating the circuit.

As an example, consider again the data of Table 9.1. If a 5 % delay increase can be tolerated (i.e., $d_{max} = 130.2 \cdot 1.05 = 136.71$ ps), then the sleep transistor must be as large as $16W$ to achieve the required speedup; this will cause the circuit to have a delay of 134.9 ps $< d_{max}$, but a power consumption due to the sleep transistor that is 16 times larger than in the nominal case. Clearly, without a dedicated architecture the use of power-gating as a M&C strategy may not be feasible.

4.3.2 Architectures

As described in Sect. 3, a key design variable in determining the area/power tradeoff of a power-gated circuit is the granularity at which the sleep transistor is inserted. This is true also when considering tunable power-gating architectures.

A first option consists of power-gating the whole circuit using a single tunable sleep transistor (left configuration in Fig. 9.8). In this case, even if the transistor is set to the proper width according to the detected delay value, its size must be the largest of the range W_{max}. Therefore, although effective in mitigating the process variation effects, large area and power overheads can not be avoided.

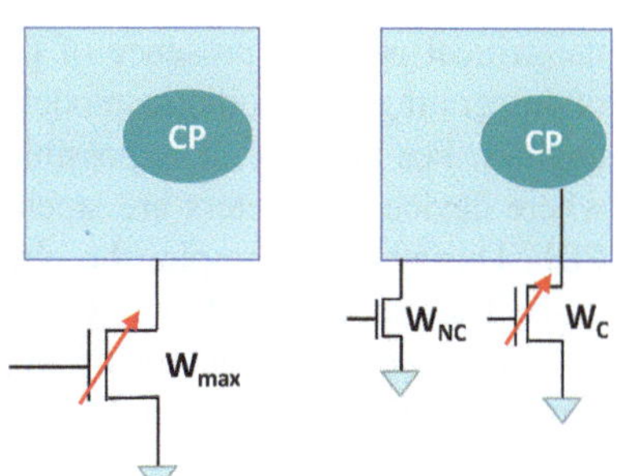

Fig. 9.8 Tunable power-gating (TPG) options: Full-TPG (*Left*) and clustered-TPG (*Right*)

A clustered sleep transistor insertion, on the other hand, may represent the most appropriate architecture. In fact, tunability is required only for the cells that determine the critical paths (CPs). On the contrary, all the other cells (i.e., cells for which the delay increase due to process variations do not slow down the critical paths) can be power gated with a regular, non-tunable, small-sized sleep transistor. We call this approach Clustered-Tunable Power-Gating (right configuration in Fig. 9.8). Under this scheme, two distinct sets of cells (i.e., clusters) use separate sleep transistors: The critical cells identify the critical cluster C, which is gated by a tunable transistor of size W_C; the rest of the cells form the non-critical cluster NC, which is gated by a regular transistor of size W_{NC}. Concentrating tunability only where needed allows to reduce the total sleep transistor width ($W_{NC} + W_C \ll W_{max}$), and guarantees power savings while keeping the same delay compensation capability.

4.3.3 Design Flow and Results

The design methodology for the automated implementation of clustered TPG does not differ from that described in Sect. 3. Starting from a (row-based) placed design, clustering, peak-current estimation, sleep transistor sizing and sleep transistors insertion are the main phases of the flow. However, working with clustered tunable power-gating architectures requires some changes and additions to the algorithms.

Design of the Tunable Sleep Transistor Cell Tunable sleep transistor cells are not available in standard CMOS libraries. Then, it is important to support design kits with new customized cells that contain parallel sleep transistors of different sizes and driven by dedicated control signals. Figure 9.9 shows the schematic of a possible architecture, as described in [14] and [32]. Each parallel sub-transistor is driven by a NAND gate that receives an external configuration bit (*b3, b2, b1, b0*, in Fig. 9.9), and whose value can enable (in case of 1-logic) or disable (in case of 0-logic) the corresponding transistor. An additional sleep signal provided by an external power-management unit is in charge of defining the operating mode of the gated circuit.

Statistical Static Timing Analysis At design time, characterizing the effects induced by process variations is key. In fact, this allows the identification of the critical paths in presence of process variations. To do so, an option consists of integrating probabilistic models into standard Static Timing Analysis (STA) engines. For instance, it is possible to resort to Monte Carlo statistical sampling, where device parameters are stochastic variables described by process-dependent PDF. During each sample, the circuit timing is computed using traditional STA tools. The output of a Monte Carlo analysis includes the path delay distributions. From that, we can extract the list of statistical critical paths, i.e., the list of paths that show a certain probability to have a slack smaller than a user defined threshold.

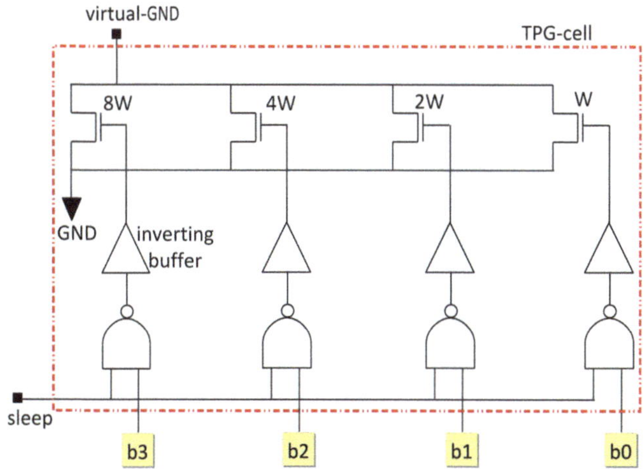

Fig. 9.9 Architecture of the tunable sleep transistor cell [32]

Clustering and Sleep Transistor Sizing The clustering phase is crucial, since it defines which portion of the circuit has to be assigned to which cluster (critical C or non-critical NC). At this stage, it is important to identify the granularity at which the sleep transistors are inserted into the layout. As described in Sect. 3, different options are available, but row-based insertion can guarantee the finer granularity at the minimum cost in terms of area and layout disruption. Hence, the clustering problem comes down to the selection of which rows have to be considered as critical or non-critical. An effective solution to this problem is to mark as critical those rows that host at least one gate belonging to a statistical critical path [32]. Once the clusters C and NC are formed, it is possible calculating their maximum active currents, I_C and I_{NC}, respectively, and performing the sleep transistor sizing which returns the actual width of the two transistors W_C and W_{NC}.

Sleep Transistor Insertion During this stage, the sleep transistor cells belonging to a customized power-gating library are placed into the layout. If a row-based approach is adopted, one can follow the strategy proposed in [31], where the sleep transistors cells are placed in dedicated layout rows, called sleep rows. It is worth mentioning that the sleep transistor of the critical cluster C is implemented by means of modular tunable sleep transistor cells connected in parallel. All the programmable sleep transistor cells are centered in their middle configuration, namely, the configuration word is set in the middle of the dynamic range (1000 for the 4-bits cell shown in Fig. 9.9). This guarantees the maximal extension of the compensation range. For the non-critical cluster one can use the same scheme, but using non-tunable sleep transistor cells of fixed size [8].

Figure 9.10 offers a quantitative analysis of the timing yield and leakage savings that a clustered tunable power-gating architecture (*CLUSTERED-TPG*) may guarantee w.r.t. standard power-gating (*FULL-PG*) and block-level tunable

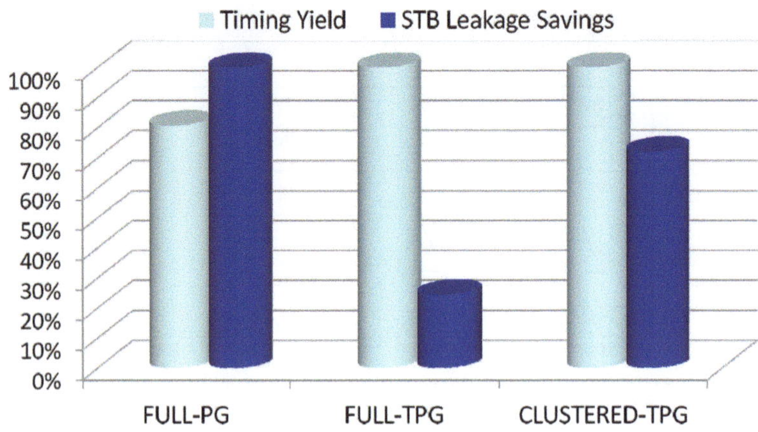

Fig. 9.10 Quantitative comparisons of various power gating options

power-gating (*FULL-TPG*). The results are obtained from the simulation of a subset of the ITC'99 benchmarks mapped onto an industrial 45 nm technology. Only the averages are reported.

The tunable approach (i.e., *CLUSTERED-TPG* and *FULL-TPG*) originates a timing yield increase from 80.67 % (*FULL-PG*) to 100 %; in other terms, tunable power-gating allows to fully recover any speed degradation induced by process variation, thus making the number of circuits that must be discarded because they incur timing violations fall to zero.

Concerning leakage power, we observe that a *FULL-TPG* architecture can successfully tackle the problem of variability compensation at the price of a drastic reduction of the achievable leakage savings; due to a larger sleep transistor (leakage savings go from almost 100 % for the *FULL-PG* to a mere 24.36 % for *FULL-TPG*). On the contrary, a clustered scheme (i.e., *CLUSTERED-TPG*) guarantees the same timing yield as *FULL-TPG*, while maintaining reasonable leakage savings (72.1 %). Clearly, applying a clustered architecture is what makes the tunable approach applicable to real-life circuits.

5 Clustered Power-Gating for NBTI-Induced Aging Minimization

Besides the non-deterministic variations described in Sect. 2, another, and possibly more insidious, type of non-ideality of scaled devices concerns *time-dependent* deviations in their operating characteristics [1].

There are two types of sources of such time-dependent variations: Bias Temperature Instability (BTI), and Hot Carrier Interface (HCI) [26]. Both phenomena cause the generation of traps at the interface between the silicon and the oxide, resulting into an increase over time of the threshold voltage of the transistors.

BTI affects both pMOS (Negative BTI—NBTI) and nMOS transistors (Positive BTI—PBTI); in current technologies, the impact of PBTI is much lower than that of NBTI, although its importance is expected to increase with the adoption of high-k dielectrics in the gate-oxide interface [24]. Conversely, HCI effects are much more significant in nMOS transistors, and are at least two orders of magnitudes larger than for pMOS devices [12].

Between NBTI and HCI, NBTI is regarded as the most significant effect, because the surface along which interface traps are created (i.e., the whole silicon-oxide interface) is much larger than that of HCI (i.e., in the neighborhood of the drain area) [1].

We can thus consider NBTI as the dominant source of aging of transistors in current sub-45 nm bulk CMOS technologies. As we will show in the rest of this section, the peculiar properties of NBTI and, in particular, its state-based manifestation will allow to use power-gating as a powerful knob for aging control while keeping the usual benefits in static power reduction.

5.1 *Background and Models*

NBTI effects occur when a pMOS is negatively biased (i.e., negative V_{gs}, or a logic '0' is applied—the stress state) and originate an increase of the threshold voltage. When a zero-bias voltage is applied (i.e., a logic '1'), NBTI stress is actually removed, resulting in a partial recovery (i.e., a decrease) of the threshold voltage (the recovery state). While there is no full consensus on the exact quantum-mechanical mechanisms that govern the NBTI effects, the reactivation-diffusion model is accepted as accurate enough for pMOS NBTI aging [1]. A simplified version of such a model is the following:

$$\textit{Stress State}: \quad \Delta V_{th} = k_s e^{\frac{-E_a}{kT}} (t - t_{str})^{\frac{1}{4}} \tag{9.5}$$

$$\textit{Recovery State}: \quad \Delta V_{th} = k_r (1 - \sqrt{\frac{t - t_{rec}}{t}}) \tag{9.6}$$

where k_s and k_r are two parameters whose magnitude depends on few technological parameters (such as channel strain and nitrogen concentration), k is the Boltzmann constant, T is the device temperature, E_a is the activation energy, and t_{str} and t_{rcv} denote stress and recovery times, respectively. t is the free variable and expresses the temporal evolution of the threshold voltage drift.

Equations (9.5) and (9.6) show the qualitative behavior of NBTI stress and recovery: V_{th} increases during stress with $t^{\frac{1}{4}}$ dependency, it decreases during recovery with $1 - \sqrt{1/t}$ dependency, and it depends exponentially on temperature during stress. The most relevant feature of the model, however, is that there are two different behaviors based on the *state* of the device. This observation, coupled with the experimental evidence that NBTI aging is *frequency-independent* [1, 22], implies that it is the total stress time that matters rather than the actual dynamics of stress-recovery.

A generic signal applied to a pMOS transistor can then be modeled as a periodic waveform with equivalent amount of stress time, paving the way to a probabilistic modeling of NBTI aging. This allows to lump the models of Eqs. (9.5) and (9.6) into a single macromodel suitable to circuit-level simulation:

$$\Delta V_{th} = K \cdot \beta \cdot t^{\frac{1}{4}} \tag{9.7}$$

where K includes all the technological and environmental (e.g.,temperature, supply voltage) constants and β denotes the stress probability of the signal connected to the gate input of the pMOS transistors, that is, the probability of a logic '0'. This macromodel is also more suitable for translating the increase of threshold voltage into a delay degradation, which better corresponds to the intuition of "aging".

Under the alpha-power law, we can write the delay of a logic gate as:

$$d = \frac{C_L \cdot V_{dd}}{(V_{gs} - V_{th})^{\alpha}} \tag{9.8}$$

where C_L is the load capacitance, V_{gs} is the gate voltage and α is a technology-related exponent that can be approximated to 1 for sub-90 nm technology.

Because of the NBTI-induced threshold voltage increase, the "aged" delay $d' > d$ then becomes:

$$d'(t) = \frac{C_L \cdot V_{dd}}{(V_{gs} - (V_{th} + \Delta V_{th}(t)))} \tag{9.9}$$

which can be expressed as a penalty with respect to d as:

$$d'(t) = d \cdot (1 + \frac{K \cdot (\beta \cdot t)^{1/4}}{V_{GT} - K \cdot (\beta \cdot t)^{1/4}}) \tag{9.10}$$

where $V_{GT} = V_{gs} - V_{th,0}$ and $V_{th,0}$ is the threshold voltage at time 0.

Equation (9.10) allows translating circuit operations (in terms of signal probabilities) to aging (i.e., delay increase over time) for the individual gates, similarly to what it is done for estimating dynamic power based on switching probabilities.

5.2 Power-Gating and Aging Reduction

Equation (9.7) clearly shows that, for a specific manufactured device and for specific operating conditions (temperature and V_{dd}), there is one single knob that can be used for mitigating the aging effects: The stress probability β.

Ideally, one would like to make β as small as possible. This would imply having as many 1's in the logic network as possible for the largest possible fraction of time. Obviously, under normal conditions, this is infeasible, because: (1) To implement meaningful functions, circuits require logic inversion, which by definition prevents achieving arbitrarily small probability for a predetermined signal value (0 or 1). (2) Circuit structure affects probability. On the other hand, information theory suggests that a signal probability of 0 or 1 carries no information—entropy is 0, and it is maximum for a 0.5 probability. Therefore, to implement a realistic function, there must be a fair distribution of 0's and 1's.

Although the ideal objective is impossible to reach, technology-independent and technology-dependent synthesis techniques can be used to *minimize* the 0-probability of internal signals. Kumar et al. [23] proposed multi-level synthesis and technology mapping algorithms that adopt a modified, NBTI-oriented metric, while Wu and Marculescu [19] used logic restructuring and pin reordering to exploit the fact that not all transistors are identically important in determining the delay of a gate in the pull-up network. These strategies can reduce the aging in the *typical* state of the circuit (i.e., determined by the implementation and by the most common input patterns) and achieve a sort of local optimum.

We propose power-gating as a knob for improving over these results, aiming at a global optimum. We suggest leveraging a well-known weakness of power-gating: The logic values of the nodes of a power-gated block are lost when the block is disconnected from the power supply or the ground nets. This poses serious problems when storing values in memory elements and when interfacing power-gated to non power-gated regions. Solutions based on the usage of special types of memory elements [3] or by proper design of the sleep transistor cell [16] do exist.

For the combinational portion of a circuit, the behavior of the internal nodes depends on the type of switch (footer or header) implementing power-gating. If a footer switch is used (and thus the block is disconnected from the ground), as described in Sect. 3, an interesting behavior occurs. Nodes that are at logic value '1' before opening the switch do keep their values, whereas nodes with value '0' become floating. Both the virtual ground line and the '0' nodes then get charged to '1' by the leakage current of the pull-up network of the cells [27]. The speed of this charge process depends on the design of the sleep transistor cell, and can be speed up by using a proper pull-up boosting mechanism, as shown in [7].

For the sake of illustration, Fig. 9.11 plots the signals of a simple two-inverter circuit connected to a footer switch after the latter has been turned off (0.0 on the timescale) [10]. We observe that the signals a and c, originally at logic value '1', stay unchanged, while signal b, at '0' when the sleep signal is activated, goes to '1' quite abruptly, and it reaches about 85 % of V_{dd} in a few tens of nanoseconds.

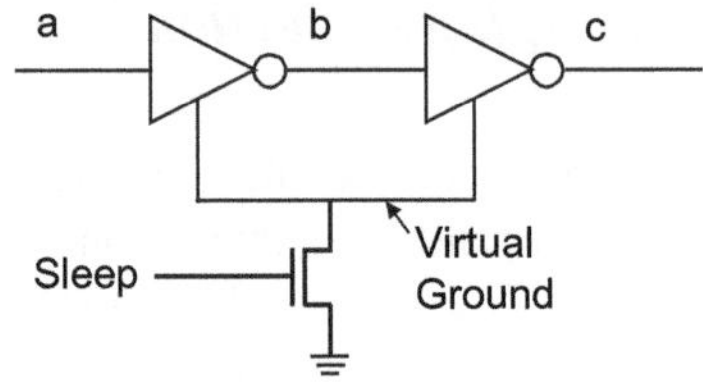

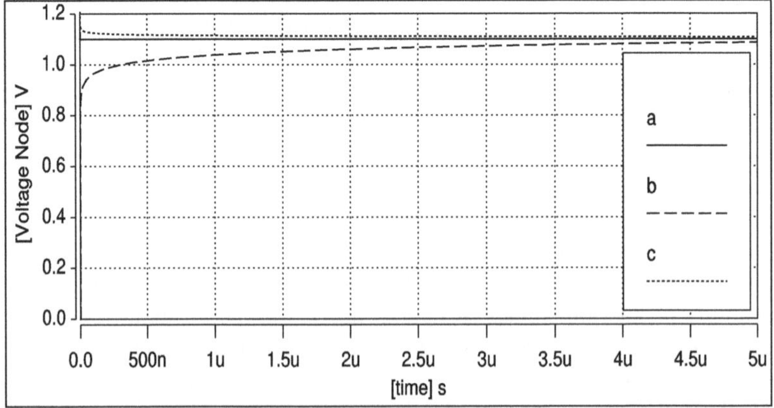

Fig. 9.11 Internal signals during standby intervals [10]

What it matters for our purposes is that all the nodes inside the gated block region (more or less quickly) reach a logic value of '1', that is, *the gated block recovers from aging during the standby intervals*. This is clearly a non-logical state that cannot be obtained by any mechanism that operates on the circuit function or the input data, and is therefore orthogonal to such design approaches.

5.3 Design Issues and Architectures

5.3.1 Design Issues

From the analysis of the previous section, two key issues deserve a deeper evaluation. First, it is evident that the benefit in terms of aging goes together with the potential leakage reductions: If the gated block is put in a standby state sporadically, the aging reduction will be marginal. Clearly, the amount of time spent in the standby state is not a quantity that is controllable by the designer and it is determined by the environment. Therefore, it is important to parameterize the aging in terms of this quantity. Second, the implementation of power-gating does not come for free. As discussed in Sect. 3, the addition of a footer switch in series with the pull-down network increases the on-resistance of each cell and results into a (nominal, i.e., at time zero) delay penalty. Since by tackling aging we are trying to compensate the

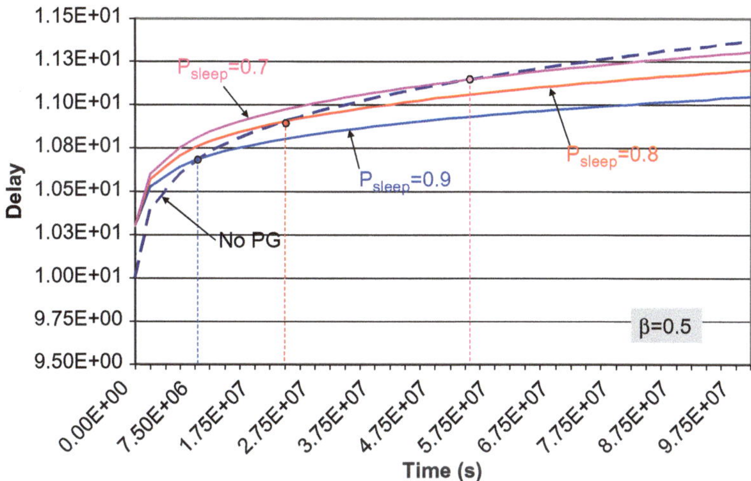

Fig. 9.12 Delay degradation as a function of sleep probability

increase of delay over time, it is fundamental considering the fact that in a power-gated circuit we start from a larger nominal delay value.

Equation (9.10) can be adapted so as to incorporate the above described effects by adding two more parameters: P_{sleep}, the probability of the sleep signal (0 means always active, 1 always in standby), and γ, the delay penalty the due to the addition of the sleep transistor (i.e., $d_s = d \cdot (1 + \gamma)$ is the time-zero delay of the gated block).

The new expression of the delay model of a gated block becomes:

$$d_s'(t) = d_s \cdot (1 + \frac{K \cdot (\beta \cdot (1 - P_{sleep}) \cdot t)^{1/4}}{V_{GT} - K \cdot (\beta \cdot (1 - P_{sleep}) \cdot t)^{1/4}}) \tag{9.11}$$

Notice that the stress probability β is now modulated by the complement of the sleep probability $(1 - P_{sleep})$; this is equivalent to an "effective" stress probability $(\beta' = \beta \cdot (1 - P_{sleep})$, with $\beta' \cdot t$ is now the effective stress time.

Figure 9.12 plots Eqs. (9.10) and (9.11); the curve for $d_s'(t)$ is parameterized with respect to P_{sleep}, and it refers to a specific value of γ. Both curves for $d(t)$ and $d_s'(t)$ refer to a value of $\beta = 0.5$. The plot is qualitative just for the sake of illustration of how the aging profiles change based on the various parameters.

The aging curves for power-gating start at a higher time-zero delay value (d_s), but they exhibit an amount of recovery that is proportional to the percentage of standby time P_{sleep}. The diamonds in correspondence to the intersections between the non power-gated curve and the power-gated ones show the breakeven points, that is, the points in time after which the power-gated circuits start aging more slowly than their non power-gated counterparts. Clearly, the intersection moves earlier in time as P_{sleep} increases, denoting better aging profiles.

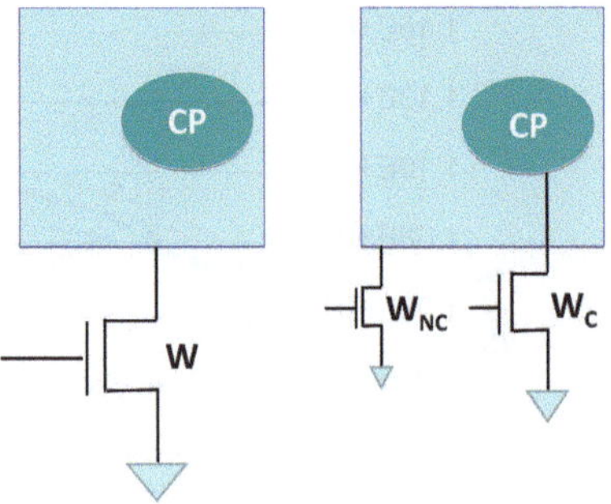

Fig. 9.13 Power-gating options: non-critical gates only (*left*) and clustered (*right*)

There are three parameters that affect the above analysis: P_{sleep}, β, and γ. The first two are not under the designer's control and depend on the functional and environmental characteristics of the circuit; for a given implementation, they assume a well defined value. On the contrary, γ is a design variable; its magnitude is related to the sleep transistor size [2–31]: A larger (smaller) γ implies a smaller (larger) sleep transistor. The ideal case of $\gamma = 0$ would correspond to a transistor of infinite size.

5.3.2 Architectures

The discussion above assumes the entire logic block being power-gated by a single sleep transistor. But, the aging benefits of power-gating are proportional to γ; a large value of d_s may result in a breakeven point which might be beyond the typical lifetime horizon of the circuit. Therefore, it is essential to keep γ as small as possible.

One way for achieving this is to selectively apply power-gating. A first option is *partial power-gating*, in which power-gating is applied only to the non-critical gates of a circuit (left schematic of Fig. 9.13). Assuming a coarse-grain implementation of power-gating, gating only a portion of the circuit requires the interaction with the physical placement of the cells [11, 31]. By not slowing down critical gates, this solution has the same delay as the nominal one ($d_s \equiv d$) and it requires a sleep transistor width, W, smaller than that required for power-gating the entire circuit. However, from the aging viewpoint, partial power-gating is almost identical to the non power-gated case. In fact, keeping the critical gates un-gated means that they will age as in the original circuit. Then, partial power-gating allows achieving leakage reductions at zero delay overhead, but it does not yield any aging benefit.

To get concurrent leakage and aging benefits, we can adopt a clustered architecture similar to the one of Sect. 4. The entire circuit is power-gated using two sleep transistors (see the right part of Fig. 9.13): One for the critical gates (of size W_C) and another one for the non-critical gates (of size $W_{NC} \ll W_C$) [10]. The idea is to decouple the sleep transistor problem by using a small transistor for the non-critical gates (with the only purpose of saving leakage) and a larger one with smaller performance penalty for critical cells. If $W_C \ll W$, with the whole circuit gated, we can achieve leakage savings comparable to the case of full power-gating.

5.3.3 Design Flow and Results

As for the case of PV-aware power-gating, the methodology for an automated implementation of an aging-aware clustered power-gating requires specialized algorithms to be integrated with the standard flow shown in Sect. 3.

Library Characterization Cell libraries are usually not characterized for aging. To this purpose, the designer has to fill look-up tables containing the delay degradation of each cell, parameterized with respect to the static 0-probabilities of the inputs, stress voltage (i.e., V_{gs}), and temperature. This task could be split into two phases by first characterizing the pMOS transistors (and modeling the corresponding ΔV_{th} variation) and then incorporating it into the cell as a negative voltage-source on the gate-terminal of pMOS transistors.

Aging-Aware Timing Analysis This step consists of a probabilistic simulation that uses the parameterized cell delay models and statistics of the input signals (static probabilities) to determine the paths with a delay that is within a given percentage of the nominal critical path d.

Clustering Entails the determination of the critical and non-critical clusters, i.e., two subset of the cells that maximize the leakage and aging benefits. The selection of which cells go in which cluster strongly interacts with the type of sleep transistor insertion strategy. In particular, the granularity of the gating unit coincides with the granularity of the cluster assignment. For instance, if the row-based strategy of Sect. 3 is adopted, clusters consist of *rows*, and therefore the assignment to clusters is done on a row-by-row basis. In [10], the clustering problem is formulated as a 0–1 nonlinear program, where the unknowns are the membership of a gating unit (in that work, layout rows) to one of the two clusters.

Sleep Transistor Sizing This phase is dictated by the timing (γ) constraints, and involves, as for clustering, the estimation of the maximum current drawn by each cluster, as discussed in Sect. 3.

The results of leakage power savings and lifetime extension achieved with the application of the clustered power-gating methodology are quite promising, and they are summarized in Fig. 9.14. Three power-gating schemes are compared: *NO-PG* (no power-gating), *PG* (the entire circuit is power-gated) and *CLUSTERED* (the two-cluster architecture in the right part of Fig. 9.13. Data are normalized with respect to the *NO-PG* case (both leakage and lifetime are assumed to be 1).

The lifetime of a circuit is defined as the time at which the circuit degrades its performance by 15 % beyond its nominal value (i.e., performance of the non

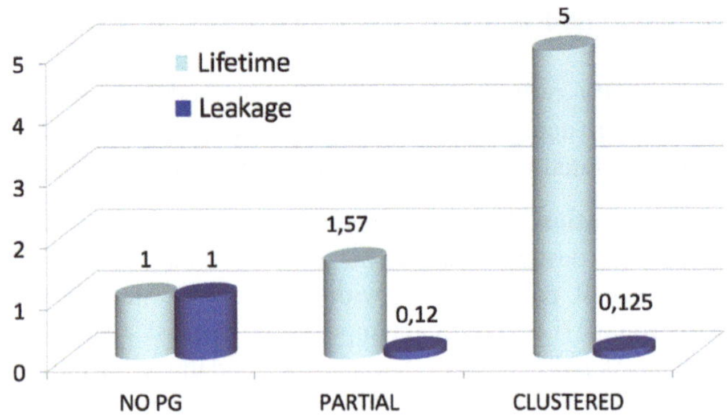

Fig. 9.14 Quantitative comparisons of various power-gating options

power-gated circuit). The results in the chart represent an average over a set of standard benchmarks and some medium-sized industrial designs, and refer to a 45 nm industrial CMOS technology.

We observe that the *PG* scheme does not fully exploit the huge leakage reduction (about 88 %) for lifetime extension; in fact, only a 57 % increase is achieved. Conversely, the *CLUSTERED* architecture yields a 5× increase with respect to the non-gated version, with a negligible loss in leakage reduction (about 0.5 % penalty with respect to *PG*). These numbers demonstrate experimentally the effectiveness of clustered power-gating as a tool for simultaneous leakage power minimization and aging effects mitigation.

Conclusions

There is wide consensus within the electronics industry that the scaling process of the CMOS technology will continue as far as the fundamental barrier of the atom size will be reached. Therefore, CMOS-based circuits and systems represent the future vehicle for digital applications of the next decade. Unfortunately, nano-scale CMOS technologies show intrinsic mechanisms, like internal leakage consumption and parametric variations, which make their use extremely challenging.

In this work, we explored the possibility of exploiting low-power techniques for concurrent leakage optimization and variability compensation, and both static (i.e., due to process variation) and dynamic (i.e., due to NBTI-induced aging mechanisms) variability have been considered. More specifically, we have shown how power-gating, when implemented through a clustered strategy, offers a suited performance control knob to compensate process variations, as well as a natural solution for reducing NBTI effects.

(continued)

New design methodologies have been implemented to support variation-aware clustered power-gating and experimental results, conducted on benchmark circuits mapped onto an industrial 45 nm technology, have highlighted their effectiveness: 100 % of timing yield in the presence of process variations have been achieved, with substantial (i.e., 5×) lifetime extensions w.r.t. non power-gated circuits.

References

1. M. Alam (2008) "Reliability- and process-variation aware design of integrated circuits," in *Microelectronics Reliability*, 48(8):1114–1122.
2. M. Anis, S. Areibi, M. Elmasry (2003) "Design and optimization of multi-threshold CMOS (MTCMOS) circuits," *IEEE Transactions on Computer-Aided Design of Integrated Circuits and Systems*, 22(10):1324–1342.
3. P. Babighian, L. Benini, A. Macii, E. Macii (2006) "Enabling fine-grain leakage management by voltage anchor insertion," *IEEE Design, Automation and Test in Europe (DATE'06)*, pp. 868–873.
4. D. Boning, S. Nassif (2000) "Models of process variations in device and interconnect," in *Design of High Performance Microprocessor Circuits*, Wiley.
5. S. Borkar (2005) "Designing reliable systems from unreliable components: The challenges of transistor variability and degradation," *IEEE Micro*, 25(6):10–16.
6. S. Borkar, T. Karnik, S. Narendra, J. Tschanz, A. Keshavarzi, V. De (2003) "Parameter variations and impact on circuits and microarchitecture," *ACM/IEEE Design Automation Conference (DAC'03)*, pp. 338–342.
7. A. Calimera, A. Pullini, A. Sathanur, L. Benini, A. Macii, E. Macii, M. Poncino (2007) "Design of a family of sleep transistor cells for a clustered power-gating flow in 65nm technology," *ACM/IEEE Great Lakes Symposium on VLSI (GLSVLSI'07)*, pp.501–504.
8. A. Calimera, L. Benini, A. Macii, E. Macii, M. Poncino (2009) "Design of a flexible reactivation cell for safe power-mode transition in power-gated circuits," *IEEE Transaction on Circuits and Systems - Part I, Regular Papers*, 56(9):1979–1993.
9. A. Calimera, E. Macii, M. Poncino (2009) "NBTI-aware power gating for concurrent leakage and aging optimization," *International Symposium on Low Power Electronics and Design (ISLPED'09)*, pp. 127–132.
10. A. Calimera, E. Macii, M. Poncino (2010) "NBTI-aware clustered power gating," *ACM Transactions on Design Automation of Electronic Systems*, 16(1):3.1–3.25.
11. L. Changbo, L. He (2004) "Distributed sleep transistor network for power reduction," *IEEE Transactions on Very Large Scale Integration (VLSI) Systems*, 12(9):937–946.
12. Z. Chen, K. Hess, J. Lee, J. Lyding, E. Rosenbaum, I. Kizilyalli, S. Chetlur, R. Huang (2000) "On the mechanism for interface trap generation in mos transistors due to channel hot carrier stressing," *IEEE Electron Device Letters*, 21(1):24–26.
13. C. Chiang, J. Kawa (2007) *Design for manufacturability and yield for nano-Scale CMOS*, Springer-Verlag.
14. H.-S. Deogun, D. Sylvester, R. Rao, K. Nowka (2005) "Adaptive MTCMOS for dynamic leakage and frequency control using variable footer strength," *IEEE International SOC Conference (SoCC'05)*, pp. 147–150.

15. D. Flynn, R. Aitken, A. Gibbons, K. Shi (2007) *Low Power Methodology Manual*, Springer-Verlag.
16. S. Henzler, G. Georgakos, M. Eireiner, T. Nirschl, C. Pacha, J. Berthold, D. Schmitt-Landsiedel (2006) "Dynamic state-retention flip-flop for fine-grained power gating with small design and power overhead," *IEEE Journal of Solid-State Circuits*, 41(7):1654–1661.
17. ITRS (2009) "Process integration, devices & structures," in *International Technology Roadmap for Semiconductors*, ITRS.
18. T. Jhaveri, V. Rovner, L. Liebmann, L. Pileggi, A. Strojwas, J. Hibbeler (2010) "Co-optimization of circuits, layout and lithography for predictive technology scaling beyond gratings," *IEEE Transactions on Computer-Aided Design of Integrated Circuits and Systems*, 29(4):509–527.
19. W. Kai-Chiang, D. Marculescu (2009) "Joint logic restructuring and pin reordering against NBTI-induced performance degradation," *IEEE Design, Automation and Test in Europe (DATE'09)*, pp. 75–80.
20. J. Kao, S. Narendra, A. Chandrakasan (1998) "MTCMOS hierarchical sizing based on mutual exclusive discharge patterns," *ACM/IEEE Design Automation Conference (DAC'98)*, pp. 495–500.
21. V. Khandelwal, S. Srivastava (2007) "Leakage control through fine-grained placement and sizing of sleep transistors," *IEEE Transactions on Computer-Aided Design of Integrated Circuits and Systems*, 26(7):1246–1255.
22. S. Kumar, C. Kim, S. Sapatnekar (2006) "An analytical model for negative bias temperature instability," *IEEE/ACM International Conference on Computer-Aided Design (ICCAD'06)*, pp. 493–496.
23. S. Kumar, C. Kim, S. Sapatnekar (2007) "NBTI-aware synthesis of digital circuits," *ACM/IEEE Design Automation Conference (DAC'07)*, pp. 370–375.
24. S. Kumar, C. Kim, S. Sapatnekar (2009) "Adaptive techniques for overcoming performance degradation due to aging in digital circuits," *IEEE Asia and South Pacific Design Automation Conference (ASPDAC'09)*, pp. 284–289.
25. M. Lavin, F. L. Heng, G. Northrop (2004) "Backend CAD flows for restrictive design rules," *IEEE/ACM International Conference on Computer-Aided Design (ICCAD'04)*, pp. 739–746.
26. S. Mahapatra, D. Saha, D. Varghese, P. Kumar (2006) "On the generation and recovery of interface traps in mosfets subjected to NBTI, FN, and HCI stress," *IEEE Transactions on Electron Devices*, 53(7):1583–1592.
27. E. Pakbaznia, F. Fallah, M. Pedram (2008) "Charge recycling in power-gated CMOS circuits," *IEEE Transactions on Computer-Aided Design of Integrated Circuits and Systems*, 27(10):1798–1811.
28. K. Roy, S. Mukhopadhyay, H. Mahmoodi-Meimand (2003) "Leakage current mechanisms and leakage reduction techniques in deep-submicrometer CMOS circuits," *Proceedings of the IEEE*, 91(2):305–327.
29. A. Sathanur, A. Pullini, L. Benini, A. Macii, E. Macii, M. Poncino (2008) "Optimal sleep transistor synthesis under timing and area constraints," *ACM/IEEE Great Lakes symposium on VLSI (GLSVLSI'08)*, pp. 177–182.
30. A. Sathanur, L. Benini, A. Macii, E. Macii, M. Poncino (2011) "Fast computation of discharge current upper bounds for clustered power-gating," *IEEE Transactions on Very Large Scale Integration (VLSI) Systems*, 19(1):146–151.
31. A. Sathanur, L. Benini, A. Macii, E. Macii, M. Poncino (2011) "Row-based power-gating: A novel sleep transistor insertion methodology for leakage power optimization in nanometer CMOS circuits," *IEEE Transactions on Very Large Scale Integration (VLSI) Systems*, 19(3):469–482.
32. L. D. Silva, A. Calimera, A. Macii, E. Macii, M. Poncino (2011) "Power efficient variability compensation through clustered tunable power-gating," *IEEE Journal of Emerging and Selected Topics in Circuits and Systems*, 1(3):242–253.

33. D. Sylvester, K. Agarwal, S. Shah (2008) "Variability in nanometer CMOS: Impact, analysis, and minimization," *Integration - The VLSI Journal*, 41(3):319–339.
34. J. Tschanz, J. Kao, S. Narendra, R. Nair, D. Antoniadis, A. Chandrakasan, V. De (2002) "Adaptive body bias for reducing impacts of die-to-die and within-die parameter variations on microprocessor frequency and leakage," *IEEE Journal of Solid-State Circuits*, 37(11):1396–1402.
35. J. Tschanz, S. Narendra, R. Nair, V. De (2003) "Effectiveness of adaptive supply voltage and body bias for reducing impact of parameter variations in low power and high performance microprocessors," *IEEE Journal of Solid-State Circuits*, 38(5):826–829.

Chapter 10
Soft Error Rate and Fault Tolerance Techniques for FPGAs

Fernanda Kastensmidt and Ricardo Reis

Abstract Different fault tolerance techniques can be applied to FPGAs according to their type of configuration technology, architecture and target operating environment. This chapter will present a set of fault mitigation techniques for SRAM, FLASH and ANTIFUSE-based FPGAs and a test methodology to characterize those FPGA under radiation. Results from neutron-induced faults will be presented and compared.

1 Introduction

Integrated circuits operating in radiation environment are sensitive to transient faults caused by the interaction of charge particles with the silicon [1]. At ground level, the main source of radiation are neutron particles that interact with the material provoking secondary particles such as alpha particles and muons that can ionize the silicon provoking transient upset in circuits fabricated in nanometer technology [2]. The interaction of the charged particles with the transistor may provoke transient and permanent effects. The effects that are caused by a single event interaction are called Single Event Effects (SEE) and they can be transient as the Single Event Upset (SEU) and Single Event Transient (SET) or permanent as single event latchup (SEL), single event gate rupture (SEGR), or single event burnout (SEB) [3]. The effect that is caused by accumulation of particle interaction is named Total Ionizing Dose (TID) and it represents degradation in the performance of the transistors as it modifies the threshold voltage and leakage current [4]. In this chapter, we focus on transient effects caused by a single event as SEU and SET.

Field Programmable Gate Array (FPGA) components are very attractive for aerospace applications [5], as well for many applications at ground level that require a high level of reliability, as automotive, bank servers, processing farms, and others.

F. Kastensmidt • R. Reis (✉)
Instituto de Informática, Universidade Federal do Rio Grande do Sul (UFRGS), Porto Alegre, Brazil
e-mail: fglima@inf.ufrgs.br; reis@inf.ufrgs.br

R. Reis et al. (eds.), *Circuit Design for Reliability*,
DOI 10.1007/978-1-4614-4078-9_10

The high amount of resources available in programmable logic devices can be applied to add flexibility to the on-board computer in satellites and to the automotive industry, for example. As FPGAs can be configured in the field, design updates can be performed until very late in the development process. In addition, new applications and features can be configured after a satellite is launched, or updated in hash environments. However, programmable devices have been found to be very sensitive to radiation. It is fundamental to experimentally measure the soft error rate of the available resources, as well as the output error rate of specific applications, to evaluate their applicability in harsh environments.

The characterization of those programmable components is mandatory to sustain its applicability under transient faults. The test methodology and characterization of FPGAs under radiation will be presented. The fault mitigation methodology to protect a design targeting those FPGAs will also be shown. Practical radiation ground testing results under neutron for SRAM-based FPGAs and Flash-based FPGAs will be presented and discussed.

2 FPGAs Under Soft Errors

Field-Programmable Gate Arrays (FPGAs) are configurable integrated circuit based on a high logic density regular structure, which can be customizable by the end user to realize different designs [6]. The FPGA architecture is based on an array of logic blocks and interconnections customizable by programmable switches. Several different programming technologies are used to implement the programmable switches. There are three types of such programmable switch technologies currently in use: SRAM, where the programmable switch is usually a pass transistor or multiplexer controlled by the state of a SRAM bit (SRAM based FPGAs); Antifuse, when an electrically programmable switch forms a low resistance path between two metal layers (Antifuse based FPGAs); and EPROM, EEPROM or FLASH cell, where the switch is a floating gate transistor that can be turned off by injecting charge onto the floating gate.

Customizations based on SRAM are volatile. This means that SRAM-based FPGAs can be reprogrammed as many times as necessary at the work site and that they loose their contents information when the memories are not connected to the power supply. The antifuse customizations are non-volatile, so they hold the customizable content even when not connected to the power supply and they can be programmed just once. Each FPGA has a particular architecture. Programmable logic companies such as Xilinx, MicroSemi, Aeroflex (licensed for Quicklogic FPGAs), Atmel and Honeywell (licensed for Atmel FPGAs) offer radiation tolerant FPGA families. Each company uses different mitigation techniques to better take into account the architecture characteristics.

2.1 Single Event Effects on SRAM-Based FPGAs

The SRAM-based FPGA is composed of an array of configurable logic blocks (CLB), a complex routing architecture, an array of embedded memories (Block RAM), an array of digital signal processing components (DSP) and a set of control and management logic. The CLBs are composed of Look-up Table (LUT) that implements the combinational logic, and flip-flops (DFF) that implements the sequential elements. The routing architecture can be very complex and composed of millions of pre-defined wires that can be configured by multiplexers and switches to build the desirable routing.

The configuration of all CLBs, routing, Block RAMs, DSP blocks and IO blocks is done by a set of configuration memory bits called bitstream. According to the size of the FPGA device, the bitstream can contain millions of bits. The memory bits that store the bitstream inside the FPGA is composed of SRAM memory cells, so they are reprogrammable and volatile. When an SEE occurs in the configuration memory bit of an SRAM-based FPGA, it can provoke a bit-flip. This bit-flip can change the configuration of a routing connection or the configuration of a LUT or flip-flop in the CLB. This can lead to a dramatic impact in the designed circuit, since an SEE may change its functionality.

An SEE in the configuration memory bits of an SRAM-based FPGA has a persistent effect and it can only be corrected when a new bitstream is loaded to the FPGA. In the combinational logic, the effect of an SEE is related to a persistent fault (zero or one) in one or more configuration bits of a LUT. Figure 10.1 exemplifies an SEU occurrence in a LUT configuration bit and in a bit controlling a routing connection. SEE in the routing architecture can connect or disconnect a wire in the matrix. This is also a persistent effect and its effect can be a modification in the mapped circuit, as a logic change or a short circuit in the combinational logic implemented by the FPGA. It can take a great number of clock cycles before the persistent error is detected and recovery actions are initiated, as the load of a faulty-free bitstream. During this time, the error can propagate to the rest of the system.

Bit-flips can also occur in the flip-flop of the CLB used to implement the user's sequential logic. In this case, the bit-flip has a transient effect and the next load of the flip-flop will correct it.

2.2 Single Event Effects on Flash-Based FPGAs

Well-known Flash-based FPGAs are from Microsemi, which presents the families PROASIC3 and SmartFusion [7]. The reconfigurable array is composed of VersaTiles and routing resources that are programmable by turning ON or OFF switches implemented by floating gate (FG) transistors (NMOS transistor with a stacked gate). The FG switch circuit is a set of two NMOS transistors: (1) a sense transistor to program the floating gate and sense the current during the threshold

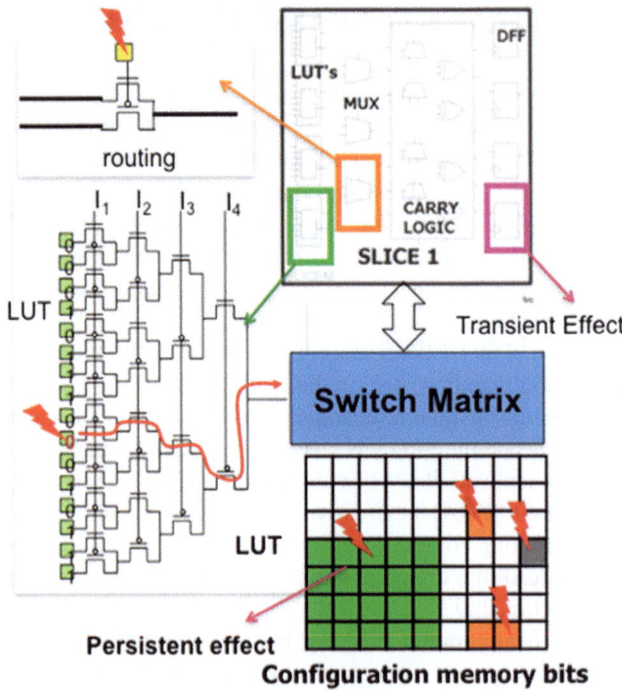

Fig. 10.1 Example of an SEU occurrence in a LUT and in the routing of an SRAM-based FPGA

Fig. 10.2 SET in the Flash-based FPGA programmable switch

voltage measurement and (2) a switch transistor to turn ON or OFF a data-path in the FPGA (Fig. 10.2). The two transistors share the same control gate and floating gate. The threshold voltage is determined by the stored charge in the FG. Figure 10.3 illustrates VersaTiles used to implement some common logic gates. The VersaTiles are connected through a four-level hierarchy of routing resources: ultra-fast local resources; efficient long-line resources; high-speed, very-long-line resources; and the high-performance VersaNet networks.

Each VersaTile can implement any 3-input logic functions, which is functionally equivalent to a 3-inputs Lookup Table (3-LUT). But it is important to highlight that the electrical implementation of the VersaTile is totally different than the electrical implementation of a Lookup Table (LUT). Hence, the VersaTile may

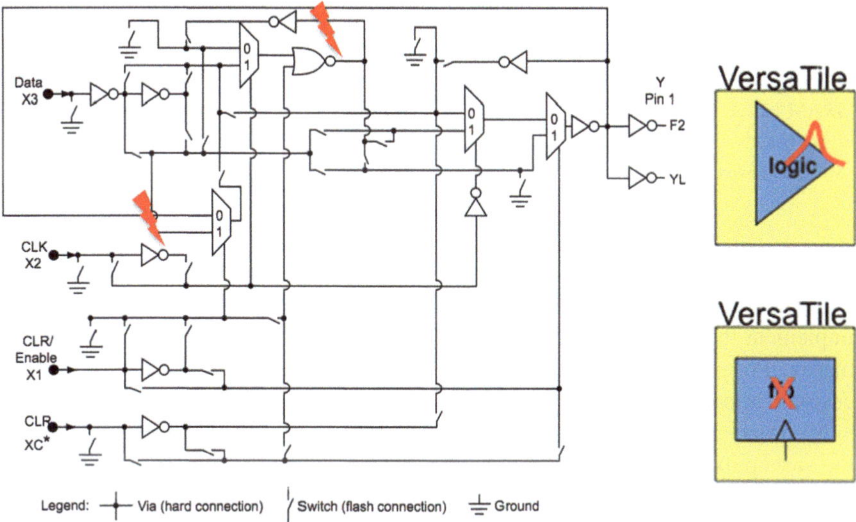

Fig. 10.3 SET and SEU in the Flash-based FPGA VersaTile

have a different electrical behavior to variability effects with respect to a 3-inputs LUT. The VersaTile can also implement a latch with clear and reset, or D flip-flop with clear or reset, or enable D flip-flop with clear and reset by using the logic gate transistors and feedback paths inside the VersaTile block. For each configuration in the VersaTile block, the number of FG switches and transistors in the critical path changes. Single Event Transient (SET) pulses can hit the drain of the transistor at OFF state as presented in Fig. 10.2 provoking a transient pulse in the configuration switches. Or it can hit the sensitive nodes of the transistors in the VersaTile provoking SET or bit-flip according to the customization of the tile (Fig. 10.3).

2.3 Single Event Effects on Antifuse-Based FPGAs

Antifuse-based FPGAs consists of a regular matrix composed of combinational (C-cells) and sequential (R-cells) surrounding by regular routing channels. One of well-known antifuse-based FPGAs is from Microsemi. All the customizations of the routing and the C-cells and R-cells are done by an antifuse element (programmable switch). Results from radiation ground testing have shown that programmable switches either based on ONO (oxide–nitride–oxide) or MIM (metal–insulator–metal) technology are tolerant to ionization and total dose effect [8]. Therefore, the customizable routing is not sensitive to SEU, only combinational logic and the flip-flops used to implement the design user sequential logic are sensitive to SEE.

Table 10.1 Summary of SEU and SET effects in FPGAs

FPGA	SEU/SET in the logic of the configuration basic block	Routing connections	Configurable switches
SRAM-based	Persistent	Persistent	Persistent
Flash-based	Transient	No	No
Antifuse-based	Transient	No	No

Another well known antifuse-based FPGA is from Aeroflex and QuickLogic. Its architecture is composed of a regular matrix of configurable logic cells used to implement the combinational logic and flip-flops, surrounding by a regular routing matrix. Programmable switches called ViaLink connector are used to do all the customizations.

In order to summarize the SEU and SET effects in FPGAs, Table 10.1 shows the susceptible parts of the architectures and classifies the effects as transient or persistent, when it is needed reconfiguration to correct the fault.

3 Fault Tolerance Techniques for FPGAs

Different fault tolerance techniques can be applied to FPGAs according to their type of configuration technology, architecture and target operating environment [6]. Techniques can be implemented by the user at hardware description language (HDL) of the design before the design is synthesized into the FPGA. Or techniques can be developed by the vendor, which provides a FPGA that is SEE robustness by layout. Figure 10.4 illustrates these two options. Here we will focus on techniques that can be applied by the user at the HDL design.

The main techniques are either based on spatial redundancy or temporal redundancy [9]. Spatial redundancy is based on the replication of n times the original module building n identical redundant modules, where outputs are merged into a voter. Usually n is an odd number. The voter decides de correct output by choosing the majority of the equal output values. The most common case of n-modular redundancy (nMR) is when n is equal to 3, where it is called Triple Modular Redundancy (TMR). In this case, a majority voter is used that is able to vote out 2 out of 3 values that are fault free. There is local TMR when only the flip-flops are triplicated or global TMR where all the combinational and sequential logic is triplicated. The TMR can be implemented in different ways by using large grain TMR, or breaking into small blocks and adding extra voters. Each one can protect SEU or SET, or both, as shown in Table 10.2.

When dealing with the routing, different techniques can be chosen to increase or decrease fan-out, delay and set of connections, which may have a different impact in the SEE sensitivity [10, 11]. Also embedded processors can use different mitigations based on software redundancy, or processor redundancy like lock-step and recomputation.

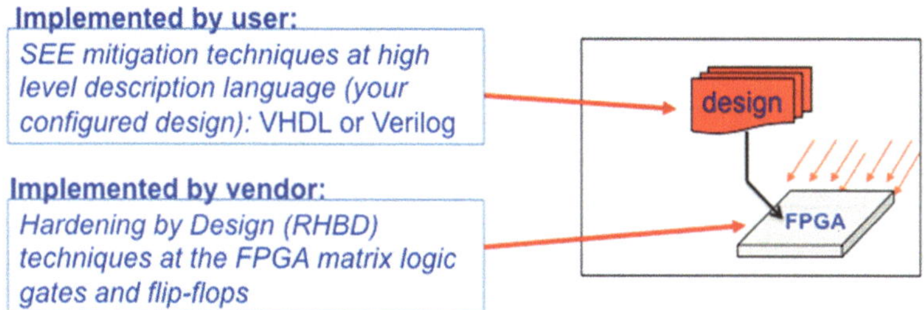

Fig. 10.4 Probability of a bit-flip as a function of the mean deposited charge by an incoming particle for a 6T-SRAM cell before and after stress

Table 10.2 List of mitigation techniques that can be applied by the User in Designs targeting FPGAs

Mitigation technique	Abstraction level	SET	SEU
Local TMR	HDL		X
Global TMR	HDL	X	X
Large grain TMR	HDL	X	X
Voter insertion	HDL	X	X
Reliability-Oriented place and Route Algorithm	FPGA Flow	X	X
Temporal redundancy	HDL	X	
Embedded processor redundancy	HDL	X	X
Scrubbing	System		X

Time redundancy is based on capturing a value twice or three times in time to vote out a transient fault. The values are shifted by a delay [9]. The idea is to be able to capture 2 out of 3 upset free values to be able to mask the fault.

3.1 *SRAM-Based FPGAs*

In SRAM-based FPGAs, radiation-induced faults have a persistent effect so spatial redundancy is needed to mask the upset combined with reconfiguration to correct the fault in the configuration memory bits (bitstream). Figure 10.5 show the flow, the design can be either protected by Global TMR (called XTMR by Xilinx) using the XTMR tool or implemented by hand, and full or partial reconfiguration, called scrubbing, must be applied to correct the upsets from time to time. Fault injection can be used to evaluate the efficiency of the fault tolerance technique and ensure its masking capability under single upsets.

The global TMR or XTMR consist on triplicating all the combinational and sequential logic, input, outputs and clock trees as illustrated in Fig. 10.6. Note that the majority voter can be applied after flip-flops in throughput logic, or after flip-flop in a feed back path. In this second case, it is mandatory to have majority voters able

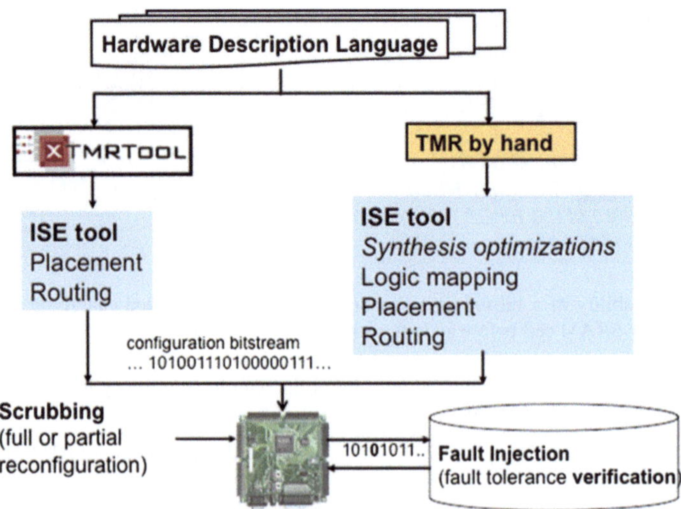

Fig. 10.5 Fault tolerance design flow for SRAM-based FPGAs

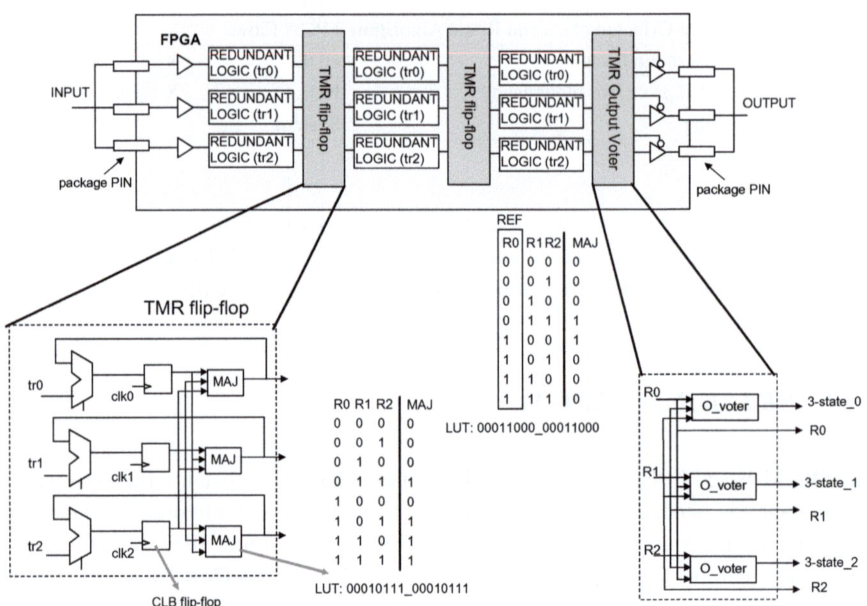

Fig. 10.6 XTMR technique example

to correct the bit-flip of the flip-flops to avoid accumulation of errors. The voter used in the output is based on a minority voter, where it blocks the output if this one is different to the other two.

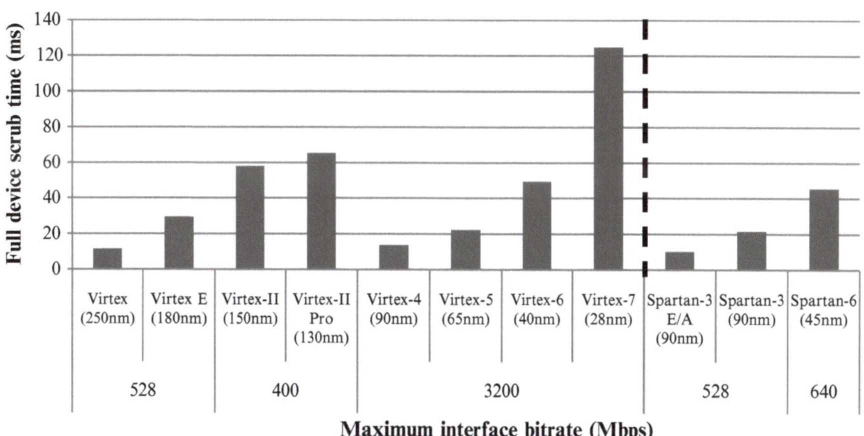

Fig. 10.7 Full configuration or scrubbing time of different SRAM-based FPGAs from Xilinx

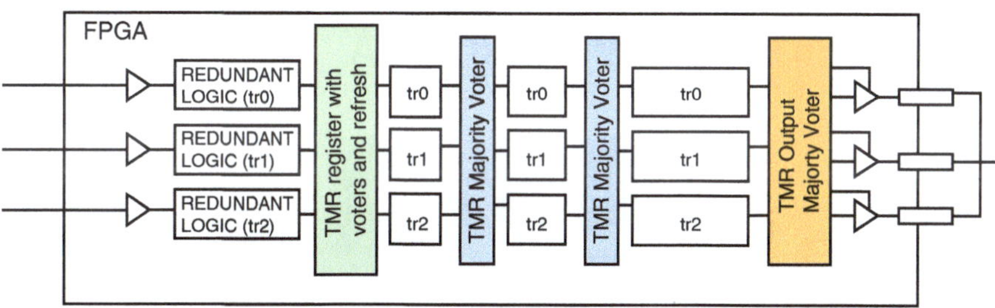

Fig. 10.8 Voting insertion in XTMR or Global TMR

Scrubbing or reconfiguration that can be full or partial can be performed by the internal block called ICAP or by the SelectMap interface. Scrubbing is mandatory to correct the upsets (bit-flips) in the configuration memory bits. Figure 10.7 shows the time for scrubbing in different FPGA families. The time for performing full scrubbing can be significant in large FPGAs, more than 50 ms. This means that some applications will run many clock cycles between two corrections of scrubbing. Consequently, spatial redundancy techniques are very important to ensure the fault masking at the output between scrubbings.

Scrubbing can be costly also in terms of power. In average, the scrubbing rate should be at least three times faster than the upset rate of the FPGA. In order to reduce the scrubbing rate, it is necessary to ensure that the spatial redundancy technique, such as XTMR for example, can tolerate more than a single fault.

Some improvements can be done as playing with different TMR granularities by voting insertion, as shown in Fig. 10.8 [12].

Another option is to use Diversity Triple Modular Redundancy (DTMR) to increase the tolerance to multiple and accumulated upsets in the configuration memory bits. A DTMR system is designed through the association of three

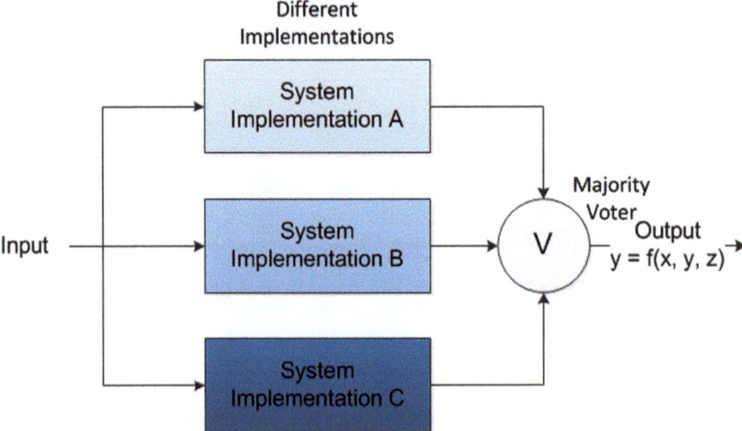

Fig. 10.9 Diversity Triple Modular Redundancy (DTMR)

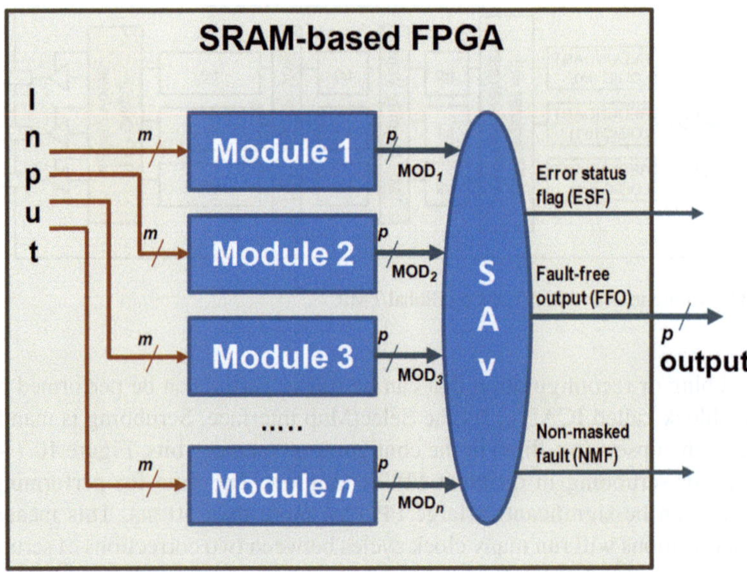

Fig. 10.10 nMR-based technique with SAV voter

diversified copies to a majority voter [13]. Figure 10.9 illustrates an example of a DTMR system. Each copy can be implemented by using a different architecture, or different processors, different logic granularities and others.

Another option to increase the masking capability to multiple upsets is to use nMR-based technique. The nMR is composed of n functionally identical modules, which receive the same m-bits input and deliver p-bits output to the Self-Adapted voter (SAv), Fig. 10.10 [14]. The SAv receives $n \times p$ bits from all modules and

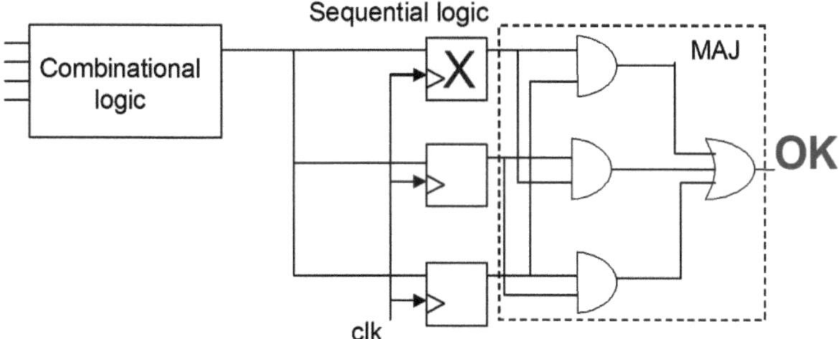

Fig. 10.11 Local TMR to mitigate SEU in the flip-flops

generates the fault-free p-output, n error status flags (ESF), and a non-masked fault signal (NMF). In this scheme, the system allows for the accumulation of defective modules, while remaining at least two modules without fault. SAv is a majority voter, considering as population fault-free modules.

3.2 *FLASH-Based and Antifuse-Based FPGAs*

The configurations of flash-based and antifuse-based FPGAs are not sensitive to radiation. So, faults as SET and SEU can only occur in the combinational and sequential logic and they have a transient effect. So, well-known techniques used in Application Specific Integrated Circuits (ASICs) can be applied in those FPGAs.

In case SEU can be observed in the user's flip-flops of the FPGA, a technique based on local TMR, where only the flip-flops are triplicated and vote, can work well, Fig. 10.11.

In case SET is an issue and also SEU in the configurable flash memory cells, a technique based on global TMR, where all the combinational logic and flip-flops are triplicated and vote, can work well, as is illustrated in Fig. 10.12. Or the designer can choose on using a temporal filtering technique, where the SET will be filtered by the added chosen delay in the clock trees or in the logic path, Fig. 10.13. In this case, the idea is that each flip-flop will capture the data at a different moment, and at least two flip-flops out or three will have the fault-free value.

4 Radiation Test Methodologies to Predict and Measure SER in FPGAs

The test of FPGAs under radiation depends on a test plan developed for each type of FPGA and design architecture. Here we will detail the radiation test for SRAM-based FPGAs. There are two types of tests: the static test and the dynamic test.

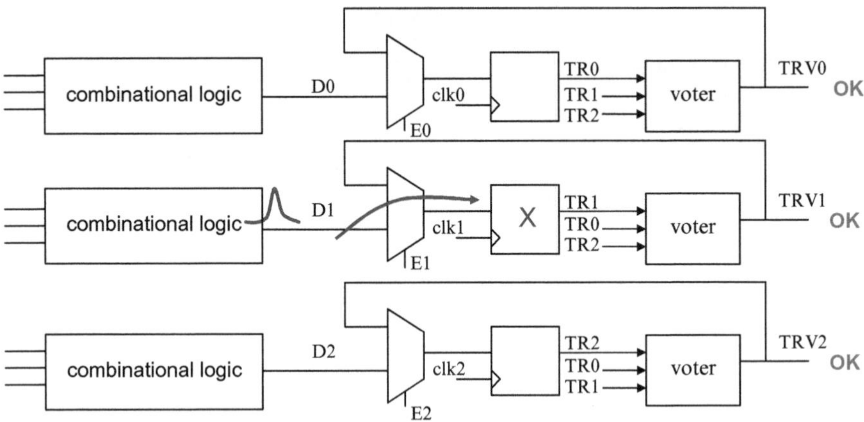

Fig. 10.12 Global TMR as XTMR

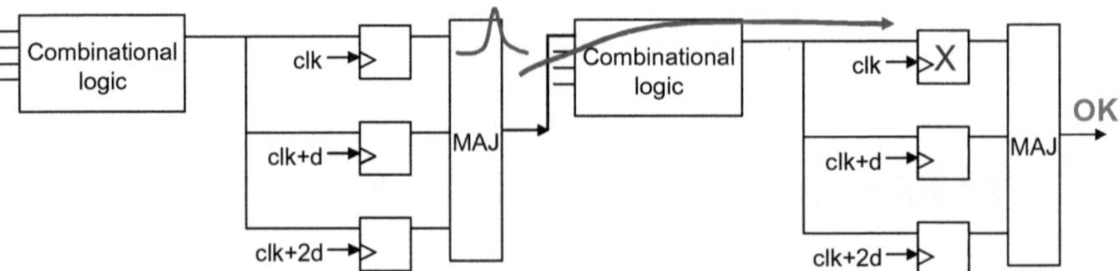

Fig. 10.13 Temporal Redundancy or SET Filtering Technique with delay d applied in the clock trees

The static test experiment consists of configuring the FPGA with a *golden* bitstream containing the test-design and then constantly read back the FPGA configuration memory with the Xilinx iMPACT tool through the JTAG interface. In the experiment control computer, the *golden* bitstream is compared against the *readback* bitstream. If differences are found, the FPGA is reconfigured with the *golden* bitstream and the differences are stored in the computer. Faults are defined as any bit-flip in the configuration memory detected by the *readback* procedure. In this case, it is possible to calculate the upset rate in the configuration memory bits for that specific particle flux, expressed by particles/s/cm^2.

The cross-sections (σ) is the sensitive area of a circuit, where one particle may cause an upset event, and it is calculated using Eq. (10.1), where fluence = flux × time, expressed by particles/cm^2. Cross-section is expressed in cm^2.

$$\sigma = \frac{\#events}{fluence} \tag{10.1}$$

There are two types of cross-section: static and dynamic cross-section. Static cross-section, also known as device cross-section (σ device) is defined as the ratio between the number of upsets in the configuration memory bits of the SRAM-based FPGA (events) and the fluence of hitting particles. Usually, it is used a normalized cross-section or bit cross-section (σ bit), where the cross-section is divided by the total amount of configuration memory bits. Static cross-section quantifies the sensitivity of the FPGA technology to a specific radiation source.

On the other hand, dynamic cross-section is defined as the ratio between the number errors observed at the output at design configured into the SRAM-based FPGA (events), divided by the fluence of hitting particles. Dynamic cross-section quantifies the sensitivity of the implemented design application to any specific radiation source. The rate at errors is defined as the soft error rate (SER). In this case, the expected error rate is much lower than the static test. Based on the Xilinx Reliability Report [15], in average it is necessary 20 upsets in the configuration memory bits to provoke one error in the design output. This relation may of course vary according to the logic density, mapping, routing and the chosen architecture for the design. This number can triplicate when XTMR is used. And the relation can increase even higher, 6 or more times, if DTMR or nMR is used.

Notice that SER is proportional to both device size sensitivity (cross-section) and flux as shown in Eq. (10.2).

$$SER = flux \times \sigma_{SEU} \tag{10.2}$$

When a charged particle (as neutron, protons or heavy ions) hits a device, part of its charge is deposited in the device. This is known as linear energy transfer (LET) and expresses the energy loss per unit length (dE/dx) of a particle and is a function of the mass and energy of the particle as well as the target material density. The units of LET are commonly expressed as MeV–cm^2/g. Radiation experiments with charged particles commonly relate the relation between cross-section and LET, which also depends on the incidence angle.

Fig. 10.14 illustrates a setup experiment under neutron at ISIS Facility in United Kingdom, where we could measure the static and dynamic cross-section of SRAM-based FPGAs and later on by knowing the sea level neutron flux, we can infer the SER of the device and circuit application running at ground level.

For example, the device was exposed for 22 min in a neutron flux of 4.11×10^4 and 70 events were counted. The fluence can be calculated and consequently by using Eq. (10.1), the cross-section obtained is 1.29×10^{-6} cm^2. And by using Eq. (10.2), the SER is 5.3×10^{-2} errors per second. By knowing particles flux at see level is 13 neutrons/cm^2/h, the expected error rate at sea level can be infer to 1.29×10^{-6} cm^2 $\times$ 13 neutrons/cm^2/h $= 1.67 \times 10^{-5}$ errors per hour.

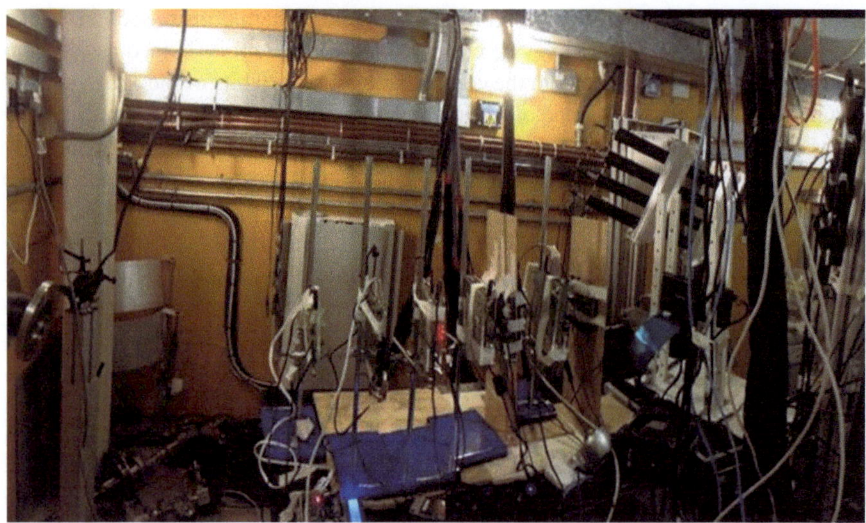

Fig. 10.14 Neutron-induced SEE in FPGAs Experiment Setup

Conclusions

Each type of FPGAs presents a different susceptibility to soft errors and according to its upset rate and architecture, distinct fault tolerance techniques may be chosen. For experiments under neutron-induced faults, it is possible to calculate the static and dynamic cross-section of SRAM-based FPGAs and to analyze its tolerance to multiple and accumulated upsets. The static cross-section can be measured for each SRAM-based FPGA device and the dynamic cross-section must be measured for each fault tolerant design implemented in the FPGA. For a certain device with a specific static cross-section, there will be many dynamic cross-sections according to the fault tolerance technique chosen.

References

1. J. Barth, "Applying Computer Simulation Tools to Radiation Effects Problems", in: IEEE Nuclear Space Radiation Effects Conference Short Course, NSREC, 1997.
2. O. Flament, J. Baggio, C. D"hose, G. Gasiot, J.L. Leray, "14 MeV neutron-induced SEU in SRAM devices," *In Nuclear Science, IEEE Transactions on*, vol. 51, no. 5, pp. 2908–2911, Oct. 2004.
3. M. Berg, "Fault tolerance implementation within SRAM based FPGA designs based upon the increased level of single event upset susceptibility," On-Line Testing Symposium, IOLTS 2006.

4. P. E. Dodd and L. W. Massengill, "Basic mechanisms and modeling of single-event upset in digital microelectronics," *IEEE Trans. Nucl. Sci.*, vol. 50, no. 3, pp. 583–602, Jun. 2003.
5. T. R. Oldham, F. B. McLean, "Total Ionizing Dose Effects in MOS Oxides and Devices," IEEE Transactions on Nuclear Science, vol. 50, no. 3, 2003. pp. 483-498.
6. F. L. Kastensmidt, R. Reis, L. Carro, Fault-Tolerance Techniques for SRAM-Based FPGAs (Frontiers in Electronic Testing), Springer, 2006.
7. Actel. ProASIC3, IGLOO and SmartFusion Flash Family FPGAs Datasheet. [Online]. Available: http://www.actel.com/documents/PA3_HB.pdf and http://www.actel.com/documents/IGLOO_HB.pdf
8. Wang, J.J., RTAXS Single Event Effects Test Rep., Aug. 2004 [available on-line at http://www.actel.com/documents/RTAXS_SEE_Report.pdf]
9. Anghel, L., Alexandrescu, D., Nicolaidis, M., Evaluation of a soft error tolerance technique based on time and/or space redundancy, in the Proceedings of Symposium on Integrated Circuits and Systems Design, SBCCI, 13, 2000. p. 237-242.
10. L. Sterpone, M. Sonza Reorda, M. Violante, RoRA: Reliability-oriented Place and Route for SRAM-based FPGAs, PRIME05: IEEE Ph.D. Research In Micro-Electronics & Electronics, 2005, pp. 147-150
11. L. Sterpone, D. Boyang, D. Merodio Codinachs, V. Ferlet-Cavrois, Accurate Mitigation of Single Event Effects on Flash-based FPGAs: A new Design Flow. RADECS 2013.
12. F. Kastensmidt, L. Sterpone, M. Sonza Reorda, L. Carro. On the Optimal Design of Triple Modular Redundancy Logic for SRAM-Based FPGAs, DATE2005: IEEE Design, Automation and Test in Europe, 2005, pp. 1290-1295
13. L. Tambara, F. Kastensmidt, J. Azambuja, E. Chielle, F. Almeida, G. Nazar, L. Carro, P. Rech, C. Frost. Evaluating the Effectiveness of a Diversity TMR Scheme under Neutrons, RADECS 2013.
14. J. Tarrillo, P. Rech, F. Kastensmidt, C. Valderrama, C. Frost, Neutron Cross-section of N-Modular Redundancy Technique in SRAM-based FPGAs. RADECS 2013.
15. Xilinx, Inc. (2013) "Device Reliability Report Third Quarter 2013" [Online]. Available: http://www.xilinx.com/support/documentation/user_guides/ug116.pdf

Chapter 11
Low Power Robust FinFET-Based SRAM Design in Scaled Technologies

Sumeet Kumar Gupta and Kaushik Roy

Abstract FinFETs have emerged as alternatives to conventional bulk MOSFETs in scaled technologies due to superior gate control of the channel, lower short channel effects and higher scalability. However, width quantization in FinFETs constrains the design space of FinFET-based circuits, especially SRAMs in which transistor sizing is critical for the circuit robustness. The adverse effects of width quantization can be mitigated by appropriate device-circuit co-design of FinFET-based memories. This chapter describes some of such techniques with an emphasis on the device-circuit interactions associated with each methodology. The impact of different technology options in FinFETs like gate-underlap, fin orientation, fin height, gate workfunction and independent control of the gates on the stability, power and performance of 6 T SRAMs is discussed.

1 Introduction

The past few decades have seen the evolution of semiconductor industry driven by technology scaling. Miniaturization of bulk Metal Oxide Semiconductor Field Effect Transistors (MOSFETs) along with scaling of power supply voltage (V_{DD}) has provided the benefits of higher performance, lower power and larger integration density. However, for the past few technology generations, scaling of transistors has been becoming increasingly difficult. Short channel effects (SCEs) like drain-induced barrier lowering [1], device threshold voltage (V_T) roll-off [1] and degradation in the ratio of ON current (I_{ON}) to OFF current (I_{OFF}) have started to manifest themselves to a larger extent in scaled MOSFETs. Another critical challenge in scaled technologies is the increasing impact of process variations on the device, circuit and system performance due to random dopant fluctuation, line edge roughness etc. [2]. Such detrimental effects in scaled technologies make the design and optimization of devices and circuits extremely challenging.

S.K. Gupta • K. Roy (✉)
School of Electrical and Computer Engineering, Purdue University,
West Lafayette, IN 47907, USA
e-mail: kaushik@ecn.purdue.edu

R. Reis et al. (eds.), *Circuit Design for Reliability*,
DOI 10.1007/978-1-4614-4078-9_11

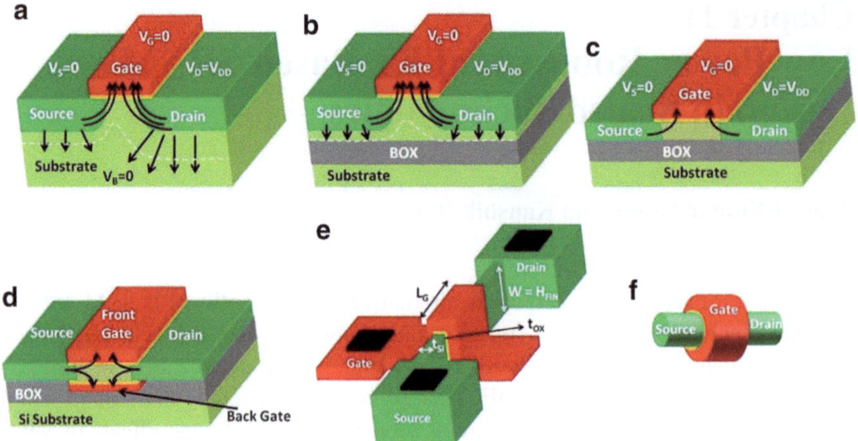

Fig. 11.1 Device structures with single gate viz. (**a**) bulk MOSFETs (**b**) partially depleted (*PD*) SOI MOSFETs (**c**) fully depeleted (*FD*) SOI MOSFETs and multiple gates viz. (**d**) planar double gate (*DG*) MOSFETs (**e**) FinFETs and (**f**) gate-all around MOSFETs. The *arrows* show the electric field lines in the structures

1.1 Alternate Device Structures

In order to meet the challenges posed by transistor scaling, different device structures have been explored [3–5] as alternates to the bulk MOSFETs. Partially depleted silicon-on-insulator (PD SOI) MOSFETs [3] reduce the drain electric fields penetrating to the source side due to the buried oxide (BOX), leading to mitigation of short channel effects (Fig. 11.1b). However, the body in PD SOI MOSFETs is floating which leads to indeterminacy in the threshold voltage (V_T) of the device due to variable body potential [3]. The undesirable floating body effects in PD SOI MOSFETs can be mitigated by reducing the thickness of the silicon body and making it fully depleted (FD-SOI MOSFETs—Fig. 11.1c) [4]. Reduced thickness of the devices also leads to lower short channel effects due to reduced penetration of drain electric fields to the source side. Short-channel effects can be further reduced in multi-gate structures (Fig. 11.1) like double-gate (DG) MOSFETs [5] and gate all-around (GAA) MOSFETs [6] which exhibit superior control of the channel electrostatics by the gate potential, compared to single-gate structures. The focus of this chapter is DG MOSFETs.

Planar DG MOSFETs, shown in Fig. 11.1d suffer from a fabrication issue, known as gate misalignment [7]. Alignment of the two gates of a DG MOSFET is required in order to effectively control the short channel effects and obtain maximum I_{ON} in the device. However, from the point of view of fabrication, gate alignment in planar DG MOSFETs is extremely challenging. FinFETs (Fig. 11.1e) in which the gate is formed by depositing the gate metal/poly-silicon around the fin-shaped body [8–9] makes it possible to easily achieve gate alignment in DG MOSFETs. Henceforth, all the discussion on DG MOSFETs will pertain to FinFETs.

1.2 Fundamentals of FinFETs

FinFETs exhibit reduced short channel effects (compared to bulk MOSFETs) viz. lower drain induced barrier lowering (DIBL), lower sub-threshold swing (SS) and higher ratio of ON current to OFF current (I_{ON}/I_{OFF}) due to ultra-thin silicon body. Lower fin thickness (t_{Si}) limits the electric field penetration from drain to source and enhances the scalability of FinFETs. However, t_{Si} cannot be scaled indefinitely. As t_{Si} decreases, quantum confinement effects start to play a bigger role, increasing the threshold voltage (V_T) of the transistors and leading to loss in the device performance. Moreover, device variability increases at lower t_{Si}.

Ultra-thin body (UTB) in FinFETs along with channel control by two gates make it feasible to use undoped (intrinsic) body, while still achieving excellent device electrostatics. Due to undoped body in FinFETs, the variations due to random dopant fluctuations (a major source of random variations in bulk MOSFETs) is eliminated. However, due to UTB, an additional source of variation viz. body thickness variation is introduced in FinFETs.

Short channel effects (SCEs) in FinFETs can also be controlled by reducing gate dielectric thickness (t_{OX}). Lower t_{OX} results in superior gate control of the channel, but leads to increase in the gate leakage (I_G). Use of high-k dielectrics expands the device design space by allowing the scaling of effective oxide thickness (EOT) with higher physical thickness of the dielectric, limiting the gate leakage current in scaled FinFETs.

Another device parameter which plays an important role in determining the characteristics of FinFETs is the gate-source/drain (S/D) overlap ($L_{OVS/D}$). Larger $L_{OVS/D}$ leads to decrease in effective channel length (L_{CH}) for the same gate length (L_G) which worsens SCEs in the device. In case the electrostatic integrity of the device cannot be maintained by scaling t_{OX} and/or t_{Si}, a negative overlap between gate and S/D, called underlap (L_{UN}) may be introduced to increase the effective channel length without increasing L_G. This achieves improved electrostatics and higher I_{ON}-I_{OFF} ratio. Gate workfunction (Φ_G) is another important design parameter which can be changed to obtain the desired V_T of the transistor [10].

The width of FinFET is along the height of the fin (H_{FIN}-Fig. 11.1e) and for ensuring mechanical stability of the device, the aspect ratio H_{FIN}:t_{Si} is limited [11]. For a particular technology, H_{FIN} is fixed and hence, the width of the device can only be increased in quanta of H_{FIN}. This effect is known as width quantization. Restricted device sizing due to width quantization in FinFETs leads to constrained design space of FinFET-based circuits.

The effect of width quantization becomes even more critical for circuits like SRAMs and flip-flops in which transistor sizing has a significant effect on the

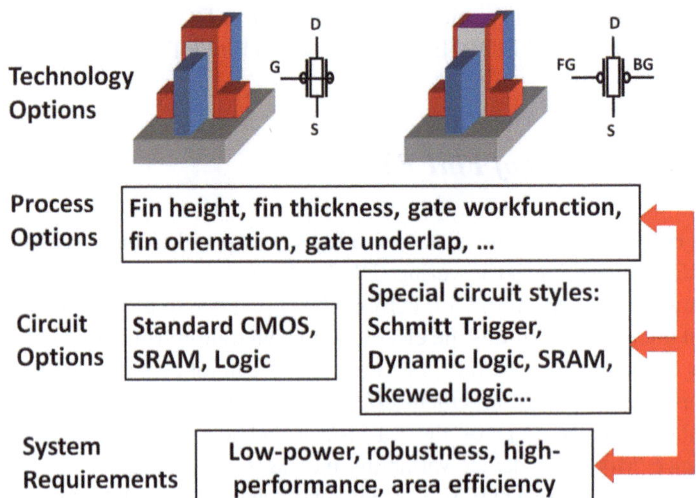

Fig. 11.2 Device-Circuit-System interactions in FinFET technology

circuit functionality [12]. Conflicting requirements for transistor sizes for stable read and write operations in SRAMs further aggravates the challenge posed by width quantization.

The adverse effects of width quantization on FinFET-based circuits can be mitigated by employing device-circuit co-design techniques [13–24]. Figure 11.2 illustrates different technology options provided by FinFETs. The two gates of FinFETs can be tied together or controlled independently to enhance the design flexibility of FinFET-based circuits. In addition, device parameters like fin thickness, gate-source/drain underlap, gate workfunction, fin orientation and fin height can be optimized considering the circuit and system requirements. This chapter explores such device-circuit interactions with a focus on FinFET-based SRAMs. Before we present the discussions on the device-circuit co-design methodologies, let us review the fundamentals of 6 T SRAMs.

1.3 Fundamentals of 6 T SRAMs

6 T SRAM cell (Fig. 11.3) comprises of 6 transistors: 2 access transistors (AXL and AXR), 2 pull-down transistors (NL and NR) and 2 pull-up transistors (PL and PR). Pull-down and pull transistors form the cross-coupled inverters which maintain the state of the cell in the hold mode. Read and write operations in the cell are performed through access transistors which are controlled by a word-line (WL), as described subsequently. Cells in a row of the array share the same word-line. The cell comprises of two bit-lines (BL and BLB) which read from or write into the cell and are shared amongst the cells in the same column.

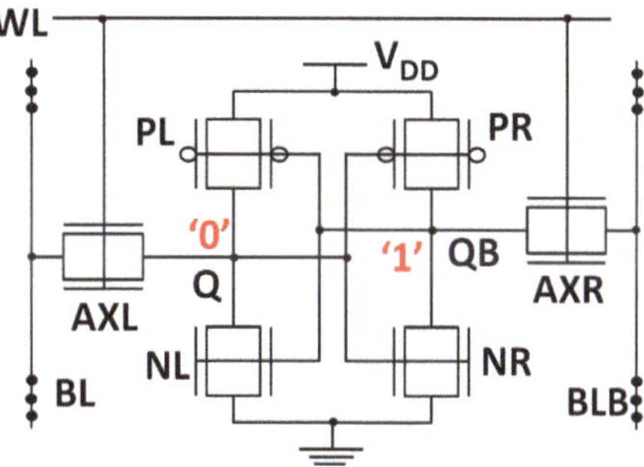

Fig. 11.3 Schematic of FinFET-based 6 T SRAM

Suppose the cell stores '0' i.e. $V_Q = 0$ and $V_{QB} = V_{DD}$. (Here, V_Q and V_{QB} are voltages at nodes Q and QB). During the read operation, bit-lines BL and BLB are pre-charged to V_{DD} and word-line WL is asserted. V_Q rises to some positive voltage V_{READ} determined by the relative strengths of AXL and NL. At the same time BL starts discharging while BLB remains at V_{DD}. The difference in the voltages of BL and BLB is amplified by the sense amplifier to obtain the value stored in the cell. Since the storage node voltages are disturbed during the read, read stability of the cell is less than the hold stability. If V_{READ} becomes greater than the trip point of the inverter formed by AXR, PR and NR during read, the contents of the cell can flip, causing a read failure [25]. During the write operation, BL is charged to V_{DD} and BLB is discharged to *GND*. On asserting WL, V_{QB} is discharged to V_{WRITE}, determined by the relative strengths of PR and AXR. Successful write occurs if V_{WRITE} is less than the trip point of the inverter formed by PL, AXL and NL.

With the basic understanding of FinFETs and cell operation of 6 T SRAMs, let us discuss the design techniques for FinFET-based SRAMs. We describe the co-optimization of fin ratio and t_{Si} in Sect. 2, joint optimization of H_{FIN}-t_{Si}-t_{OX}-V_{DD}-V_T in Sect. 3, analysis of fin orientation in Sect. 4, spacer thickness optimization in Sect. 5 and independent gate FinFETs in Sect. 6.

2 Co-optimization of Fin Ratio and Fin Thickness

Width quantization in FinFETs restricts the ratio of the transistor widths in 6 T SRAM to integer values and results in constrained design space. As a result, the conflict between read stability, write-ability, hold stability, performance and power of 6 T SRAM is aggravated. The design space of FinFET-based 6 T SRAMs can be expanded by co-optimizing the number of fins with the device design parameters. In this section, we will explore the joint effect of the fin ratio and fin thickness (t_{Si}) on the cell stability, performance and leakage of FinFET-based 6 T SRAMs.

Let us start by explaining the effect of t_{Si} on the device characteristics. Higher t_{Si} results in aggravated short channel effects leading to increase in I_{OFF} and lower I_{ON}-I_{OFF} ratio. As in [13], let us perform the comparison of devices with different

Table 11.1 Comparison of FinFETs with different fin thickness (t_{Si}) at iso-I_{OFF} [13]. ($I_{OFF} = 25$ nA/μm)

t_{Si} (nm)	Φ_G (eV)	I_{DSAT} (mA/μm)	I_{DLIN} (mA/μm)	DIBL (mV/V)
7	4.61	2.18	0.88	51
9	4.73	1.95	0.92	115
11	4.80	1.56	0.8	194

t_{Si} under iso-I_{OFF} conditions, which is achieved by appropriately adjusting the gate work-function (Φ_G). At iso-I_{OFF}, lower short channel effects (SCE) in devices with a reduced t_{Si} tend to increase drain current (I_D). However, lower t_{Si} results in mobility degradation [26] and increase in source/drain (S/D) resistance, which tends to decrease I_D. At high V_{DS} (saturation region), the effect of lower SCE has a dominant effect on I_D due to which saturation current (I_{DSAT}) increases with decreasing t_{Si} (Table 11.1 [13]). However, at low V_{DS} (linear region), the impact of SCE, mobility degradation and S/D resistance on I_D are comparable. Due to this, maximum value of linear current (I_{DLIN}) is obtained for $t_{Si} = 9$ nm in Table 11.1. With this discussion in mind, let us describe the effect of t_{Si} on the cell stability, performance and leakage of 6 T SRAMs.

During read, NL (Fig. 11.3) operates in the linear region while AXL is in saturation. Due to increase in I_{DSAT} at lower t_{Si} and comparable I_{DLIN} for different t_{Si} at iso-I_{OFF}, the relative strength of AXL with respect to NL increases as t_{Si} is reduced. This results in higher V_{READ} which tends to lower the read static noise margin (SNM). However, lower t_{Si} also leads to higher I_{ON}-I_{OFF} ratio, which tends to increase the read SNM. Depending on the relative impact of the two effects on read stability, read SNM might increase or decrease with t_{Si} at iso-I_{OFF}. The access time of SRAM reduces with decreasing t_{Si} due to increase in the strength of the access transistor. Hold SNM increases for lower t_{Si} due to higher I_{ON}-I_{OFF} ratio of the transistors. Write-ability of the cell reduces with decreasing t_{Si} due to reduction in the relative strength of AXR with respect to PR. Moreover, higher I_{ON}-I_{OFF} ratio at reduced t_{Si} makes it difficult to flip the contents of the cell and lowers the write-ability. Since the comparison is performed under iso-I_{OFF} conditions, the cell leakage is similar for different t_{Si}.

Let us, now, discuss the impact of fin ratio on SRAM characteristics. We define N_{AC}, N_{PD} and N_{PU} to be the number of fins in the access, pull-down and pull-up FinFETs, respectively. Increasing N_{PD}:N_{AC} results in lower V_{READ} and higher read stability. Improvement in read stability can also be achieved by increasing N_{PU}:N_{PD} which leads to higher logic threshold voltage (V_M) of the inverter formed by PR and NR (see Fig. 11.3). Write-ability of the cell can be increased by increasing N_{AC}:N_{PU}, which results in lower V_{WRITE} or by increasing N_{PU}:N_{PD} which results in higher V_M of the inverter formed by PL and NL. Hold stability improves with increasing N_{PU}:N_{PD}. Note, the conflicting requirements for the fin ratios for achieving high read stability, write-ability and hold stability. As an example, if N_{AC} is decreased to increase N_{PD}:N_{AC} and improve read stability, then N_{AC}:N_{PU} decreases, resulting

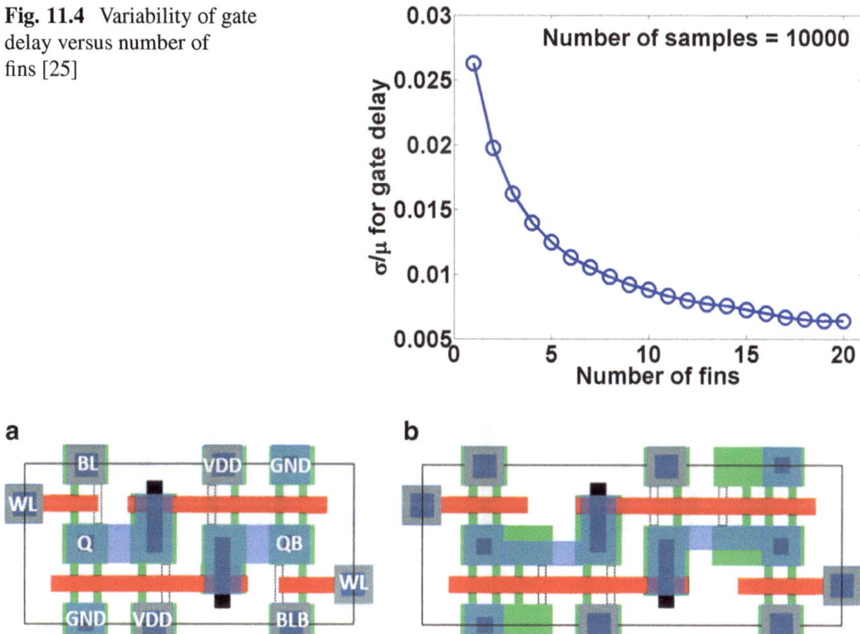

Fig. 11.4 Variability of gate delay versus number of fins [25]

Fig. 11.5 Layouts of FinFET-based 6 T SRAM showing increase in area with increasing number of fins: (**a**) layout with using one/two fins in the transistors (**b**) layout for a high performance cell

in degradation in write-ability. Decreasing N_{PU} results in improved write-ability; however, this leads to lower N_{PU}:N_{PD} and reduced read and hold stabilities. Access time decreases with (a) higher N_{AC} due to increase in the strength of AXL and (b) higher N_{PD} due to lower V_{READ} which increases the gate-to-source voltage (V_{GS}) and drain to source voltage (V_{DS}) of AXL, increasing its strength. Larger number of fins results in higher cell leakage. However, at the same time, more fins result in lower parameter variations [27] as the variation effects get averaged out for larger number of fins. This is illustrated in Fig. 11.4 which shows a decrease in the variability of gate delay as the number of fins increases [27].

Increase in the number of fins may also lead to increase in the cell area. Considering the spacer lithography technology for the fabrication of FinFETs [28], FinFETs with 2*i*-1 and 2*i* fins can be fabricated in the same lithographic pitch (Here, *i* is a natural number). For minimum area of the layout, all transistors with one fin each may be used. To increase the cell stability, number of fins in the pull down transistors can be increased to 2, as shown in Fig. 11.5. For high performance cells, number of fins in the access transistors and the pull down transistors may be increased at the cost of area (Fig. 11.5).

Considering the trade-offs discussed above, the design space for FinFET-based 6 T SRAM comprising of the fin ratio and fin thickness can be explored. As an example, the joint impact of fin ratio and t_{Si} on the read stability and write-ability

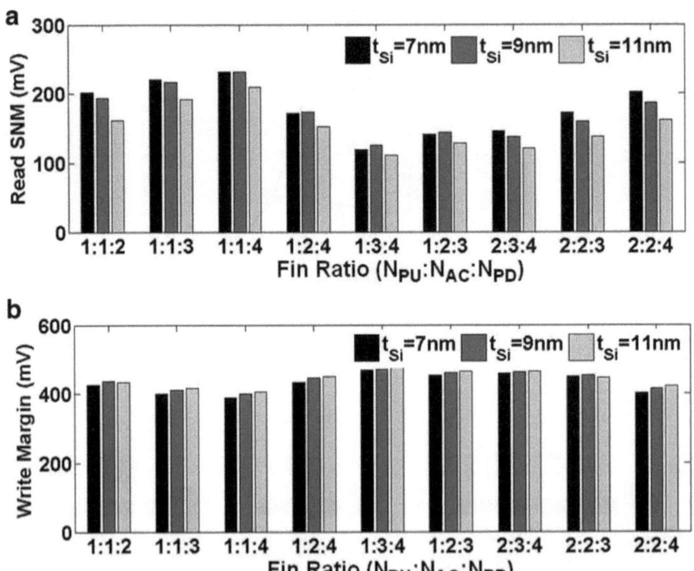

Fig. 11.6 (**a**) Read SNM and (**b**) Write margin as a function of fin ratio and fin thickness [13]

is shown in Fig. 11.6 [13] under iso-I_{OFF} conditions. Increase in read SNM with decreasing t_{Si} can be observed and is attributed to improved I_{ON}-I_{OFF} ratio for low t_{Si}, as explained before. However, higher I_{ON}-I_{OFF} ratio for low t_{Si} results in degradation in write margin (WM). Read SNM and WM for different fin ratios are also shown in Fig. 11.6. Fin ratio $N_{PU} : N_{AC} : N_{PD} = 1 : 1 : 2$ provides the optimal balance between read stability, write-ability and cell area. The plots in Fig. 11.6 can be used for the co-optimization of t_{Si} and fin ratio to achieve the optimal cell stability of FinFET-based SRAMs.

3 Joint Optimization of Fin Height, Fin Thickness, Oxide Thickness, Supply Voltage and Threshold Voltage

Due to quasi-planar structure of FinFETs, the device current can be increased by increasing the fin height (H_{FIN}) with a minimal increase in the device footprint. This is in contrast with the planar MOSFETs, in which the increase in the device footprint is similar to that in the device width. The minimal increase in the device footprint with increasing H_{FIN} in FinFETs is due to the fact that in order to maintain the mechanical stability, the aspect ratio H_{FIN}:t_{Si} has to be below a critical value. Hence, increase in H_{FIN} is accompanied by a proportional increase in t_{Si} and consequently, a small increase in the device footprint. Increase in t_{Si} leads to aggravation of short channel effects in FinFETs. In order to maintain the electrostatic integrity of the device, gate dielectric thickness (t_{OX}) can be reduced. However, t_{OX} scaling is limited by the gate current. As a result, t_{Si} and hence, H_{FIN} cannot be increased beyond a certain value. In order to increase the device strength further, one has to increase the number of fins, which leads to width quantization in FinFETs, as explained earlier. However, for small increase in the device current, the optimization of H_{FIN} can be a potentially useful technique, which we discuss in this section.

Increase in the device strength with the increasing H_{FIN} allows (a) increase in threshold voltage (V_T) to achieve exponential reduction in sub-threshold leakage and (b) V_{DD} scaling to achieve lower leakage and dynamic power with a similar read access time. Considering these options, two kinds of co-design techniques have been explored in [14]: (a) joint optimization of H_{FIN}-t_{Si}-t_{OX}-V_T (or device-V_T) design space and (b) joint optimization of H_{FIN}-t_{Si}-t_{OX}-V_T-V_{DD} (or device-V_T-V_{DD}) design space. The desired V_T is obtained by adjusting the gate workfunction appropriately.

In the device-V_T co-optimization, H_{FIN} is increased along with increase in t_{Si} and reduction in t_{OX}, keeping V_{DD} constant. The gate workfunction is adjusted to increase V_T and achieve similar I_{ON} per fin. On the other hand, device-V_T-V_{DD} co-optimization involves increasing H_{FIN} and t_{Si}, reducing t_{OX}, scaling V_{DD} and increasing V_T to achieve iso-I_{ON} per fin. V_{DD} scaling results in lower short channel effects, which allows higher t_{OX} to be used in device-V_T-V_{DD} co-optimization compared to device-V_T co-optimization, leading to lower gate leakage (Fig. 11.7a). Exponential reduction in sub-threshold current due to increase in V_T and consequently, higher I_{ON}-I_{OFF} ratio for larger H_{FIN} in device-V_T co-optimization leads to increase in read (Fig. 11.7b) and hold stabilities. For device-V_T-V_{DD} co-optimization, higher I_{ON}-I_{OFF} ratio for larger H_{FIN} tends to increase read and hold stabilities. However, operation at lower V_{DD} for devices with larger H_{FIN} tends to decrease the read and hold stabilities. As a result, an optimal point in Fig. 11.7 is observed for design point D8. The optimal design point can be chosen by considering the impact of the co-optimization techniques on cell leakage and SNM. In Fig. 11.7a, the design point D9 (see Table 11.2) yields minimum leakage along with a moderate improvement in read SNM for both device-V_T and device-V_T-V_{DD} co-optimizations. The read access time remains similar across different design points (Fig. 11.7c) since the devices are designed under iso-I_{ON}. Increasing H_{FIN} leads to increase in capacitance at the storage nodes which leads to increased immunity to soft errors due to increase in the cell charge (Fig. 11.7d). However, increase storage node capacitance results in higher write time (Fig. 11.7c).

In summary, joint optimization of H_{FIN}, t_{Si}, t_{OX}, V_T and V_{DD} leads to increase in static noise margin, lower cell leakage and improved immunity to soft errors at iso-read access speed at the cost of increase in write access time.

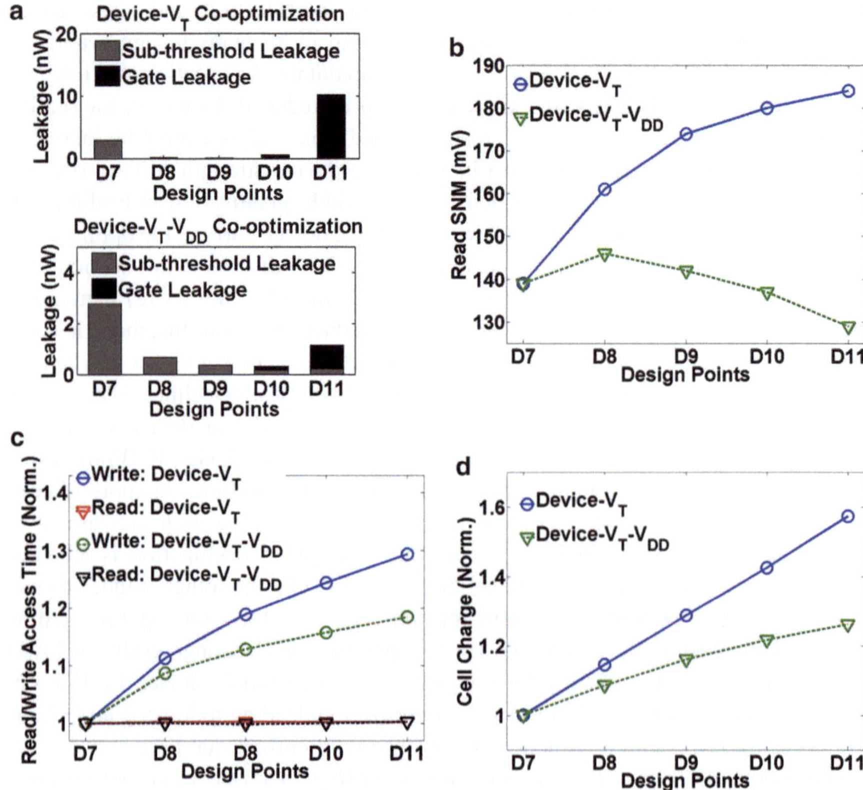

Fig. 11.7 (**a**) Cell leakage (**b**) read SNM (**c**) read and write access time and (**d**) cell charge for different design points for Device-V_T and Device-V_T-V_{DD} co-optimizations [14]

Table 11.2 Device parameters for design points at iso-I_{ON} per fin [14]

Design name	D7	D8	D9	D10	D11
T_{Si} (nm)	7	8	9	10	11
H_{FIN} (nm)	35	40	45	50	55
H_{FIN}-V_TCo-optimization					
V_{DD} (V)	1.0	1.0	1.0	1.0	1.0
T_{OX} (nm)	2.4	2.1	1.8	1.5	1.2
NMOS Φ_G (eV)	4.6	4.72	4.82	4.89	4.95
PMOS Φ_G (eV)	4.8	4.68	4.58	4.51	4.45
H_{FIN}-V_T-V_{DD}Co-optimization					
V_{DD}(V)	1.0	0.95	0.9	0.85	0.8
t_{OX} (nm)	2.4	2.15	1.9	1.65	1.4
NMOS Φ_G (eV)	4.6	4.67	4.71	4.73	4.74
PMOS Φ_G (eV)	4.8	4.73	4.69	4.67	4.66

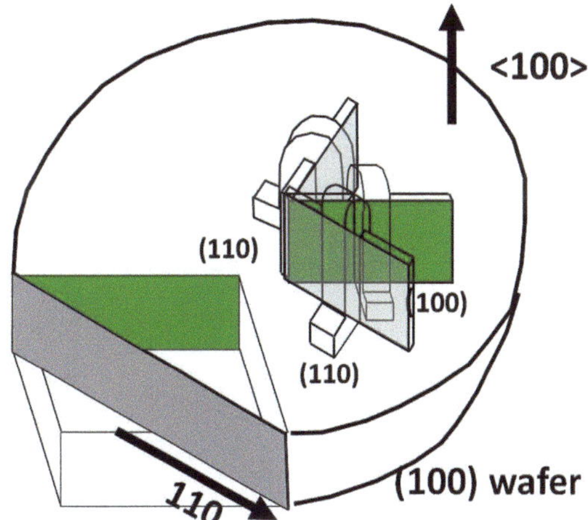

Fig. 11.8 Fins aligned along different crystal orientations using fin rotation [15]

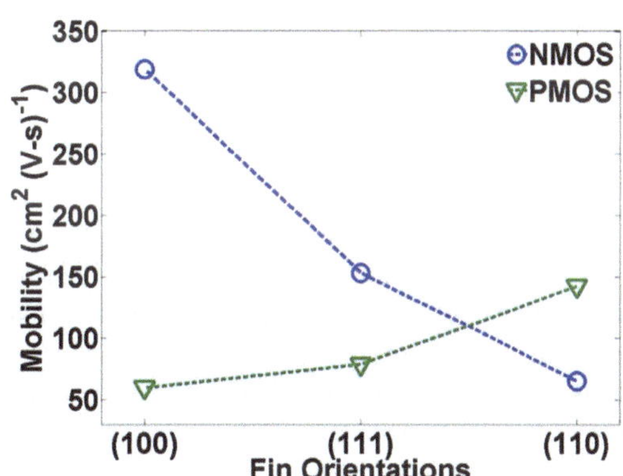

Fig. 11.9 Electron (NMOS) and hole (PMOS) mobilities along different crystal orientations [15]

4 Fin Rotation and Orientation

Quasi-planar structure of FinFETs enables the alignment of transistors along a particular crystal orientation by fin rotation (Fig. 11.8). Due to the difference in the carrier effective masses, FinFETs with different fin orientations exhibit difference in mobilities and scattering rates [15]. Figure 11.9 shows the mobility for different crystal orientations for electrons and holes [15]. NMOS shows a higher mobility along (100) direction while for PMOS, the highest mobility is along (110) direction. Hence, the drive strength of FinFETs can be optimized by aligning them along appropriate crystal orientations. In this section, we investigate the impact of fin rotation and orientation on the characteristics of 6 T SRAMs.

Figure 11.10a shows read SNM for different fin orientations of the access (AX), pull-down (PD) and pull-up (PU) transistors. Maximum read SNM is obtained when AX is along (110) direction and PD is along (100) direction due to reduced

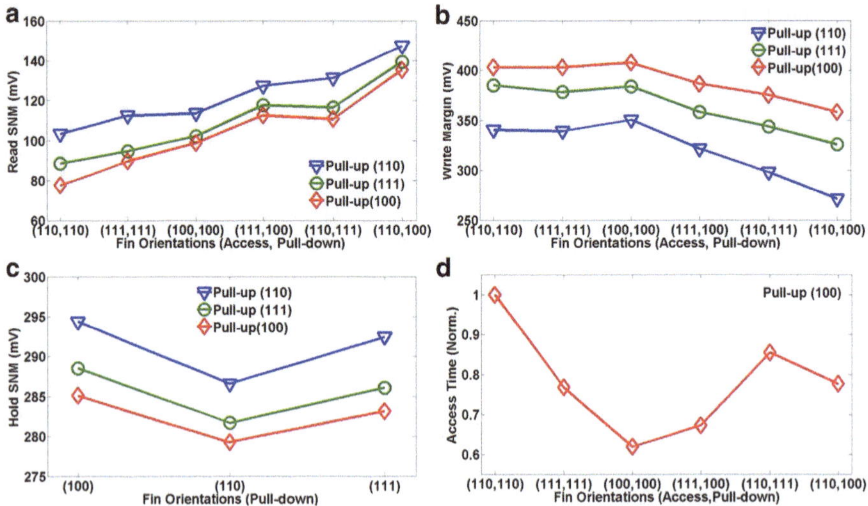

Fig. 11.10 (**a**) Read SNM (**b**) write margin (**c**) hold SNM and (**d**) access time for different fin orientations [15]

strength of AX relative to PD resulting in lower V_{READ}. The optimal orientation of PU for maximum read stability is (110) which increases the strength of PU leading to higher V_M of the cross-coupled inverters. However, from the point of view of write margin, (100) orientation of PU is the most superior (Fig. 11.10b). This is because of reduced strength of PU relative to AX leading to lower V_{WRITE}. For AX, (100) orientation yields maximum write-ability due to increase in the strength of access transistor. For hold stability, stronger PU ((110) orientation) as well as PD ((100) orientation) leads to maximum benefits (Fig. 11.10c) due to increase in the coupling between the two inverters. For access time, strong AX and PD (both (100) orientations) yield the minimum access time (Fig. 11.10d). Fin rotation and orientation enhances the design flexibility of FinFET based SRAMs by mitigating the effect of width quantization and allowing finer modulation of the transistor strengths. Compared with all (110) transistors (conventional orientation for (100) wafer), multi-oriented FinFETs in a 6 T SRAM show benefits in terms of increased read stability with a negligible effect on the write margin. However, when compared with all (100) FinFETs (conventional orientation for (110) wafer), optimization of fin orientation shows less improvements in the cell stability. Thus, the effectiveness of fin orientation optimization depends on the wafer orientation and yields maximum benefits for (100) wafers. However, on a (100) wafer, rotation of fins along (111) direction is not feasible, which restricts the combinations of fin orientations that can be used in a 6 T SRAM. Fin orientation also leads to area penalty and is expected to be more susceptible to process variations.

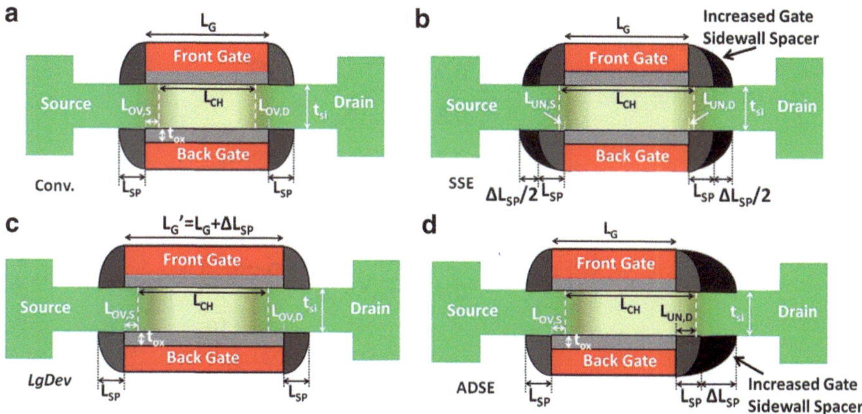

Fig. 11.11 (**a**) Conventional FinFET (**b**) Symmetric Spacer Extension (*SSE*) FinFET (**c**) *LgDev* FinFET with larger gate length and (**d**) Asymmetric Drain Spacer Extension (*ADSE*) FinFET

5 Spacer Thickness Optimization

FinFETs exhibit significant reduction in the junction capacitance compared to bulk MOSFETs due to lower source/drain (S/D) junction cross-section. As a result, overlap capacitance becomes the dominant component of the parasitic capacitance in FinFETs. In order to reduce the parasitic capacitance, a negative gate-S/D overlap (called underlap) can be introduced in FinFETs by optimizing the spacer thickness. In addition to the capacitance, gate underlap has a strong impact on the device electrostatics, and in turn, on the device characteristics like I_{ON}, I_{OFF}, sub-threshold swing, gate leakage and drain-induced barrier lowering (DIBL). In this section, we discuss the optimization of spacer thickness and evaluate the impact of introducing a gate underlap in FinFETs on 6 T SRAM characteristics. We perform the analysis for FinFET structures with both symmetric and asymmetric gate underlap (Fig. 11.11).

5.1 Symmetric Spacer Optimization

Figure 11.11b shows the FinFET structure with symmetric gate underlap (Symmetric Spacer Extension (SSE) FinFET) [16]. Gate sidewall spacer thickness is increased by ΔL_{SP} on both sides of the gate to introduce an equal gate underlap on the source and drain sides. Gate underlap results in (a) increase in the effective channel length of the device and (b) change in the device electrostatics due to weaker gate control in the underlapped portion of the channel. Due to these two effects, I_{ON} degrades with increasing ΔL_{SP}. However, at the same time, FinFETs with higher ΔL_{SP} exhibit lower capacitance due to the reduction in the gate overlap. I_{ON} and

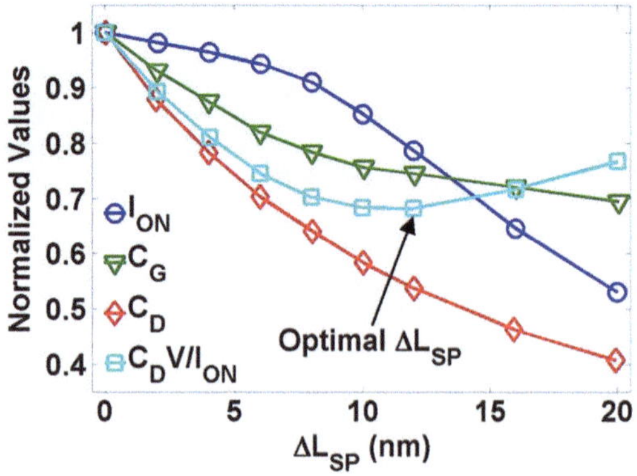

Fig. 11.12 Dependence of ON current, capacitance and intrinsic device delay ($C_D V/I_{ON}$) on ΔL_{SP}[16]

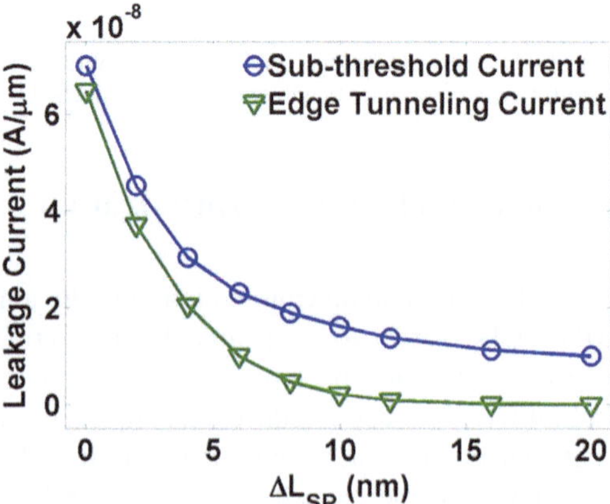

Fig. 11.13 Sub-threshold and edge tunneling leakage as a function of ΔL_{SP} [16]

capacitance versus ΔL_{SP} is shown in Fig. 11.12. It can be observed that for small ΔL_{SP}, I_{ON} degradation is small and is due to increase in the effective channel length. However, for large ΔL_{SP}, sharp decrease in I_{ON} is observed and is attributed to a significant loss of gate control over the channel, especially on the source side. This effect is explained in a greater detail in the next sub-section. Figure 11.12 also shows a decrease in gate and drain capacitance (C_G and C_D) with increasing ΔL_{SP}. The net effect of decreasing I_{ON} and capacitance on intrinsic device delay ($C_D V/I_{ON}$) can be observed in Fig. 11.12 which shows a minimum in $C_D V/I_{ON}$ for an optimal ΔL_{SP}. Thus, symmetric spacer optimization can lead to lower intrinsic device delay. Moreover, increase in the effective channel length leads to lower short channel effects viz. lower DIBL, lower sub-threshold swing and reduced sub-threshold leakage with increasing ΔL_{SP} (Fig. 11.13). Reduced gate overlap also results in lower edge tunneling current (I_{ET}), as shown in Fig. 11.13.

Let us now explain the impact of symmetric underlap in FinFETs on SRAM stability, leakage and performance. Improvement in short channel effects of the transistors with increasing ΔL_{SP} results in higher read and hold SNM, leading to reduction in the data retention voltage (DRV). At the same time, lower intrinsic

Table 11.3 Comparison of SSE FinFET SRAM with conventional (Conv.) and *LgDev* FinFET SRAMs [16]

Metric	Conv. ($\Delta L_{SP} = 0$)	SSE ($\Delta L_{SP} = 12$ nm)	*LgDev* ($\Delta L_{SP} = 12$ nm)
Read SNM (mV)	129	132	137
Write Time (ps) : Nominal	25.4	25.7	23.8
:Worst Case	54.2	42.3	52.9
Worst Case Data Retention Voltage (mV)	220	160	150
Leakage Power (nW): P_{SUB}	13.05	2.52	2.10
(T = 27 °C) : P_{GATE}	34.05	13.78	60.9
:P_{TOT}	47.1	16.3	63

delay of the optimized devices (as explained earlier) and lower storage node capacitance result in lower write time. Moreover, lower sub-threshold leakage leads to significant reduction in the cell leakage. This comes at the cost of (a) increase in cell area due to increase in the device footprint and (b) higher access time due to lower strength of the access transistors. Table 11.3 shows the comparison of conventional FinFETs and SSE FinFETs [16]. Comparison is also performed with a device which has the same spacer thickness as conventional FinFETs but with channel length equal to that of SSE FinFET (see *LgDev* in Fig. 11.11c). Note, *LgDev* has a larger gate length compared to SSE FinFET. Larger read SNM and lower DRV is observed in *LgDev* based SRAM, compared to SSE FinFET SRAM because I_{ON}-I_{OFF} ratio in *LgDev* is higher due to superior gate control. However, since the overlap capacitance is higher in *LgDev*, the worst case write time is larger compared to SSE FinFET based SRAM. Moreover, large gate length results in higher gate current in the ON state due to larger cross-sectional area of the gate. This leads to higher cell leakage in *LgDev* FinFET SRAM.

5.2 Asymmetric Drain Spacer Extension FinFETs

In the previous sub-section, we discussed the pros and cons of SSE FinFETs. One of the drawbacks of introducing gate underlap on the source side is that carrier injection from the source to the channel is significantly affected by the underlap, which results in a large degradation in I_{ON} for large ΔL_{SP}. Instead of introducing underlap on both sides, if the underlap is introduced only on the drain side, the degradation in I_{ON} can be mitigated. Figure 11.11d shows the FinFET structure with asymmetric gate underlap (Asymmetric Drain Spacer Extension (ADSE) FinFET) [17]. Asymmetry in the structure of ADSE FinFETs results in unequal I_D for $V_{DS} > 0$ and $V_{DS} < 0$. This asymmetry is used to mitigate the read-write conflict in 6 T SRAMs. In order to explain the benefits of ADSE FinFET based SRAM, let us discuss the impact of asymmetric gate underlap on the device characteristics.

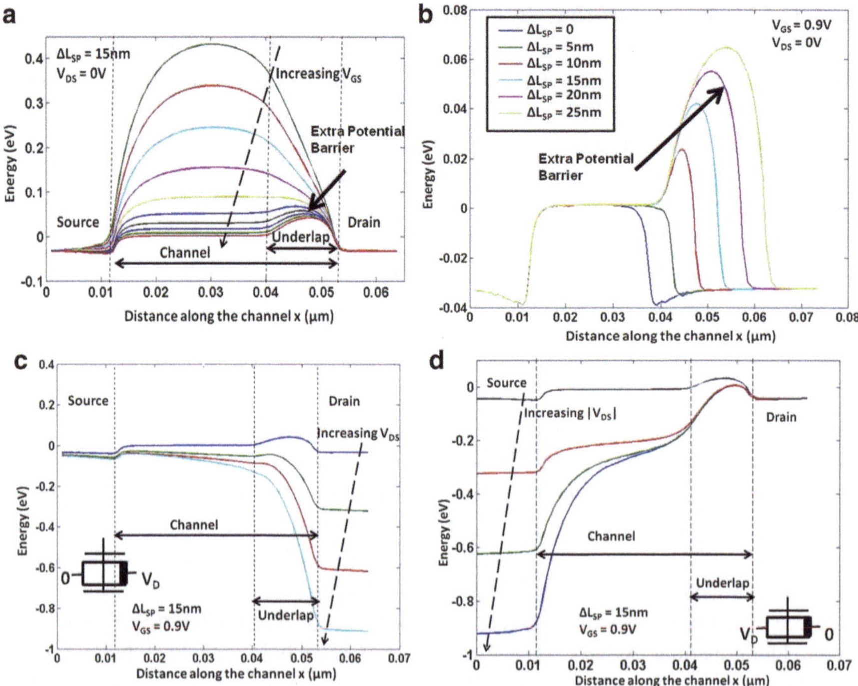

Fig. 11.14 Conduction band profile of ADSE FinFETs (**a**) at different V_{GS} (**b**) for different ΔL_{SP} (**c**) at different positive V_{DS} and (**d**) at different negative V_{DS} [17]. The thick line on the device symbol indicates the underlapped terminal

As in SSE FinFETs, gate underlap in ADSE FinFETs leads to increase in the channel length and change in the device electrostatics. Due to weaker gate control in the underlapped part of the channel, an extra potential barrier is introduced on the drain side at high V_{GS} (Fig. 11.14a). The height of the extra potential barrier increases with increasing ΔL_{SP} (Fig. 11.14b). Figure 11.14c shows that the positive drain bias (V_{DS}) pulls down the extra potential barrier (Fig. 11.14c). On the other hand, negative V_{DS} has a small effect on the extra potential barrier since positive bias is applied on the non-underlapped terminal (Fig. 11.14d).

Increase in the channel length and extra potential barrier in the underlapped part of the channel leads to reduction in I_{ON} (Fig. 11.15), as in SSE FinFETs. A knee point (L_{KP}) can be observed in Fig. 11.15. For $\Delta L_{SP} < L_{KP}$, decrease in I_D of ADSE FinFETs is gradual and matches with that of *LgDev*. This implies that slow decrease in I_D is due to increase in channel length. For $\Delta L_{SP} > L_{KP}$, sharp decrease in I_D is due to increase in extra potential barrier. For $V_{DS} > 0$, an increase in L_{KP} with increasing V_{DS} can be observed (see Fig. 11.15). This is due to the fact that positive V_{DS} lowers the extra potential barrier (Fig. 11.14c), which results in the onset of sharp decrease in I_D at a higher ΔL_{SP}. For $V_{DS} < 0$, L_{KP} is relatively insensitive to V_{DS} (Fig. 11.15) because of weak dependence of the extra potential barrier on

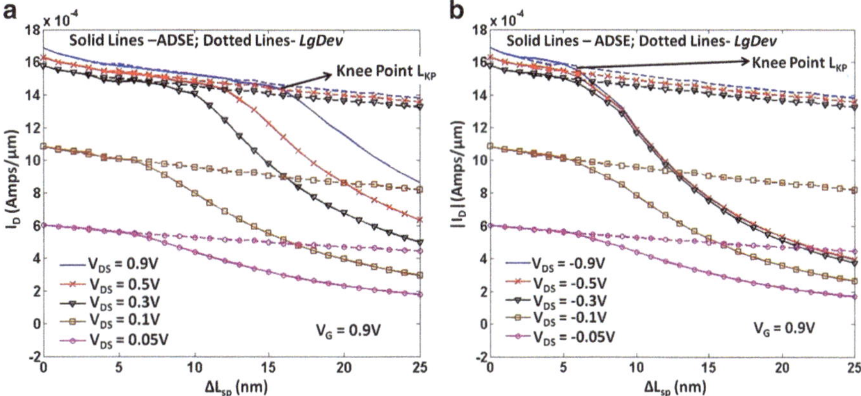

Fig. 11.15 Drain current versus ΔL_{SP} of ADSE and *LgDev* FinFETs for (**a**) $V_{DS} > 0$ and (**b**) $V_{DS} < 0$ [17]

Table 11.4 Comparison of ADSE FinFET SRAM with conventional and SSE FinFET SRAMs

Parameter	Improvement over conventional	Improvement over SSE (iso-device-footprint)
I_{ON}/I_{OFF}	16X	2.3X
DIBL	51 %	20 %
SS	12.6 %	3.5 %
C_{GD}	29 %	27 %

negative V_{DS} (Fig. 11.14d). The difference in the dependence of extra potential barrier and hence, L_{KP}, on the polarity of drain bias creates an asymmetry in current for positive and negative V_{DS}.

In addition, ADSE FinFETs show improved short channel effects due to increase in the channel length. Reduction in I_{ET} is also observed with increasing ΔL_{SP} for positive V_{DS} [17]. The comparison of ADSE FinFETs with conventional FinFETs and iso-device footprint SSE FinFETs is summarized in Table 11.4. ADSE FinFETs show improvement in short channel characteristics and lower gate-drain capacitance compared to iso-area SSE FinFETs. In addition to superior device characteristics, ADSE FinFETs exhibit asymmetry in I_D for $V_{DS} > 0$ and $V_{DS} < 0$. This unique feature of ADSE FinFETs is exploited to mitigate the read-write conflict in 6 T SRAMs, which we discuss next.

Figure 11.16 shows the ADSE FinFET based 6 T SRAM. The underlapped terminal is indicated by the thick lines. During the read operation, V_{DS} across AXR is negative, which results in lower I_D due to extra potential and hence, lower strength of AXR relative to NR. During the write operation, V_{DS} across AXL is positive which reduces the effect of the extra potential barrier on I_D and increases the strength of AXL relative to PL. This leads to the mitigation of the read-write conflict (Fig. 11.17). Moreover, improved short channel effects in ADSE FinFETs results in lower cell leakage (I_{LEAK}) and hold stability. At the same time, lower storage node capacitance results in lower write time. However, at large ΔL_{SP}, degradation in I_{ON}

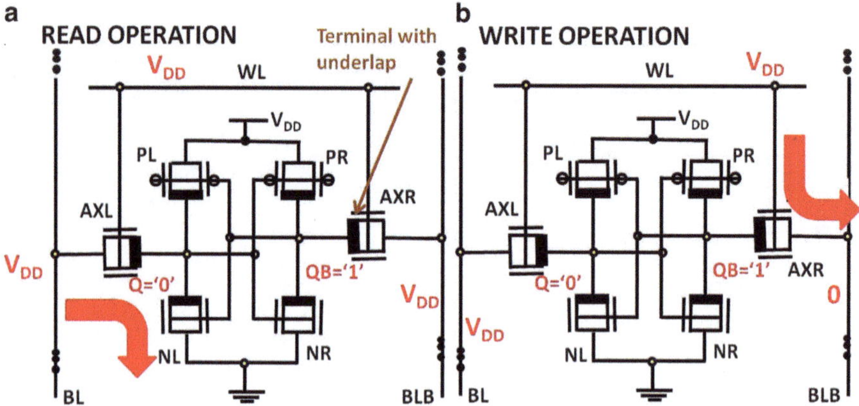

Fig. 11.16 Schematic of ADSE FinFET based 6 T SRAM showing current direction during (**a**) read and (**b**) write [17]

becomes dominant over the reduction in capacitance which leads to increase in write time. Weak access transistors during read leads to increased access time. ADSE FinFETs also have larger cell area due to increased device footprint. Figure 11.17 also shows comparison between ADSE and *LgDev* FinFET SRAMs. ADSE FinFET based SRAM exhibits superior read and write stability compared to *LgDev* based SRAM due to asymmetry in ADSE FinFETs which mitigates the read-write conflict. Moreover, ADSE FinFET SRAM shows lower cell leakage. I_{GATE} (gate leakage current) decreases with ΔL_{SP} for ADSE FinFET based SRAM. This is because of reduced edge tunneling current and negligible change in direct tunneling current (since gate length L_G does not change with ΔL_{SP}). On the other hand, *LgDev* based SRAM shows an increase in I_{GATE} due to increase in L_G which results in larger direct tunneling current. Also, larger storage node capacitance in *LgDev* leads to larger write time. However, *LgDev* FinFET SRAMs show lower access time than ADSE FinFET SRAMs due to higher strength of the access transistors. Moreover, superior short channel control in *LgDev* results in larger hold stability.

6 SRAMs based on Asymmetrically Doped (AD) FinFETs

In the previous section, we discussed that transistors with asymmetric gate underlap leads to the mitigation of the read-write conflict in SRAMs. In this section, we discuss another technique of introducing asymmetry in the transistor by unequally doping the source and drain of the transistors, as shown in Fig. 11.18. Similar to the ADSE FinFETs, AD FinFETs exhibit unequal currents for positive and negative V_{DS} [23]. The current flowing from the terminal with lower doping (i.e. when the terminal with larger doping acts as the source) is higher than the current flowing in the other direction (i.e. when the terminal with lower doping acts as the source).

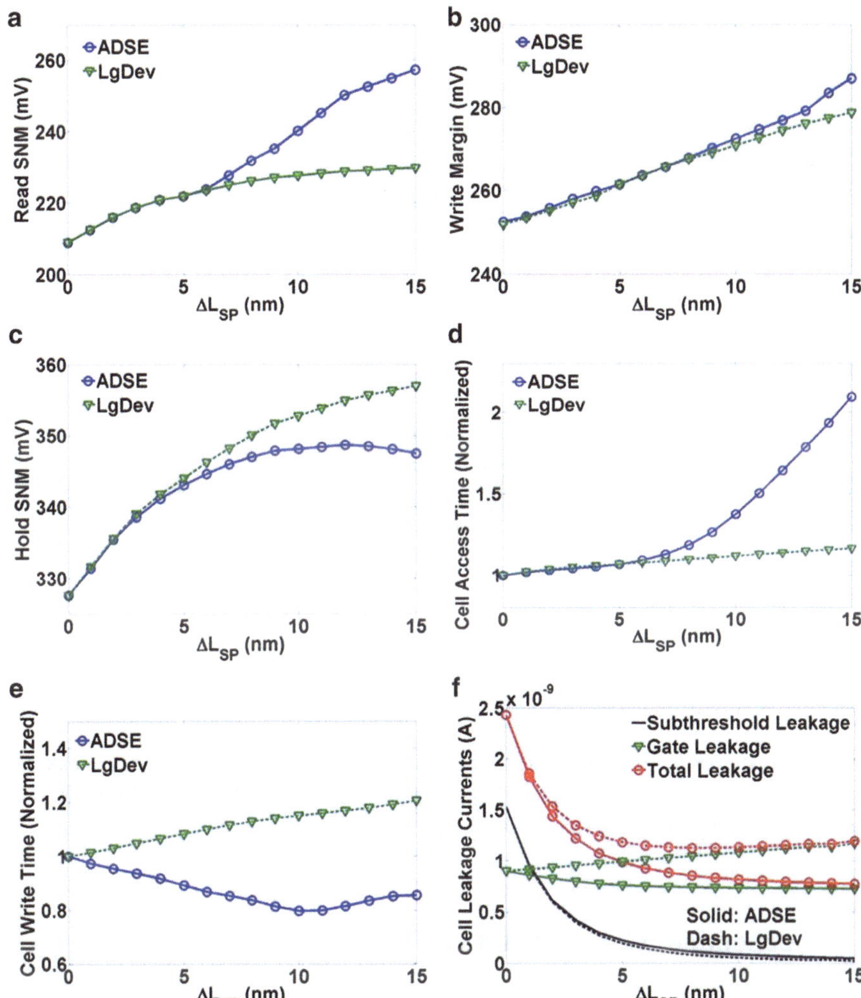

Fig. 11.17 (**a**) Read SNM (**b**) write margin (**c**) hold SNM (**d**) access time (**e**) write time and (**f**) cell leakage versus ΔL_{SP} for ADSE FinFET and LgDev FinFET SRAMs [17]

The asymmetry in current increases as the doping concentration of the drain terminal is reduced, with the source doping kept fixed. In order to achieve the mitigation of the read-write conflict, the terminal with lower doping is connected to the storage nodes (Fig. 11.18). During the read operation, the strength of the access transistor reduces, increasing the read stability. During the write operation, higher drive of the

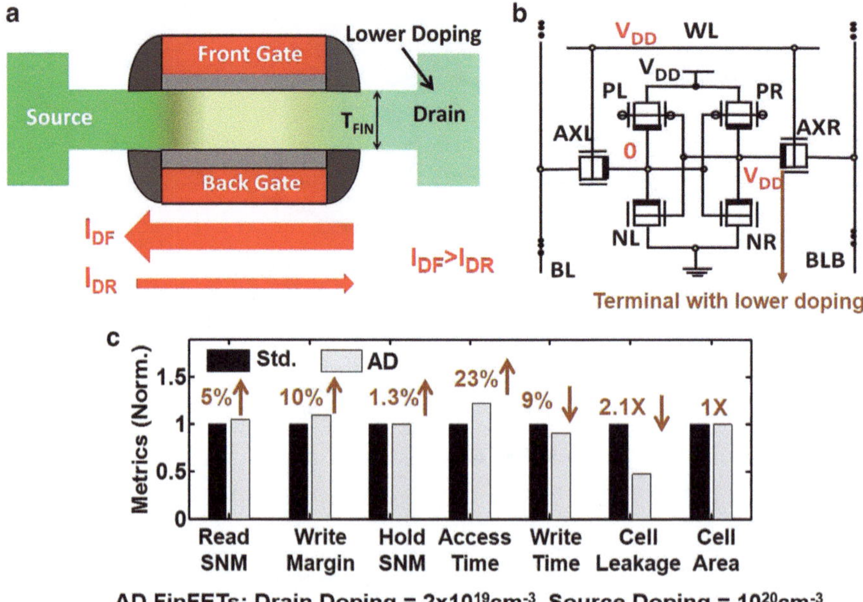

Fig. 11.18 (**a**) Asymmetrically Doped (*AD*) FinFETs (**b**) AD FinFET based SRAM and (**c**) comparison of the metrics of AD FinFET SRAM with standard FinFET SRAM (Std.) [23]

access transistors leads to increase in the write-ability. In addition lower cell leakage is achieved at the cost of increase in the access time. Unlike ADSE FinFET SRAM, there is no area penalty associated with this technique.

7 Independent Gate FinFETs

Two gates in FinFETs offer an option to modulate the transistor strength by independently controlling the front and back gates (FG and BG, respectively). Drain current in independent gate (IG) FinFETs is strongly dependent on the coupling between FG and BG. Due to ultra-thin body, FinFETs exhibit strong FG-BG coupling which increases as t_{Si} is reduced in scaled technologies. As an example, Fig. 11.19 shows the I_D-V_{GS} characteristics of tied-gate (TG) FinFETs and IG-FinFETs (with back gate voltage $(V_{BG}) = 0$) for $L_G = 32$ nm and 10.8 nm with corresponding $t_{Si} = 9$ nm and 4.5 nm, respectively. The electric fields emanating from BG reduce the inversion charge in IG FinFET formed due to FG. This leads to the ratio of I_D of IG and TG FinFETs less than 0.5. FinFETs with $t_{Si} = 4.5$ nm exhibits stronger FG-BG coupling compared to the FinFET with $t_{Si} = 9$ nm due to lower body thickness and increase in the quantum mechanical effects. As a result, the ratio of the currents of IG and TG FinFETs is smaller for FinFET with $t_{Si} = 4.5$ nm.

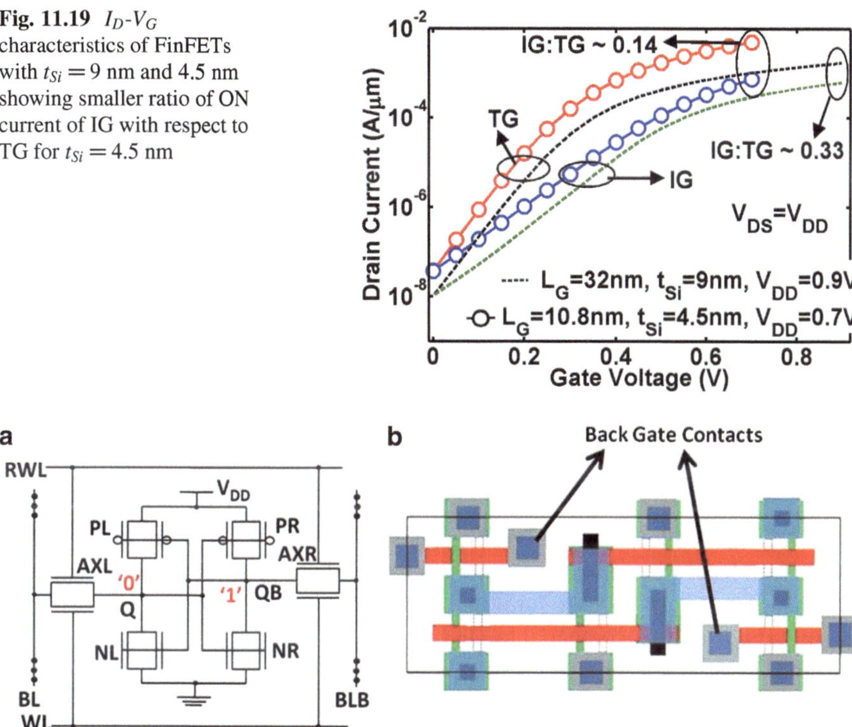

Fig. 11.19 I_D-V_G characteristics of FinFETs with $t_{Si} = 9$ nm and 4.5 nm showing smaller ratio of ON current of IG with respect to TG for $t_{Si} = 4.5$ nm

Fig. 11.20 (**a**) Schematic and (**b**) thin-cell layout of BMIG FinFET SRAM [20]

Independent control of two gates in IG FinFETs can be exploited to enhance the design flexibility of FinFET based SRAMs and mitigate the read-write conflict. In this section, we discuss some of the design techniques for SRAM using IG FinFETs and compare each with standard tied gate (TG) FinFET SRAM.

7.1 Bi-mode Independent Gate FinFET SRAMs

Figure 11.20 shows the IG FinFET SRAM proposed in [20], in which the two gates of the access transistors are controlled by two word-lines. The strength of the access transistor can be modulated between two values in the ON state by independently asserting/de-asserting the two wordlines. (Henceforth, the IG FinFET SRAM in Fig. 11.20 is called bi-mode independent gate (BMIG) FinFET SRAM). The two ON state modes for BMIG FinFETs are, henceforth, called high strength (HS) and low strength (LS) modes. During the read operation, only RWL is asserted (LS mode). Due to weak access transistor, read stability increases. This allows one fin to be used in the pull down transistors (instead of two, as in standard FinFET—see Sect. 2), thus reducing cell leakage. During the write operation,

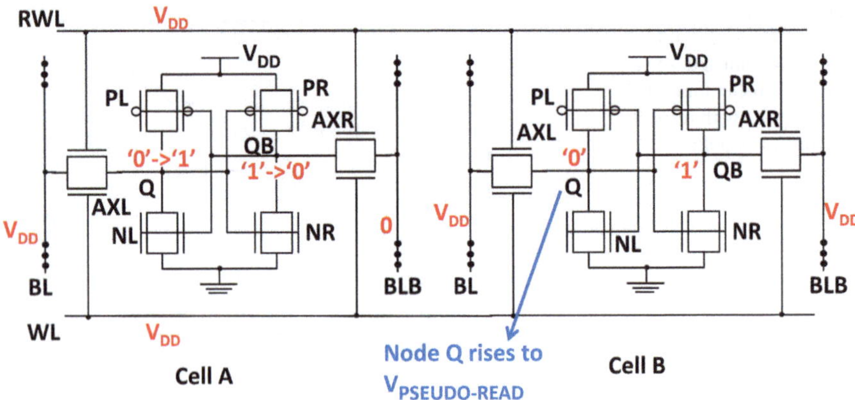

Fig. 11.21 Two BMIG SRAM cell showing pseudo-read problem during write

both WL and RWL are asserted (HS mode), increasing the strength of the access transistor. Hence, the conflicting requirements for the access transistor for high read stability and write-ability are mitigated. However, this technique incurs a cell area penalty (Fig. 11.20b), compared to TG FinFET SRAM due to the back gate contact of the access transistors. Moreover, weak access transistor results in significant degradation in the access time and therefore, limits the application of BMIG FinFET SRAM, as proposed in [20], to low power low throughput applications. Further, decreasing the number of fins in the pull-down transistors to 1 while using the access transistor in the HS mode during write worsens the pseudo-read issue [29, 30] in BMIG FinFET SRAM array, as illustrated in Fig. 11.21. Suppose write operation needs to be performed on cell A for which the bitlines are charged to the appropriate voltages. For another cell B in the same row, belonging to a different word, the bitlines are precharged to V_{DD} so that no new value is written on that cell. On asserting the wordlines, the node of cell B storing '0' gets charged to some positive voltage $V_{PSEUDO\text{-}READ}$ depending on the resistive divider action of AXL and NL, similar to the read operation. In the presence of process variations, $V_{PSEUDOREAD}$ may become larger than the trip point of the inverter formed by AXR, PR and NR causing the contents of cell B to flip. This is known as the pseudo-read or half-select problem. In BMIG FinFETs, the relative strength of access transistors with respect to pull-down transistor is higher than TG FinFET SRAM since a single fin is used in pull-down FinFET. This worsens the pseudo-read problem in BMIG FinFET SRAM compared to TG FinFET SRAMs. Mitigation of the pseudo-read problem can be achieved by architectural techniques like write-back scheme [31] in which the whole row is written simultaneously. Table 11.5 summarizes the differences between TG and BMIG FinFET SRAMs [20].

In addition to low throughput applications, BMIG FinFETs are also suitable for dynamic voltage frequency (DVF) scalable SRAMs [22]. During high workload conditions (high voltage - V_{DDH} operation), both the wordlines are asserted (HS mode) during read and write to achieve fast operations. At low workload conditions,

Table 11.5 Comparison of BMIG FinFET SRAM [20] with TG FinFET SRAM

Metric	TG	BMIG
Read SNM (mV)	210	240
Access time (Norm.)	1	2.2
Write time (Norm.)	1	~1

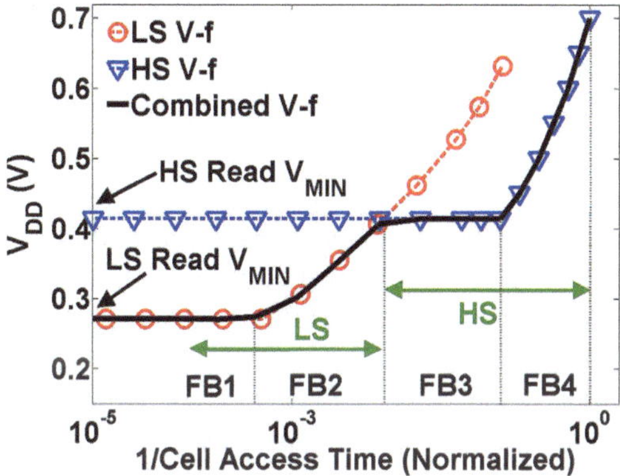

Fig. 11.22 V_{DD} versus the inverse of cell access time for BMIG FinFET SRAM showing HS and LS modes of operation and switching between the two modes. HS mode is switched to LS mode as transition is made from FB3 to FB2 and vice versa. The plot corresponds to the worst case process corner for read stability [22]

the cell is operated at low voltage (V_{DDL}). Although the frequency requirements are low, V_{DD} scaling may not be possible in conventional TG FinFET SRAM due to increase in read and write failures. However, in BMIG FinFET SRAM, only one wordline is asserted (LS Mode) during read at V_{DDL} which increases the read stability and allows low voltage operation. During write at V_{DDL}, both the wordlines are asserted (HS mode) to increase the strength of the access transistors and achieve low write failures. Since both the wordlines are asserted during read at V_{DDH}, two fins need to be used in the pull-down FinFETs to achieve high read stability. This makes the operation of BMIG FinFET SRAM at V_{DDH} identical to TG FinFET SRAM. Also, the cell leakage of BMIG FinFET SRAM is the same as TG FinFET SRAM. Moreover, the pseudo-read issue in BMIG FinFET SRAM with a single fin in the pull-down FinFET discussed above is mitigated in DVF scalable BMIG FinFET SRAM since the write operation of TG and BMIG FinFET SRAMs are identical.

Figure 11.22 shows V_{DD} versus the inverse of access time for BMIG FinFET at worst case process corner for read stability, illustrating the switching between LS and HS modes during the read operation. In the frequency band FB4, access FinFET is operated in the HS mode since the frequency requirements are high. In FB3, V_{DD} scaling is precluded by read failures in the HS mode. The cell is still operated in the HS mode since LS mode of operation needs higher V_{DD} to meet the frequency requirements in FB3. As transition is made from FB3 to FB2, the access transistor is switched to the LS mode, which allows further V_{DD} scaling. Finally, V_{DD} scaling is limited by read failures in the LS mode in FB1. Note, in TG FinFET SRAM, V_{DD} scaling is not possible at low frequencies due to read failures; however due to bi-modal operation during read, BMIG FinFET SRAM exhibits significantly lower minimum supply voltage (V_{MIN}) at low frequencies. The write operation of BMIG FinFET is identical to TG FinFET SRAM, hence write-ability of the cell is not affected. Thus, read-write conflict is mitigated in BMIG FinFET SRAM at low frequencies, at the cost of larger cell area due to back gate contacts of the access transistors [22].

7.2 Tri-Mode Independent Gate FinFET SRAMs

In Sect. 7.1, we discussed the benefits of BMIG FinFETs for DVF scalable SRAMs. The bi-modal operation of BMIG FinFET SRAMs results in significant reduction in V_{MIN} at low frequencies, compared to TG FinFET SRAM. However, no improvement in write-ability is achieved. In this sub-section, we discuss tri-mode independent gate (TMIG) FinFET SRAM which achieves improvement in write-ability along with increase in read stability at low frequencies and similar access time for high frequency operation [22].

This technique exploits the fact that spacer lithography technology used for fabrication of FinFETs [26] offers a single-fin and two-fin transistors in the same lithographic pitch. Thus, depending on the contact size and the fin pitch, the difference between the device footprint of FinFETs with one and two fins may be minimal. In TG and BMIG FinFET SRAMs, one fin is used in the access transistor, which requires etching of one of the fins after the two fins are fabricated using spacer lithography technique. However, in TMIG FinFET SRAM, the second fin is not etched away (Fig. 11.23). Instead, the gate metal is etched from the top of Fin2 (Fig. 11.23) to allow independent control of the two gates of Fin2. One of the gates of Fin2 is connected to the two gates of Fin1 (FG) while the other gate of Fin2 is controlled by BG. This structure allows three-way modulation of the strength of the access transistor, as shown in Fig. 11.24. Let us define V_{FG} and V_{BG} to be the voltages applied at the FG and BG of TMIG FinFET. Therefore, $(V_{FG},V_{BG}) = (V_{DD},V_{DD})$ leads to the highest strength of TMIG FinFET (high strength (HS) mode) followed by $(V_{FG},V_{BG}) = (V_{DD},0)$ (medium strength (MS) mode) and $(V_{FG},V_{BG}) = (0,V_{DD})$ (low strength (LS) mode).

Figure 11.25 shows the schematic and layout of TMIG FinFET based 6 T SRAM. An increase in area in TMIG FinFET SRAM compared to TG FinFET SRAM is

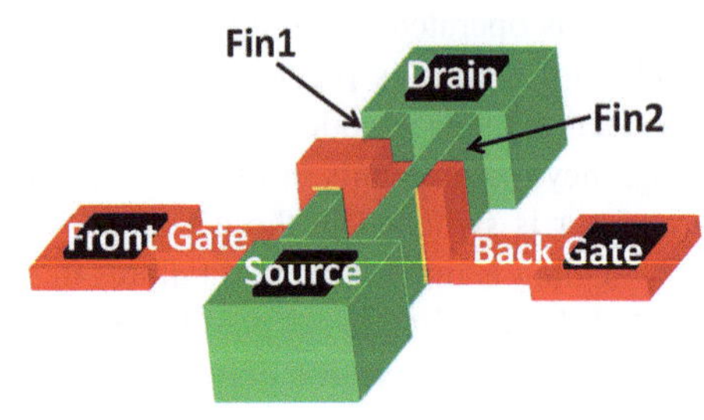

Fig. 11.23 TMIG (tri-mode independent gate) FinFET [22]

Fig. 11.24 Four configurations of TMIG FinFET [22]

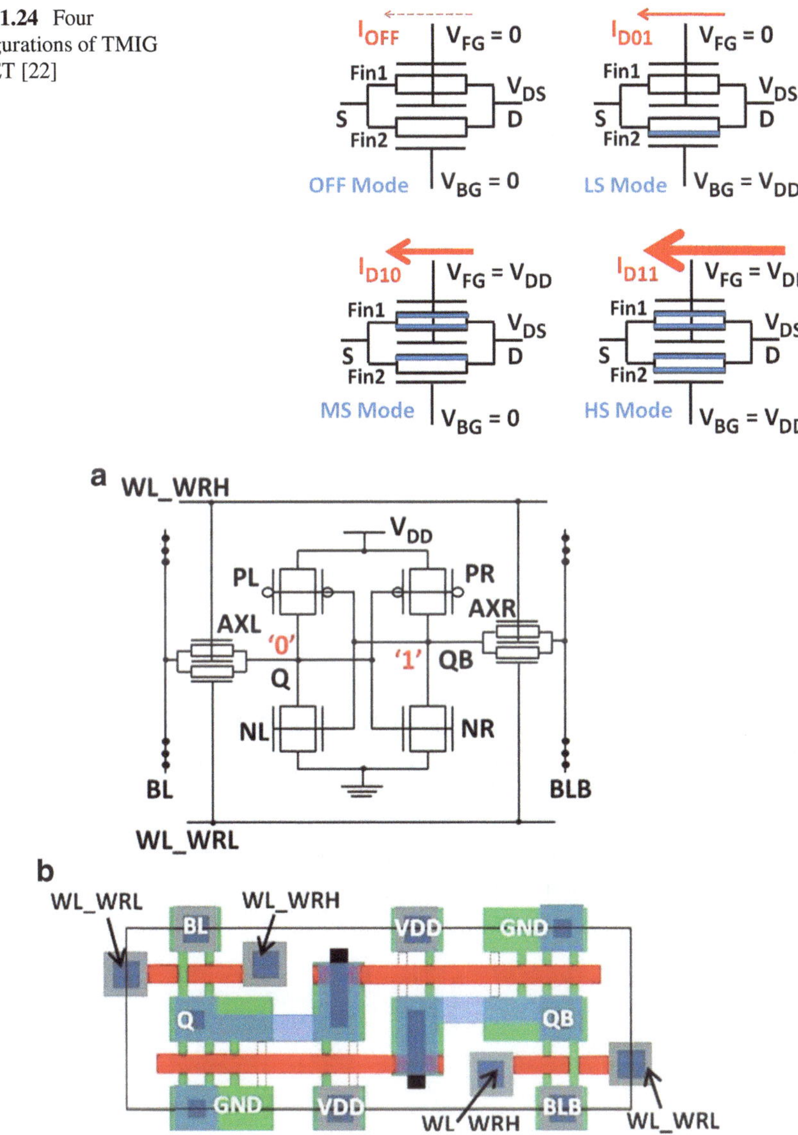

Fig. 11.25 (**a**) Schematic and (**b**) thin-cell layout of TMIG FinFET SRAM

observed due to the back gate contacts of the access transistors, similar to BMIG FinFET SRAM. Also note, an increase in x-dimension of TMIG FinFET SRAM cell allows three fins to be used in the pull down transistor without further increase in the area. The two wordlines modulate the strength of the access transistor between three values and achieve reduction in cell V_{MIN} along with the mitigation of read-write conflict, as explained next.

During write operation, TMIG FinFETs are operated in the HS mode by asserting both the wordlines. This leads to increase in the write-ability of the cell. During the read operation at high workload conditions, the access transistors are operated in the MS mode by asserting only WL_WRH to achieve similar access time and read stability as TG FinFET SRAM. Read operation during low workload conditions is performed by operating TMIG FinFETs in the LS mode to increase read stability and allow low voltage operation. The switching between MS mode and LS modes of operation during read in TMIG FinFET SRAM is similar to that between HS and LS modes in BMIG FinFET SRAM, as described in Sect. 7.1 (see Fig. 11.22). Due to larger number of fins the TMIG FinFET SRAM, cell leakage increases. However, note, MS mode in TMIG FinFET SRAM has a larger strength compared to a single fin access transistor in TG FinFET SRAM. Therefore, V_T of TMIG FinFET SRAM can be increased to achieve reduction in cell leakage while still achieving similar cell access time at high V_{DD} compared to TG FinFET SRAM. Hold stability in TMIG FinFET SRAM is slightly degraded compared to TG and BMIG FinFET SRAM due to increase in fin ratio of pull-down and pull-up FinFETs (N_{PD}:N_{PU}).

Figure 11.26 shows the iso-leakage comparisons of V_{DD} versus frequency curves for TMIG and TG FinFET SRAMs at the process corners that are worst from the point of view of read stability and write-ability. It can be observed that dynamic switching between MS and LS modes of operation during read achieves significant reduction in V_{MIN} at low frequencies while still maintaining similar performance at high V_{DD}. The plot in Fig. 11.26 is obtained by considering the four frequency bands similar to Fig. 11.22 in case of BMIG FinFET SRAM. However, unlike BMIG FinFET SRAM, TMIG FinFET SRAM achieves significant improvement in write-ability and reduction in V_{MIN} for the process corner that is the worst for write operation (Fig. 11.26). Hence, tri-modal operation of TMIG FinFET SRAM allows significant reduction in active V_{MIN} of the cell across different process corners at the cost of slight degradation in hold V_{MIN} [22].

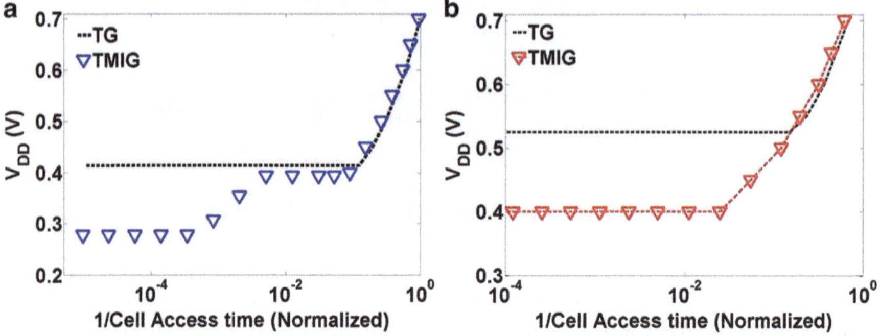

Fig. 11.26 Comparison of VDD versus 1/cell access time of TG and TMIG FinFET SRAMs for process corners that are the worst for (**a**) read stability and (**b**) write-ability [22]

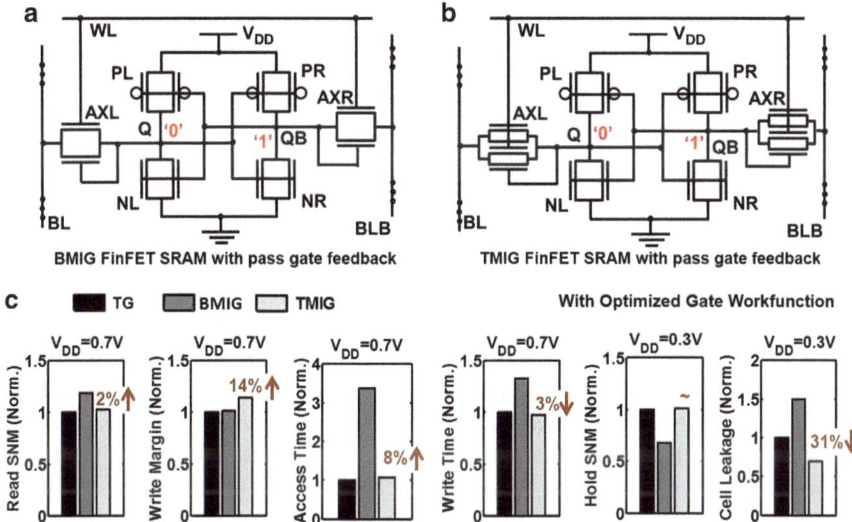

Fig. 11.27 Schematics of (**a**) BMIG [19] and (**b**) TMIG FinFET SRAMs with pass gate feedback (PGFB) FinFET 6 T SRAM [24] and (**c**) comparison with TG FinFET SRAM

7.3 Pass Gate Feedback in FinFET SRAMs

Pass-gate feedback is an interesting technique to mitigate the read-write conflict in SRAMs by providing the feedback to the back gate of the access transistor from the respective storage nodes. 6-T SRAM cell with BMIG FinFET employing pass gate feedback (PGFB) [19] is shown in Fig. 11.27a. The back gate of the access transistor is connected to the storage node. During the read operation, the access transistor corresponding to the node storing zero (AXL) is rendered weak, thus increasing read stability. The access transistor corresponding to the node storing '1' (AXR) is strong because both of its gates are ON. This further contributes to increase in the read stability. However, the access time increases significantly due to weak access transistor AXL. Moreover, during the write operation, the node QB discharges and decreases the strength of AXR. Depending on the coupling between the front gate and the back gate, decrease in the drive of AXR may lead to lower write-ability. This is especially critical in scaled technologies, in which quantum confinement effects lead to increase in the coupling between the front and back gates, as discussed before. However, pass-gate feedback is an interesting technique to mitigate the design conflicts in SRAMs. Coupled with write-assist techniques, pass-gate feedback can offer a useful design option for SRAMs in scaled technologies. It may be noted that since the back gate of the transistor is connected to the storage node, there is no area penalty with respect to TG FinFET SRAM.

It has been argued by the authors in [19] that write-ability of the cell can be further improved by connecting the back gate of PR and PL to a write word-line

(WWL) which is charged to V_{DD} during write. This weakens the pull-up transistors, making it easier to flip the contents of the cell. During the read operation, WWL is de-asserted which increases the strength of the pull-up transistors and raises the logic threshold voltage of the inverter formed by NR, PR and AXR, thereby increasing the read stability of the cell. However, the extra back gate contact can lead to significant increase in the area of the bit-cell. Other techniques like gate-workfunction engineering can be used to increase the write-ability. However, this leads to lower hold stability and increase in the cell leakage [24].

Pass-gate feedback can also be employed in SRAMs using TMIG FinFETs as the access transistors, as shown in Fig. 11.27b. During the read operation, the strength of AXL (with BG connected to Q = '0') is comparable to a single-fin FinFET [24]. In other words, AXL in TMIG FinFET SRAM is slightly stronger than AXL in TG FinFET SRAM (since single-fin FinFETs are typically used as access transistors in TG FinFET SRAM, as discussed previously). This tends to mildly decrease the read stability of the cell. On the other hand, the strength of AXR (with BG connected to QB = '1') is twice of AXR in a TG FinFET SRAM. This increases the trip point of the inverter formed by PR. NR and AXR and therefore, tends to increase the read stability. The overall effect is that read stability of TMIG FinFET SRAM is comparable to that of TG FinFET SRAM.

During the write operation, the strength of AXR (with BG connected to QB = '1') in TMIG FinFET SRAM is higher than that in TG FinFET SRAM. As a result, large improvement in write-ability is achieved. As the voltage at node QB discharges during write, the strength of AXR decreases, but is still sufficient to provide high write-ability.

Hence, TMIG FinFET SRAM with pass-gate feedback achieves significant improvement in the write-ability at comparable read-stability. In addition, co-optimization of the gate workfunction leads to lower cell leakage along with comparable hold stability at the cost of a mild increase in the access time. The cell-level comparison of BMIG and TMIG FinFET SRAMs employing pass-gate feedback is shown in Fig. 11.27c.

Connecting the back gate of the access transistor to the storage node may have issues with the leakage and stability of the unaccessed cells during read and write operations [19]. During the read operation, cell leakage and leakage through AXL may increase significantly for the bit-cell belonging to an unaccessed row. Considering the worst case, suppose the accessed bit-cell stores '0' and all the unaccessed bit-cells in the same column store '1' (value opposite to that of the accessed bit-cell), as shown in Fig. 11.28a. During the read operation, BL discharges to some voltage (V_{DD}-ΔV). Hence, V_{GS} for the back gate of AXL of all the unaccessed cells becomes positive (ΔV), leading to an increase in cell leakage and bitline leakage through AXL. Increase in bitline leakage delays the development of voltage differential ΔV between BL and BLB, resulting in increased access time.

During the write operation, the bit-cells belonging to the unaccessed rows may be disturbed if they store the same value as the accessed node (see Fig. 11.28b—note both the accessed and unaccessed cells store '0' i.e. $V_Q = 0$). Since the back gate of AXR is turned ON, the storage node V_{QB} of the unaccessed cells is disturbed

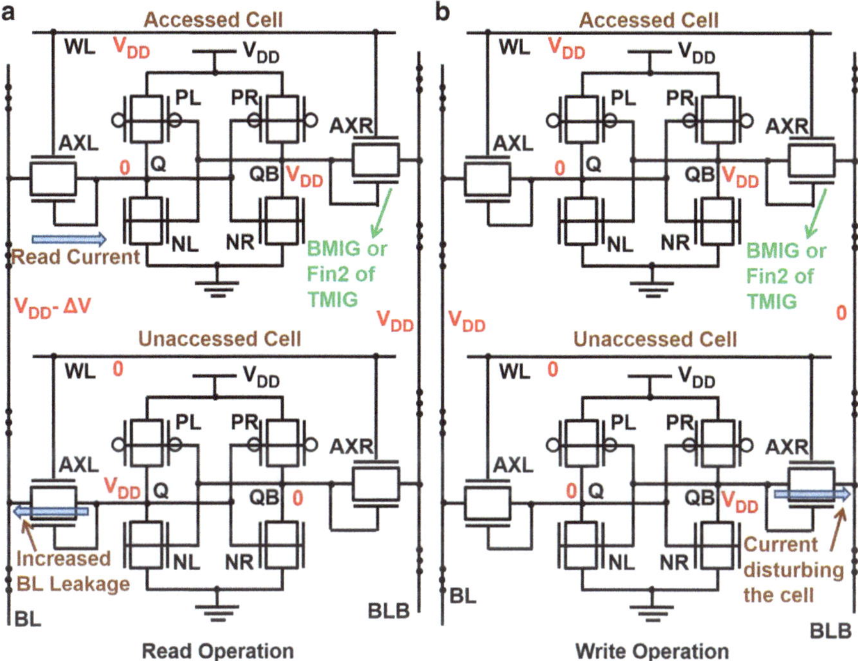

Fig. 11.28 Array-level implications of using pass gate feedback in SRAMs during (**a**) read and (**b**) write. The access transistors shown in the figure are BMIG FinFETs or Fin2 of TMIG FinFETs (Fin1 of TMIG FinFETs not shown)

due to current flowing through AXR, thus introducing an additional source of failure under process variations. These issues are not just with 6 T PGFB SRAM [19] but with other cells that uses PGFB configuration like 4 T SRAM cells in [32]. It is interesting to note that increase in the coupling between front gate and back gate may lead to the mitigation of these issues, as shown in the analysis in [24]. Moreover, in TMIG FinFET SRAMs, these issues are significantly reduced compared to BMIG FinFET SRAMs. Nevertheless, one needs to be aware of these issues while designing SRAMs with pass-gate feedback.

8 Summary

FinFETs have emerged as promising alternatives to bulk MOSFETs in scaled technologies because of lower short channel effects and superior scalability. However, quasi-planar structure of FinFET results in width quantization which reduces the design flexibility of FinFET based circuits. This effect has an aggravated impact on SRAMs, in which transistor sizing is critical for cell stability. Therefore, device-circuit co-design techniques are required which mitigate the effect of width quantization in FinFET based SRAMs. In this chapter, we discussed such techniques and presented the implications of each on SRAM design. We showed that techniques like fin ratio-t_{Si} co-optimization, joint optimization of H_{FIN}-t_{Si}-t_{OX}-V_{DD}-V_T, fin orientation and spacer thickness optimization alleviate the effect of width quantization and expand the design space of FinFET-based SRAMs. We also described asymmetric drain spacer extension (ADSE) FinFETs and asymmetrically doped (AD) FinFETs in which the asymmetry in the drain current for positive and negative V_{DS} was exploited to mitigate the read-write conflict in 6 T SRAMs. Finally, we described various IG FinFET based SRAMs and discussed their benefits in mitigating the design conflicts and the trade-offs associated with each technique.

References

1. Yuan Taur and Tak. H Ning, "Fundamentals of Modern VLSI Devices", *Cambridge University Press*, pp: 140–144.
2. Shekhar Borkar, Tanay Karnik, Siva Narendra, Jim Schantz, Ali Keshavarzi and Vivek De, "Parameter variations and impact on circuits and microarchitecture", *Proc. IEEE Design Automation Conference*, 2003, pp: 338–342.
3. G.G. Shahidi, T.H. Ning, T.I. Chappell, J. H. Comfort, B. A. Chappell, R. Franch, C. J. Anderson, P. W. Cook, S. E. Schuster, M.G. Rosenfield, M.R. Polcari, R.H. Dennard and B. Davari, "SOI for a 1-volt CMOS technology and application to a 512 Kb SRAM with 3.5 ns access time", *Int. Electron Device Meeting*, Dec 1993, pp: 813–816.
4. K. K. Young, "Short-Channel Effect in Fully Depleted SOI MOSFETs", *IEEE Trans. Electron Devices*, vol. 36, no. 2, Feb. 1989, pp: 399–402.
5. M. Vinet, T. Poiroux, J. Widiez, J. Lolivier, B. Previtali, C. Vizioz, B. Guillaumot, Y. Le Tiec, P. Besson, B. Biasse, F. Allain, M. Casse, D. Lafond, J.-M. Hartmann, Y. Morand, J. Chiaroni and S. Deleonibus, "Bonded planar double-metal-gate NMOS transistors down to 10 nm", *IEEE Electron Device Letters*, vol. 26, no. 5, May 2005, pp: 317–319.
6. J.P. Colinge, M.H. Gao, A. Romano-Rodriguez, H. Maes, C. Claeys, "Silicon-on-insulator 'gate-all-around device'",*Int. Electron Device Meeting*, Dec. 1990, pp: 595–598.
7. J. Widiez, F. Dauge, M. Vinet, T. Poiroux, B. Previtali, M. Mouis, S. Deleonibus, "Experimental gate misalignment analysis on double gate SOI MOSFETs", *Proc. Int. SOI Conf.*, Oct. 2004, pp:185–186.
8. D. Hisamoto, W. –C. Lee, J. Kedzierski, H. Takeuchi, K. Asano, C. Kuo, E. Anderson, T-J King, J. and C. Hu;, "FINFET- a self-aligned double-gate MOSFET scalable to 20 nm", *IEEE Trans. Electron Devices*, vol. 47, no. 12, pp: 2320–2325, Dec 2000.
9. B.S. Doyle, S. Datta, M. Doczy, S. Hareland, B. Jin, J. Kavalieros, T. Linton, A. Murthy, R. Rios, R. Chau, "High performance fully-depleted tri-gate CMOS transistors," *IEEE Electron Device Letters*, vol. 24, no. 4, April, 2003, pp: 263–265.
10. P. Ranade, H. Takeuchi, T. –J King and C. Hu, "Work Function Engineering of Molybdenum Gate Electrodes by Nitrogen Implantation," *Electrochemical and Solid State Letters*, vol. 4, no. 11, 2001, pp: G85-G87.
11. B. Yu, L. Chang, S. Ahmed, W. Haihong, S. Bell, Y. Chih-Yuh, C. Tabery, H. Chau, Q. Xiang, T. –J. King, J. Bokor, C. Hu, M. –R. Lin and D. Kyser, "FinFET scaling to 10 nm gate length", *Int. Electron Device Meeting*, Dec 2002, pp: 251–254.
12. T. Ludwig, I. Aller, V. Gernhofer, J. Keinert, E. Nowak, R. V. Joshi, A. Mueller and S. Tomaschko, "FinFET Technology for Future Microprocessors", *IEEE SOI Conference*, October 2003, pp. 33–34.

13. D. Lekshmanan, A. Bansal and K. Roy, "FinFET SRAM: Optimizing Silicon Fin Thickness and Fin Ratio to Improve Stability at iso Area," *IEEE Custom Integrated Circuits Conference*, Sept. 2007, pp. 623–626.
14. H. Ananthan and K. Roy, "Technology and circuit design considerations in quasi-planar double-gate SRAM,"*IEEE Transactions on Electron Devices*, vol. 52, no. 2, Feb. 2006, pp. 242–250.
15. S. Gangwal, S. Mukhopadhyay and K. Roy, "Optimization of Surface Orientation for High-Performance, Low-Power and Robust FinFET SRAM," *IEEE Custom Integrated Circuits Conference*, Sept 2006, pp. 433–436.
16. A. Bansal, S. Mukhopadhyay and K. Roy, "Device-Optimization Technique for Robust and Low-Power FinFET SRAM Design in NanoScale Era," *IEEE Transactions on Electron Devices*, vol. 54, no. 6, June 2007, pp. 1409–1419.
17. A. Goel, S. K. Gupta and K. Roy, "Asymmetric Drain Spacer Extension (ADSE) FinFETs for Low Power and Robust SRAMs", *IEEE Trans. Electron Devices*, vol. 58, no. 2,Feb 2011, pp:296–308.
18. K. Endo, S. –I. O'uchi, Y. Ishikawa, Y. Liu, T. Matsukawa, K. Sakamoto, J. Tsukada, K. Ishii, H. Yamauchi, E. Suzuki, M. Masahara," Enhancing SRAM cell performance by using independent double-gate FinFET", *Int. Electron Device Meeting*, Dec. 2008.
19. A. Carlson, Z. Guo, S. Balasubramanian, R. Zlatanovici, T. J. King Liu, B, Nikolic, "SRAM Read/Write Margin Enhancements using FinFETs", *IEEE Trans. VLSI*, vol. 18, no. 6, June 2010, pp:887–900.
20. Z. Liu, S.A. Tawfik, V. Kursun, "An independent-gate FinFET SRAM cell for high data stability and enhanced integration density", *IEEE Int. SOI Conf.*, 2007, pp: 63–66.
21. S. A. Tawfik, Z. Liu, and V. Kursun, "Independent-Gate and Tied-Gate FinFET SRAM Circuits: Design Guidelines for Reduced Area and Enhanced Stability", *Int. Conf. Microelectronics*, 2007, pp: 171–174.
22. S. K. Gupta, S. P. Park and K. Roy, "Tri-mode Independent Gate (TMIG) FinFETs for dynamic voltage/frequency scalable 6 T SRAMs," to appear in *IEEE Trans. Electron Devices*.
23. F. Moradi, S. K. Gupta, G. Panagopoulos, D. T. Wisland, H. Mahmoodi and K. Roy, "Asymmetrically Doped FinFETs for Low-Power Robust SRAMs," *IEEE TED*, vol.58, no.12, pp.4241,4249, Dec. 2011.
24. S. K. Gupta, J.P. Kulkarni, and K. Roy, "Tri-Mode Independent Gate FinFET-Based SRAM With Pass-Gate Feedback: Technology–Circuit Co-Design for Enhanced Cell Stability," *IEEE TED,* vol.60, no.11, pp.3696,3704, Nov. 2013
25. Jan M. Rabaey, Anantha Chandrakasan and Borivoje Nikolic, "Digital Integrated Circuits, A Design Perspective", *Prentice Hall Electronics and VLSI series*, pp:657–661.
26. D. Esseni, A. Abramo, L. Selmi, E. Sangiorgi, "Physically based modeling of low field electron mobility in ultrathin single- and double-gate SOI n-MOSFETs", *IEEE Trans. Electron Devices*, vol. 50, no. 12, Dec. 2003, pp:2445–2455.
27. D. Lekshmanan, A. Bansal, K. Roy, "Body Thickness Optimization and Sensitivity Analysis for High Performance FinFETs," *Device Research Conf.*, June 2007, pp: 91–92.
28. Y. –K.Choi, T. –J. King, C. Hu, "A Spacer Patterning Technology for Nanoscale CMOS", *IEEE Trans. Electron Devices*, vol. 49, no. 3, March 2002, pp:436–441.
29. F. Bauer. G. Georgakos, D. Schmitt-Landseidel, "A Design Space Comparison of 6 T and 8 T SRAM Core-Cells", *Lecture Notes in Computer Science*, vol. 5349, 2009, pp: 116–125.
30. M. –L. Fan, Y. –S Wu, V P –H Hu, C –Y Hsieh, P. Su, C. –T. Chuang, "Comparison of 4T and 6T FinFET SRAM Cell for Subthreshold Operation Considering Variability – A Model-Based Approach", *IEEE Trans. Electron Devices*, vol. 58, no. 3, March 2011, pp:609–616.
31. K. C. Chun, P. Jain, J. H. Lee and C. H. Kim, "A 3 T Gain Cell Embedded DRAM Utilizing Preferential Boosting for High Density and Low Power On-Die Caches", *IEEE, Journal Solid State Circuits*, vol. 46, no. 6. June 2011, pp: 1495–1505.
32. M –L Fan et al, "Comparison of 4T and 6T FinFET SRAM Cell for Subthreshold Operation Considering Variability – A Model-Based Approach", *IEEE Trans. Electron Devices*, vol. 58, no. 3, March 2011, pp: 609–616.

Chapter 12
Variability-Aware Clock Design

Matthew R. Guthaus and Gustavo Wilke

Abstract High-performance clock network design has been a challenge for many years due to the drastically increasing effect of process variability. In addition, tight power budgets have lowered supply voltage levels which make designs more sensitive to noise. Together, variability and noise present a colossal challenge to clock designers in order to meet timing, yield, and power simultaneously. This chapter discusses the different strategies that designers use to ameliorate variability and noise problems in clock network design.

1 Introduction

The clock is the most important signal in any synchronous design, because it controls all synchronous communication. A failure in the clock signal can lead to errors in one or more sequential elements. Strict constraints on clock timing must be used to guarantee the correct behavior of synchronous digital blocks. The clock period must be defined in such a way that data will always be ready and stable before clock edge arrives at the clock pin of the sequential elements which are also called clock sinks.

Figure 12.1 shows the timing parameters that must be considered to safely determine clock frequency and ensure correct operation. Assume that $T_{CK}(n)'$ is clock arrival time at flip-flop A, at clock cycle n and $T_{CK}(n+1)''$ is clock arrival time at flip-flop B during clock cycle $(n+1)$. Data propagation time through flip-flop A is represented by TP_{FFA}. The minimum (maximum) delay associated with the combinational logic is represented by $T_{C,MIN}$ ($T_{C,MAX}$). Flip-flop B setup and hold times are represented respectively by TS_{FFB} and TH_{FFB}. Given that T_{CLOCK}

M.R. Guthaus (✉)
University of California Santa Cruz, 1156 High St. MS SOE3,
Santa Cruz, CA 95064, USA
e-mail: mrg@ucsc.edu

G. Wilke
UFRGS, Av. Bento Gonçalves, 9500 - Campus do Vale - Bloco IV, Porto Alegre,
RS 91509-900, Brazil
e-mail: wilke@inf.ufrgs.br

R. Reis et al. (eds.), *Circuit Design for Reliability*,
DOI 10.1007/978-1-4614-4078-9_12

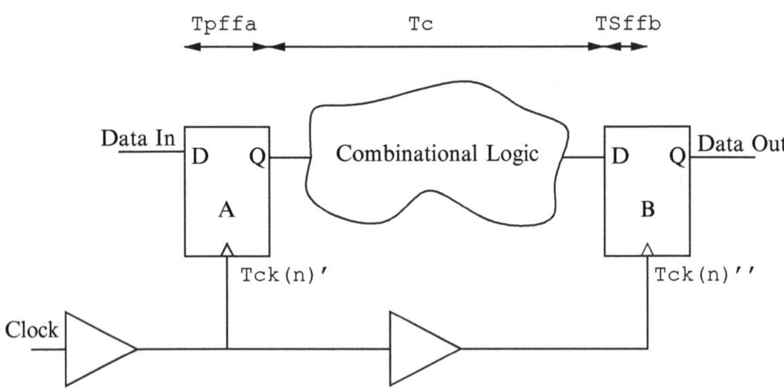

Fig. 12.1 Clock period definition

is the clock period, Eqs. (12.1) and (12.2) represent the timing constraints that must be satisfied to guarantee the correct data storage for the case illustrated by Fig. 12.1.

$$T_{CLOCK} \geq TP_{FFA} + T_{C,MAX} + TS_{FFB} + (Tck(n)' - T_{CK}(n+1)'') \quad (12.1)$$

$$TH_{FFB} \geq TP_{FFA} + T_{C,MIN} + (T_{CK}(n)' - T_{CK}(n)'') \quad (12.2)$$

Equation (12.1) represents a lower bound to the clock period. The clock period has to be chosen such that Eq. (12.1) is going to be respected for any path connecting any two flip-flops in the design. Equation (12.2) limits the minimum delay $T_{C,MIN}$ allowed for any combinational path. Both equations have to be simultaneously respected to guarantee the correct circuit behavior.

The main challenge faced by clock designers is how to design robust clock distribution networks respecting Eqs. (12.1) and (12.2) within the allowed power budget. Traditional worst-case analysis is known to be robust but is usually too pessimistic and leads to over-designed circuits that waste power. In the last decade statistical techniques have begun to aid designers to safely reduce the amount of pessimism in IC design. Statistical design, however, is taking longer than expected to be applied to clock design, because the importance of clock timing on the circuit robustness makes designers conservative.

An alternative to traditional clock tree based architectures began to appear in microprocessor designs in the late 1990s. To decrease the effect of process and environmental variability, microprocessor designers adopted clock meshes and clock spines in many designs [3, 27, 35, 40]. Due to the redundant structure of meshes and spines, fast and slow paths are able to compensate each other and reduce the clock arrival time variability [18]. The main disadvantage presented by these architectures, however, is the large power penalty caused by the extra wiring resources and short circuit currents.

The remainder of this chapter discusses the impact of variation sources on the clock design and presents different techniques to make the clock network less sensitive to these variations. In the remainder of this section, we will present some important definitions and discuss the parameters that make a clock network robust. In Sect. 2, the different sources of variability and their impact on clock networks are discussed. Sections 3 and 4 present design strategies to create clock networks that are less sensitive to variability.

1.1 Definitions

The clock network is designed so that the time it takes for a clock edge to travel from the clock root to any sink is the same. This delay is defined as the clock *arrival time* (At). Due to design mismatch, the clock arrival time at the clock sinks is often not exactly the same. The difference between the clock arrival time at different sinks is defined as *clock skew*. Given clock sinks $s_i, s_j \in S$ and $At(s_i)$ which represents the clock arrival time at the clock sink s_i, the clock skew (Sk) for a set of clock sinks S can be defined in a more formal manner as follows:

$$Sk(S) = max_{(i,j)}|At(s_i) - At(s_j)|. \tag{12.3}$$

In addition to design mismatch clock skew, environmental variations such as temperature, crosstalk, and power supply noise are also responsible for degrading circuit performance along with the clock skew. Many of these variations can occur over time resulting in uncertainty in the clock arrival time at each clock sink which is called *clock jitter*.

When defining the clock period, clock skew and clock jitter must be budgeted. The sum of both values represents an upper bound for the difference in the clock arrival time at any two flip-flops. In Eq. (12.1), the effect of clock skew and clock jitter is illustrated by the $T_{CK}(n)' - T_{CK}(n+1)''$ term. For sake of simplicity, all future references to clock skew will refer to the combined uncertainty due to both clock skew and clock jitter.

1.2 Robustness

Clock robustness is key for the correct operation of any circuit. Unlike the other signal lines, a fault in the clock signal can easily propagate to hundreds of sequential elements generating errors in multiple logic bits. These logic errors can occur due to two reasons: delay violations and glitches. Both types of faults can in turn create logic errors.

Clock delay violations are caused by the non-robust budgeting of clock margins. When the clock skew violates the budgeted margins due to process or environmental variations, Eqs. (12.1) and (12.2) are also violated. This situation leads to the storage of invalid values in some sequential elements. If the clock arrival time at the destination flip-flop is delayed or the clock edge at the transmitter flip-flop is expedited, the data hold time constraints can be violated by the fastest paths in combinational logic (Eq. (12.2)). When the inverse situation happens, i.e., transmitter clock is late or receiver clock is early, the time for data to propagate through combinational logic is shortened leading to setup time violations (Eq. (12.1)).

Clock glitches are the another potential cause of logic errors. It can be cause by two main issues, radiation [28] and poor synchronization of clock gating control signals on latch-free clock gaters. A clock glitch is a spurious transition in the clock signal that can lead to invalid values stored in affected sequential elements. Although invalid data can be stored in memory elements due to a clock glitch, it is possible that the failure is corrected when the next clock edge arrives. In case the invalid data is not corrected by the following clock edge (e.g., write enable signal was turned off) the error may propagate to multiple sequential elements in the following clock cycles. Even when the invalid data is quickly corrected it can still cause errors. During the time that a flip-flop stores an invalid value a data glitch is generated. The data glitch will cause errors if it has enough time to propagate to a memory element.

A robust clock network has to minimize the probability that a delay fault will occur and minimize the chances of a clock glitch reaching several clock sinks. In the next sections, techniques to design robust clock networks are described.

2 Variability

Understanding the different variability sources and properly modeling them is a very important task during the clock design. When process and environmental effects were not the main cause of clock skew, the analysis of clock network performance could be done using a corner-based approach. Corner-based analysis bounds all the parameters in the design by minimum and maximum values. Corner analysis is robust if the worst case corner can be identified, but it also became overly pessimistic as the impact of variability increases.

Statistical analysis resurged in the early 2000s as an alternative to reduce pessimism under scenarios where the impact of variation sources is large. The use of statistical methods do not guarantee by itself an accurate estimation of clock skew values. The way that each source of variation is modeled strongly affects the accuracy in the clock skew estimation. The correct modeling of each variability source is fundamental for the proper estimation of the delay margins. In this section, we discuss the characteristics of the different types of variation and how they should be modeled.

2.1 Types of Variation

Variability in clock networks can be due to manufacturing process variation (P), on-chip supply voltage variation (V), on-chip temperature (T), and cross-talk between the clock and signal wires (X). These are often referred to as PVT variations (ignoring the X). PVTX variations, however, are a significant challenge in the synthesis of clock networks, because they can add skew and jitter which reduces timing margins.

Each form of variation can be modeled as either a statistical distribution or as a bounded corner value. The distribution and, more importantly, the correlation of parameters must be considered in order to model a source of variation [29]. Variation that is observed from die-to-die (D2D) is called *inter-die variation* while variation that occurs within a die (WID) is called *intra-die variation.* Often the WID variation exhibits significant *spatial correlation* so that nearby locations are correlated but distant locations are not. Many sources of variation are *systematic* which can be predicted in terms of design or process details. Systematic variation is distinctly different from correlated variation because it is not a random variable when the cause of variation is considered [15].

2.1.1 Process

Process variation can be either front-end (device) variation or back-end (interconnect) variation. Some front-end variation such as V_{th} are uncorrelated and do not exhibit spatial correlation or systematic behavior [31]. Other device parameters such as L_{eff} are mostly systematic [15] with high spatial correlation except for a small portion due to line-edge roughness (LER) [26, 37]. Interconnect variation is primarily due to systematic chemical mechanical polishing (CMP) variation which affects the metal height and therefore interconnect parasitics with high correlation [38]. However, line-edge roughness (LER) can also affect interconnect as a highly uncorrelated behavior.

2.1.2 Voltage

On-chip power supply variation is one of the most significant sources of variation in a clock network [39, 41]. Due to the high power consumption during peak switching, large $L\frac{di}{dt}$ voltage drops can occur in the power grid. Chip-level clock distributions include decoupling capacitors with the large clock drivers to reduce supply-related variation. The decoupling capacitors low-pass filter the supply variation and remove sharp current peaks. The remaining supply variation has high spatial correlation [41].

2.1.3 Temperature

On-chip temperature fluctuations are also very significant for both device and interconnect variations. Temperature directly affects the mobility of silicon and the resistance of metal wires [2, 22]. The effect on interconnect capacitance, however, is relatively small. The spatial correlation of temperature is highly dependent on the package and heat sink model. For high-performance designs with large heat sinks, there can be temperature gradients of 30–40 °C on-chip. Whereas, for GPUs (with high activity in all regions) or low-performance designs without active heat removal, the ambient chip temperature will be higher yet gradients will be lower [41]. Despite the high spatial correlation, significant temperature-induced skew can be present between the initial (cool) chip and a steady-state thermal map due to mismatch in clock sensitivities to temperature. The temperature map can also dynamically change depending on activity patterns.

2.1.4 Crosstalk

The clock network is the largest cross-talk aggressor due to its large capacitance and span. However, signal nets also affect the jitter of individual clock branches in trees. Adjacent layers have different routing orientations and thicker oxides, so within-layer coupling is the most significant problem. Clock wires are usually shielded to reduce these crosstalk effects [39].

2.2 Impact of Variation Models

The law of large numbers suggests that independent WID variation can be reduced simply by increasing the number of random variables. For example, the mean and standard deviation of a sum of independent Gaussian distributions is

$$\mu = \sum_{\forall i} \mu_i \qquad \sigma_{ind} = \sqrt{\sum_{\forall i} \sigma_i^2} \tag{12.4}$$

which implies that increasing $|i|$ will also decrease the total $\frac{\sigma_{ind}}{\mu}$. As a limit, this will approach zero with infinite Gaussians. As an example, consider a single transistor which is known to have V_{th} variation due to random dopant fluctuation. Pelgrom's model [31] demonstrates this with the relation

$$\sigma_{V_{th}} \propto \frac{1}{\sqrt{WL}} \tag{12.5}$$

that implies that larger transistors will have less variation. Clock buffers in general are large devices ranging from 10 to 64× minimum size for tree-type distributions to huge 29 cm drivers for clock grids [4]. The previous models correctly imply that $\sigma_{V_{th}}$ is in fact not very significant for clock buffers due to the averaging over very large device sizes.

The problem with the previous observation is that the law of large numbers only holds true if the variations are uncorrelated. In the case of full correlation, the mean and standard deviation become

$$\mu = \sum_{\forall i} \mu_i \qquad \sigma_{corr} = \sum_{\forall i} \sigma_i \tag{12.6}$$

which will not decrease the total $\frac{\sigma_{corr}}{\mu}$ as $|i|$ increases. As an example, assume that every clock buffer has an uncorrelated power supply variation of $V_{dd} = 1.0 \pm 0.15$. According to the independence assumption, it is therefore better to implement many instances of a smaller device as opposed to fewer instances of the larger device to reduce the total variation while having the same total drive strength. By simply doing this, one can achieve a relative reduction in variability of

$$\frac{\sigma_{ind}}{\sigma_{corr}} = \frac{\sqrt{\sum_{\forall i} \sigma_i^2}}{\sum_{\forall i} \sigma_i}. \tag{12.7}$$

If a simple long-channel transistor model is used, a single 8× library buffer will have 12 % variation in delay due to supply variations while eight 1× library buffers in parallel will have a variation of 4.2 %. In reality, however, all of these buffers will be adjacently placed and will have the same power supply rail so this is an artifact of using the wrong correlation model.

Similarly, assume a wire model with an uncorrelated 10 % wire variation which affects both resistance and capacitance. It is better to segment a 1 mm wire into 100 segments as opposed to 2 segments, because this will reduce the delay variation by 7×. However, if this variation is due to CMP, it will have high spatial correlation at distances less than 0.5–1 mm, so the above independent assumption is not valid and leads to an incorrect observation. Conversely, if the variation is due to uncorrelated LER, then segmenting the wires into large 500 μm intervals is not correct and will overestimate the variation.

It is also often not accurate to assume that a variation is a percentage of a nominal value. As an example, consider an uncorrelated 10 % wire variation due to line-edge roughness (LER). A wide wire will have less LER delay variation compared to a minimum width wire because LER is an absolute number and not a percentage of the nominal resistance or capacitance.

One overlooked approach for reducing variation is to improve the correlation rather than minimize variation. This is especially true in clock distributions since we are concerned with the relative difference in delays for skew. If you consider two insertion delays with $\sigma = 0.1$, the skew variation is $3\sigma = 3\sqrt{0.1^2 + 0.1^2} = 0.42$ if they are independent whereas it is reduced to zero if they are perfectly correlated. Correlation can be improved by using the same gate types/sizes, using the same metal layers for D2D variation, placing items closer together when spatial variation is present, etc.

2.3 Common Industry Models

There are several approaches used in industry to model variation. The oldest approach is to increase the t_{budget} with a timing margin to "guard band" the setup and hold times without specifically analyzing the variation. Most frequently, timing sign-off will use worst-case corners that analyze the worst local skew when portions of a clock distribution are "fast" and others are "slow". The number of corners in such approaches were traditionally only for PVT variations, but additional corners have been added for individual metal layers, Negative Bias Temperature Instability (NBTI), and other sources of variation. This is to prevent the process (P) term from becoming overly pessimistic when all sources are assumed worst-case simultaneously.

Despite an exponentially increasing number of corners [44], multi-corner timing analysis is the current industry standard practice. Incremental algorithms are used to perform common path pessimism removal (CPPR) [20, 49] due to shared clock buffers that cannot be simultaneously fast and slow in a single corner. Most CPPR algorithms are iterative and worst-case exponential time complexity. Other methods have used data from multiple corners and normalized delay spreads to quantify the effects of variation on clocks [33].

Statistical static timing analysis (SSTA) solves the previous problems [1, 7, 45], but has a number of daunting challenges of its own such as obtaining accurate variation models, efficiently considering non-Gaussian and non-linear sources of variation, and gaining designer acceptance. There are several surveys and books that cover the recent advances in SSTA [5, 13, 36] and are applicable to clock researchers.

2.4 Requirements for Variation Models

We now summarize the requirements that are needed to accurately assess the robustness of a clock distribution with respect to variation.

Both device and interconnect models must be considered. Ignoring device variation implies that one can over-buffer a design making interconnect variation less significant. Similarly, without interconnect variation, one can under-buffer the design which makes the delay of the devices less significant. The relative importance of device and interconnect variability also depends on whether the synthesis algorithm is considering local, regional, or global distributions. Local distributions are affected more significantly by device variability whereas global distributions are affected more significantly by interconnect variability.

The physical manifestation of variability must be considered in the model. Blindly selecting variation as a percentage of delay or a parameter is not correct. Most variations (V_{th}, LER, etc.) are not a percentage. The correct model requires careful consideration of the specific source of variation.

Assuming that only WID variation matters is not correct. D2D variation also matters because parameter mismatch (between layers, implants, etc.) can lead to different insertion delay sensitivities and therefore skew.

Ignoring correlation for WID variation is not correct. If WID variation is assumed to be completely uncorrelated the law of large numbers can always be applied, but this may not be appropriate. It is necessary to consider the right correlation model for each source of variability.

Both skew and jitter must be simultaneously considered. Otherwise, insertion delay and jitter sensitivity can be increased to reduce skew or skew can be increased to reduce jitter. In high-performance designs, skew and jitter are on the same order of magnitude.

Variability must be analyzed statistically. Simple corner-based approaches are too pessimistic and do not allow for the cancellation of correlated variation or the reduction of variation due to the law of large numbers (when appropriate) [44]. Multi-corner methods do not scale since the number of corners will become too large [44]. Therefore, statistical methods should be used with the appropriate variation models.

2.5 Analysis Methodologies

When variability impact is considered, the clock network cannot be analyzed in a single spice run. Statistical methods have to be applied for an accurate evaluation of the variability effect on the clock performance. Many of the methods developed for statistical timing analysis can be adapted to evaluate the clock network. Monte Carlo analysis however is still the most common analysis methodology due to its simplicity and accuracy.

The main limitation of Monte Carlo simulation is the number of samples required for an accurate analysis. A single flat spice run on a full clock network can take many hours. To enable the characterization of large clock trees the analysis can be done level-by-level. This methodology however cannot be applied to clock meshes and clock spines. Whenever clock buffers are shorted they have be analyzed as a whole. In a clock mesh or a clock spine the simulation of the clock tree driving the mesh/spine can still be separated from the stages in which the buffers are shorted but every shorted set of buffers has to be handled in a single simulation. Large clock meshes may be prohibitively large to be fit in a single spice run, so several strategies have been proposed [9,48] to enable the analysis of these large meshes.

Frequency domain analysis [46,50] can achieve speed ups in the order of 1,000× but these techniques require that the buffer model be linearized which introduces error. Model order reduction techniques are more accurate but achieve speed ups of at most 10× [48]. The method of [9] is the most flexible as it can be applied using any electrical simulation, but it can only reduce the clock mesh characterization time if the different runs are parallelized.

3 Clock Trees

The simplest style of clock distribution network is an H-tree. The H-tree, as shown in Fig. 12.2, uses a recursive "H" routing pattern and inserts buffers at regular intervals in the hierarchy. H-trees are advantageous because they are simple and fairly power efficient. They are also symmetric and therefore resistant to inter-die sources of variation. However, H-trees can still be plagued by intra-die sources of variation as shown in [19]. Besides their robustness, H-trees are only typically useful for the top levels of clock distribution since the sink loads and locations are never uniform. Loads can be balanced with dummy gates, but this can increase the clock tree power consumption considerably. In addition, individual sinks may have differing clock edge requirements according to a useful skew budget.

In response to the non-uniformity of clock sinks, balanced clock trees offer a low-power solution with equal delay to clock sinks as illustrated by Fig. 12.3. The diagonal routes are later embedded into traditional rectilinear layers. Several researchers have presented exactly zero skew [6, 8, 11, 43], bounded skew [10, 21] and heuristic methods [23, 24] to construct these trees. Most balancing techniques rely on simplified delay models such as Elmore delay [12], which are known to be only approximations of more complex delays [32, 34]. However, they are still useful in generating initial solutions that can be refined with more accurate models later.

The main disadvantage of balanced trees is that they often assume process parameters are nominally equal for different interconnect layers and do not consider

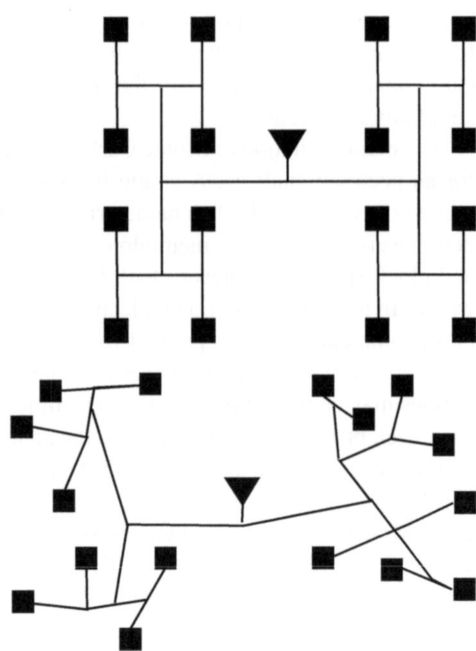

Fig. 12.2 H-Tree clock distribution

Fig. 12.3 Balanced clock tree distribution

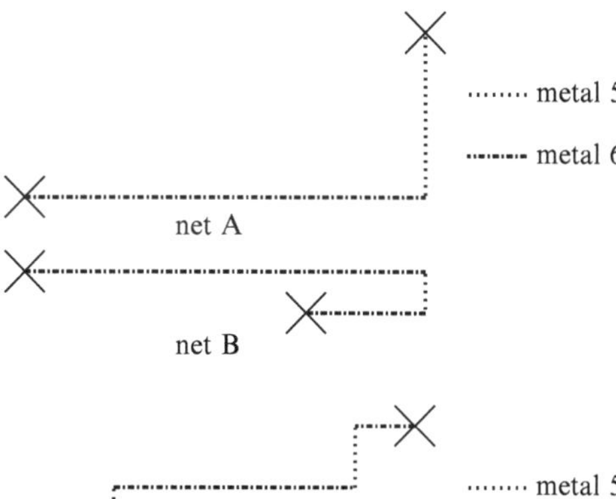

Fig. 12.4 Non symmetrical equal wirelength routing example

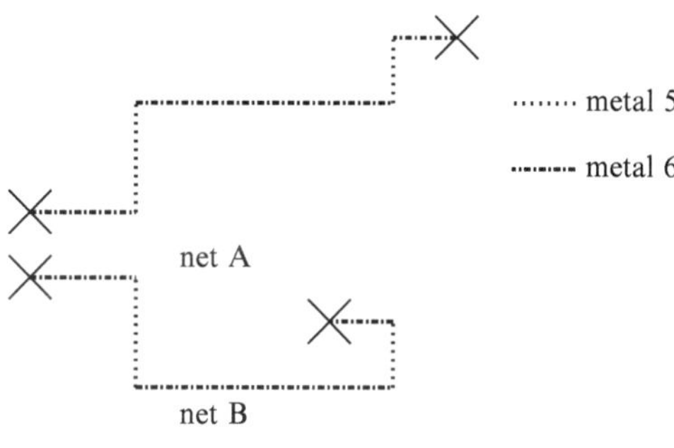

Fig. 12.5 Symmetrical routing example

the impact of process variation on skew. The different variability sources described in Sect. 2 can introduce skew in the clock tree even when it has zero-skew under the nominal case. The effect of variations sources is very significant in today's technologies and its effect has to be account on every design.

One way to make a clock network more robust is by equalizing the way it affects the clock network. Die-to-die variability effects on the clock network, for instance, can be completely eliminated by equalizing the sensitivity of each parameter on the delay on each path. Consider the two nets shown in Fig. 12.4, net A and B that are routed using metal 5 (vertical) and metal 6 (horizontal) wires. Assuming that the delay of nets A and B must be equalized, it is important that both nets have exactly the same wire length. The routing shown in Fig. 12.4 guarantees nominal zero skew between them, but if during fabrication metal 5 is faster while metal 6 is slower, net B would become faster than net A. This situation can be avoided if both nets are routed using the same amount of routing in each layer in the same order as shown in Fig. 12.5. By doing so, if one metal layer gets faster while the other gets slower both nets are affected in the same fashion. Sensitivity matching algorithms have attempted to do this by sizing buffers and wires to equalize the sensitivities of different process parameters using statistical timing analysis [17].

On the other hand, symmetrical clock trees are very efficient to reduce the effect of die-to-die variations. Custom symmetrical clock trees can be built using symmetrical routing algorithms such as [16, 30]. In general symmetrical clock trees are more robust to die-to-die variations at the expense of higher power consumption due to the longer total wirelength.

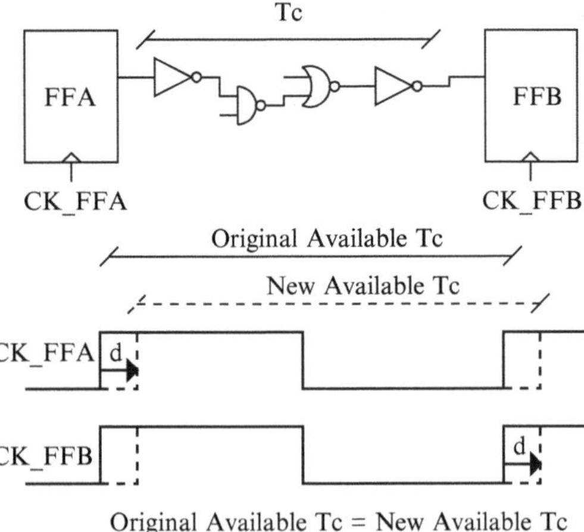

Fig. 12.6 Time available for signal to propagate between flip-flops is unaltered if both clock edges are modified by the same amount

Unlike die-to-die variation, within-die variation is not easily addressed. The effect of within-die variation on skew can be minimized by increasing the correlation between the clock signals. As illustrated in Fig. 12.6, if the variability affecting the clock skew is correlated, a delay in the launching clock edge will coincide with a delay in the capturing clock edge. This results in an identical clock period length. This concept was used in [17] for a modified version of the DME algorithm in which the correlation between clock sinks that share timing paths is increased which reduces the variability of those paths.

Besides the techniques described above, simple design principles can be followed to make clock networks more robust. Power supply noise is one of the main contributors to clock skew, therefore inserting decoupling caps in the power supply close to the clock buffers helps to decrease clock skew due to power supply noise. Another strategy to minimize clock skew variation is to reduce the number of buffers in the clock network as buffers are typically more sensitive to variation than wires. Aggressively reducing the number of buffer stages is not always good, however, as it can increase slew rates. Excessively large slew rates make the clock network more sensitive to threshold voltage variations as buffer delays are more dependent on the threshold voltages when the input signal slew is high. It should be noticed that those techniques are general and can be applied to non-tree clock network topologies.

In many microprocessor designs the clock network uses active compensation through de-skew buffers to reduce the effect of process variation on clock skew. Active deskewing uses programmable delay buffers and a control feedback. The control feedback is responsible for detecting and minimizing the clock skew. De-skew buffers can dynamically monitor the clock network delays [14, 25] or be configured only once after fabrication [42]. Active deskewing requires extra circuitry to monitor the clock skew and adjust the delay at the variable delay buffer. Static deskewing is much simpler as delays at the variable delay buffer are set only once and no skew monitoring is required. Due to the simpler and more robust design, static deskewing is more commonly found than dynamic deskewing in industry.

4 Clock Meshes

A clock mesh is a grid of orthogonal wires to which clock buffers are connected. The grid structure allows mesh buffer outputs to interact to each other. Whenever a buffer switches before the others, it is slowed down by the late switching buffers. This behavior reduces the overall difference between early and late arrival time. The capability to compensate for any variability source makes clock meshes popular in high-performance designs. Meshes and spines are widely used in the design of the clock distribution for microprocessors [3, 27, 35, 40].

The buffers that drive clock meshes are typically driven by a clock tree. Non-homogeneous clock sink distribution and obstructions may make the clock mesh irregular requiring the clock tree to be custom designed for it. Clock sinks are usually not directly connected to the clock mesh. Local buffered clock trees are designed to cluster clock sinks and decouple the actual sink capacitance from the clock mesh. The total clock mesh power consumption is reduced by decoupling clock sink capacitance.

The high performance offered by clock meshes comes at the cost of high power consumption. The redundant routing that shorts all mesh buffers represents extra capacitance that must be charged/discharged every clock cycle. Clock gating techniques are not as effective on clock meshes as they are on clock trees. Clock gaters must be placed in the local cock buffers after the clock mesh. The mesh must be divided into smaller sub-meshes in order to gate the mesh itself.

Short circuit current is another important contributor to total mesh power. As usual, internal short circuit current flows between PMOS and NMOS devices when the clock signal is switching. However, the clock mesh also has significant short circuit currents flowing between different mesh buffers. Moreover, during the time that mesh buffers are compensating for variation, short circuit currents flow between buffers as illustrated in Fig. 12.7.

The contribution of short circuit current to total power consumption is dependent on two design parameters: the clock skew at the input of the mesh buffers and the total load driven by the clock mesh. When the input clock signal is poorly synchronized at the input of the mesh buffers, short circuit current can have a long duration which increases the power consumption. The contribution of short circuit current to the clock mesh power consumption as a function of the input clock skew is shown in Fig. 12.8. Figure 12.8 shows power consumption on the y-axis while the x-axis shows the maximum clock skew at the input of mesh buffers normalized to a 1 GHz clock period. A roughly linear increase in the power consumption is observed when the input skew increases.

The other main contributor to the short circuit power consumption are the mesh buffer sizes. Larger mesh buffers produce larger short circuit currents. Mesh buffer sizes are determined by the load connected to the clock mesh, the higher the load the larger mesh buffers are required to be so timing constraints can be met. The load driven by the clock mesh can be reduced by adding more level of buffers between

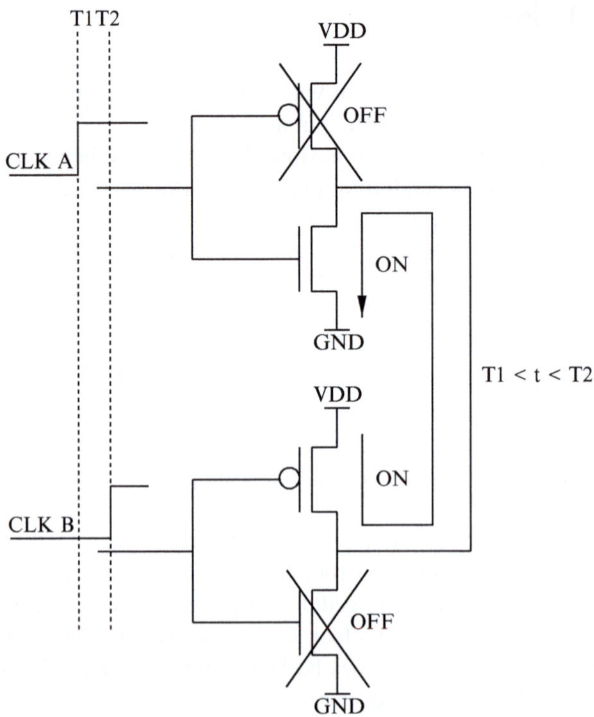

Fig. 12.7 Short-circuit current caused by different arrival time at mesh buffers

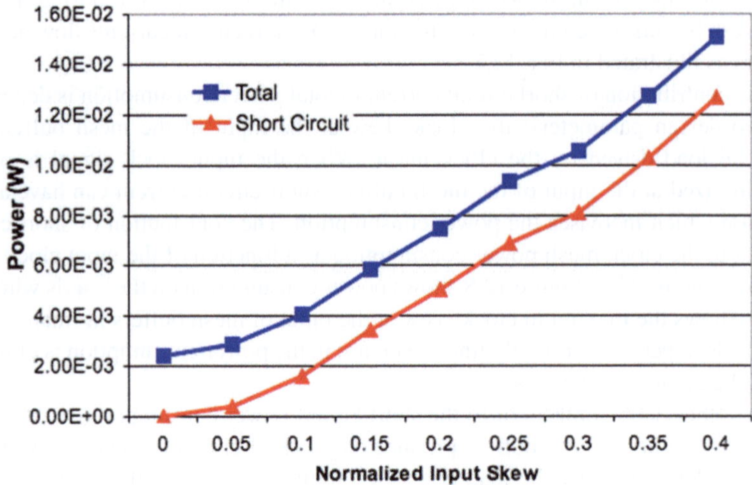

Fig. 12.8 Short-circuit current caused by different arrival time at mesh buffers [47]

the clock mesh and the clock sinks. By decoupling clock sink capacitances from the clock mesh the contribution of the clock mesh power to the total power of the clock network is minimized.

Besides the clear performance advantage offered by clock meshes, they also facilitate fast clock design turn-around. Due to the tight skew requirements, clock trees must be redesigned every time changes are made at the clock sink level. Clock meshes, on the other hand, are less sensitive since every sink is driven by a large set of nearby buffers. Therefore, the effect of changes at the clock sinks is shared among many buffers. If many random changes are performed the overall effect tends to average out. This property allows the design of the clock network to be kept unchanged when minor changes happen at the clock sink level.

The clock mesh conveniently shields the top-level clock tree from any changes in the clock sink distribution. However, the clock mesh requires that mesh buffers be spread over the entire region covered by the clock mesh. Any large macro within this region must accommodate space for mesh buffers. Leaving buffer-less regions in the clock mesh drastically reduces variability tolerance and increases clock slew. Clock spines are frequently used when macros cannot accommodate mesh buffers.

Clock spines are essentially one dimensional clock meshes with buffers placed along the line so that the outputs are shorted and the inputs are driven by a clock tree. Figure 12.9 illustrates a clock spine. Although the clock spine also allows faster switching buffers to compensate for late switching ones, its performance is significantly inferior than a clock mesh. The two dimensional structure of clock meshes reduces the equivalent resistance between the mesh buffers and enables more effective compensation for variability. Compensation between the spine buffers is almost exclusively adjacent and leads to high skew between distant points.

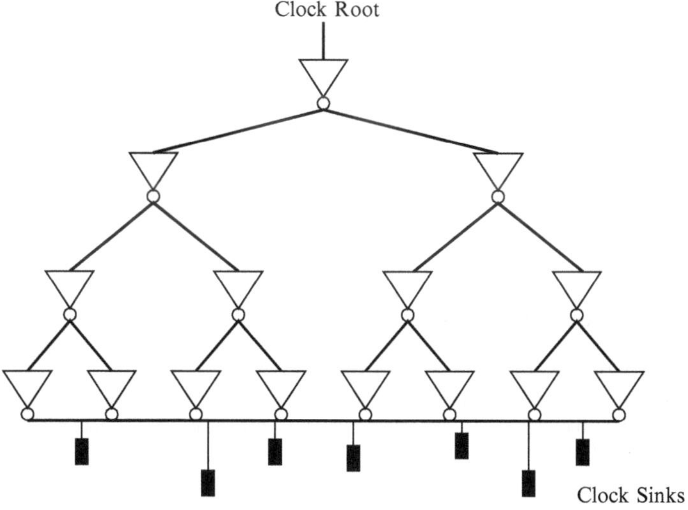

Fig. 12.9 Clock spine example

In addition, high skew compared to clock meshes is observable between different spines. Despite this, clock spines are popular in high performance designs due to the floorplanning benefits and can be found in microprocessors such as in [27] and [40].

Conclusion

Clock robustness is a critical part of chip design as any timing fault can propagate and produce multiple logic errors. A robust clock network must properly account for the effect of the process and environmental variation to ensure that timing margins are not violated. The sensitivity of the clock network becomes even more evident as technology scales. Shrinking devices amplify the effect of process variability due to fabrication challenges and random dopant fluctuation in transistor threshold implants. Reduced supply voltages are critical to reduce power consumption but also lead to reduced noise margins which make designs more susceptible to power supply noise, crosstalk and radiation effects. As discussed in the previous chapter, it is clear that clock networks must be designed with robustness in mind. Variation-tolerant clock trees, clock meshes and clock spines will continue to gain importance in high performance designs.

References

1. Agarwal, A., Zolotov, V., Blaauw, D.T.: Statistical timing analysis using bounds and selective enumeration. IEEE Transactions on Computer-Aided Design **22**(9), 1243–1260 (2003)
2. Ajami, A., Banerjee, K., Pedram, M.: Modeling and analysis of nonuniform substrate temperature effects on global ULSI interconnects. IEEE Transactions on Computer-Aided Design of Integrated Circuits and Systems **24**(6), 849–861 (2005)
3. Bailey, D., Benschneider, B.: Clocking design and analysis for a 600-MHz Alpha microprocessor. Journal of Solid-State Circuits (JSSC) **33**(11), 1627–1633 (1998)
4. Benschneider, B.J., Black, A.J., Bowhill, W.J., Britton, S.M., Dever, D.E., Donchin, D.R., Dupcak, R.J., Fromm, R.M., Gowan, M.K., Gronowski, P.E., Kantrowitz, M., Lamere, M.E., Mehta, S., Meyer, J.E., Mueller, R.O., Olesin, A., Preston, R.P., Priore, D.A., Santhanam, S.: A 300-MHz 64-b quad-issue CMOS RISC microprocessor. Journal of Solid-State Circuits (JSSC) **30**(11), 1203–1214 (1995)
5. Blaauw, D., Chopra, K., Srivastava, A., Scheffer, L.: Statistical timing analysis: From basic principles to state of the art. IEEE Transactions on Computer-Aided Design of Integrated Circuits and Systems **27**(4), 589–607 (2008)
6. Boese, K., Kahng, A.: Zero-skew clock routing trees with minimum wirelength. In: ASIC Conf., pp. 1.1.1–1.1.5 (1992)
7. Chang, H., Sapatnekar, S.: Statistical timing analysis considering spatial correlations using a single PERT-like traversal. In: International Conference on Computer-Aided Design, pp. 621–625 (2003)
8. Chao, T.H., Hsu, Y.C., Ho, J.: Zero skew clock net routing. In: Design Automation Conference (DAC), pp. 518–523 (1992)
9. Chen, H., Yeh, C., Wilke, G., Reddy, S., Nguyen, H., Walker, W., Murgai, R.: A sliding window scheme for accurate clock mesh analysis. In: IEEE/ACM International Conference on Computer-Aided Design, ICCAD, pp. 939–946 (2005)

10. Cong, J., Koh, C.K.: Minimum-cost bounded-skew clock routing. In: International Symposium on Circuits and Systems (ISCAS), pp. 215–218 (1995)
11. Edahiro, M.: Minimum path-length equi-distant routing. In: Asia-Pacific Conf. on Circuits and Systems, pp. 41–46 (1992)
12. Elmore, W.C.: The transient response of damped linear networks. Journal of Applied Physics **19**, 55–63 (1948)
13. Forzan, C., Pandini, D.: Statistical static timing analysis: A survey. Integration, the VLSI Journal **42**, 409–435 (2009)
14. Geannopoulos, G., Dai, X.: An adaptive digital deskewing circuit for distribution networks. International Solid-State Circuits Conference (1998)
15. Gupta, P., Heng, F.L.: Toward a systematic-variation aware timing methodology. In: Design Automation Conference, pp. 321–326 (2004)
16. Guthaus, M., Sylvester, D., Brown, R.: Clock tree synthesis with data-path sensitivity matching. In: Design Automation Conference, 2008. ASPDAC 2008. Asia and South Pacific, pp. 498–503 (2008)
17. Guthaus, M.R., Sylvester, D., Brown, R.B.: Process-induced skew reduction in deterministic zero-skew clock trees. In: Asia and South Pacific Design Automation Conference (ASPDAC), pp. 84–89 (2006)
18. Guthaus, M.R., Wilke, G., Reis, R.: Revisiting automated physical synthesis of high-performance clock networks. ACM Trans. Des. Autom. Electron. Syst. **18**(2), 31:1–31:27 (2013)
19. Hashimoto, M., Yamamoto, T., Onodera, H.: Statistical analysis of clock skew variation in H-tree structure. In: International Symposium on Quality Electronic Design (ISQED), pp. 402–407 (2005)
20. Hathaway, D., Alvarez, J.P., Belkbale, K.P.: Network timing analysis between signals traversing a common circuit path. United States Patent 5,636,372 (1997)
21. Huang, D.J.H., Kahng, A.B., Tsao, C.W.A.: On the bounded-skew clock and Steiner routing problems. In: Design Automation Conference (DAC), pp. 508–513 (1995)
22. Im, S., Srivastava, N., Banerjee, K., Goodson, K.E.: Scaling analysis of multilevel interconnect temperatures for high performance ICs. IEEE Transactions on Electron Devices **52**(12), 2710–2719 (2005)
23. Jackson, M.A.B., Srinivasan, A., Kuh, E.S.: Clock routing for high performance ICs. In: Design Automation Conference (DAC), pp. 573–579 (1990)
24. Kahng, A.B., Cong, J., Robins, G.: High-performance clock routing based on recursive geometric matching. In: Design Automation Conference (DAC), pp. 322–327 (1991)
25. Kapoor, A., Jayakumar, N., Khatri, S.P.: A novel clock distribution and dynamic de-skewing methodology. In: International Conference on Computer-Aided Design (ICCAD), pp. 626–631 (2004)
26. Kim, S.D., Wada, H., Woo, J.C.S.: TCAD-based statistical analysis and modeling of gate line-edge roughness effect on nanoscale MOS transistor performance and scaling. IEEE Transactions on Semiconductor Manufacturing **17**(2), 192–200 (2004)
27. Kurd, N.A., Barktullah, J.S., Dizon, R.O., Fletcher, T.D., Madland, P.D.: A multigigahertz clocking scheme for the pentium 4 microprocessor. IEEE Journal of Solid-State Circuits **36**(11), 1647–1653 (2001)
28. Mallajosyula, A., Zarkesh-Ha, P.: A robust single event upset hardened clock distribution network. In: Integrated Reliability Workshop Final Report, 2008. IRW 2008. IEEE International, pp. 121–124 (2008). DOI 10.1109/IRWS.2008.4796101
29. Nassif, S.R.: Modeling and forecasting of manufacturing variations (embedded tutorial). In: Asia and South Pacific Design Automation Conference (ASP-DAC), pp. 145–150 (2001)
30. Ozdal, M.M., Hentschke, R.F.: Exact route matching algorithms for analog and mixed signal integrated circuits. In: Proceedings of the 2009 International Conference on Computer-Aided Design, ICCAD '09, pp. 231–238. ACM, New York, NY, USA (2009)
31. Pelgrom, M., Duinmaijer, A., Welbers, A.: Matching properties of MOS transistors. Journal of Solid State Circuits (JSSC) **24**(5), 1433–1439 (1989)

32. Pillage, L.T., Rohrer, R.: Asymptotic waveform evaluation for timing analysis. IEEE Transactions on Computer-Aided Design **9**, 352–366 (1990)
33. Rajaram, A., Damodaran, R., Rajagopal, A.: Practical clock tree robustness signoff metrics. International Symposium on Quality Electronic Design (ISQED) pp. 676–679 (2008)
34. Ratzlaff, C.L., Gopal, N., Pillage, L.T.: RICE: Rapid interconnect circuit evaluator. In: Proceedings of the 28th conference on ACM/IEEE design automation, pp. 555–560 (1991)
35. Restle, P., Carter, C., Eckhardt, J., Krauter, B., McCredie, B., Jenkins, K., Weger, A., Mule, A.: The clock distribution of the POWER4 microprocessor. In: International Solid-State Circuits Conference (ISSCC), pp. 144–145 (2002)
36. Sapatnekar, S.: Timing. Springer (2004)
37. Steinhogl, W., Schindler, G., Steinlesberger, G., Traving, M., Engelhardt, M.: Impact of line edge roughness on the resistivity of nanometer-scale interconnects. Microelectronic Engineering **76**, 126–130 (2004)
38. Stine, B.E., et al., D.S.B.: The physical and electrical effects of metal-fill patterning practices for oxide chemical-mechanical polishing processes. IEEE Transactions on Electron Devices **45**(3), 665–679 (1998)
39. Sze, C.: Personal communication (2010). IBM Austin Research Lab, Austin, TX
40. Tam, S., Rusu, S., Desai, U.N., Kim, R., Zhang, J., Young, I.: Clock generation and distribution for the first ia-64 microprocessor. IEEE Journal of Solid-State Circuits **35**(11), 1545–1552 (2000)
41. Teodorescu, T.: Personal communication (2010). ATI Radeon 5870 Clock Designer, Sunnyvale, CA
42. Tsai, J.L., Zhang, L., Chen, C.C.P.: Statistical timing analysis driven post-silicon tunable clock-tree synthesis. In: International Conference on Computer-Aided Design (ICCAD), pp. 575–581 (2005)
43. Tsay, R.S.: Exact zero skew. In: International Conference on Computer-Aided Design (ICCAD), pp. 336–339 (1991)
44. Visweswariah, C.: Death, taxes and failing chips. In: Design Automation Conference, pp. 343–347 (2003)
45. Visweswariah, C., Ravindran, K., Kalafala, K., Walker, S.G., Narayan, S.: First-order incremental block-based statistical timing analysis. In: Design Automation Conference, pp. 331–336 (2004)
46. Wang, R., Koh, C.K.: A frequency-domain technique for statistical timing analysis of clock meshes. In: IEEE/ACM international conference on Computer-aided design, ICCAD, pp. 334–339. Piscataway, IEEE Press, San Jose, CA (2007)
47. Wilke, G., Fonseca, R., Mezzomo, C., Reis, R.: A novel scheme to reduce short-circuit power in mesh-based clock architectures. In: Symposium on Integrated Circuits and System Design (SBCCI), pp. 117–122 (2008)
48. Ye, X., Li, P., Zhao, M., Panda, R., Hu, J.: Analysis of large clock meshes via harmonic-weighted model order reduction and port sliding. In: IEEE/ACM international conference on Computer-aided design, ICCAD, pp. 627–631. Piscataway, IEEE Press, San Jose, CA (2007)
49. Zejda, J., Frain, P.: General framework for removal of clock network pessimism. In: International Conference on Computer-Aided Design (ICCAD), pp. 632–639 (2002)
50. Zhang, L., Yu, W., Zhu, H., Zhang, W., Cheng, C.K.: Clock skew analysis via vector fitting in frequency domain. In: International Symposium on Quality Electronic Design, ISQED, 9., pp. 476–479. Los Alamitos, IEEE Computer Society, San Jose, CA (2008). DOI 10.1109/ISQED.2008.4479780

MIX
Papier aus verantwortungsvollen Quellen
Paper from responsible sources
FSC® C105338

If you have any concerns about our products,
you can contact us on
ProductSafety@springernature.com

In case Publisher is established outside the EU,
the EU authorized representative is:
Springer Nature Customer Service Center GmbH
Europaplatz 3, 69115 Heidelberg, Germany

Printed by Libri Plureos GmbH
in Hamburg, Germany